D0947179
LIBRARY

THE GENUS
CYMBIDIUM

THE GENUS

CYMBIDIUM

DAVID DU PUY & PHILLIP CRIBB

CHRISTOPHER HELM
London

TIMBER PRESS
Portland, Oregon

Illustrations by Claire Smith
Maps by David Du Puy

Christopher Helm (Publishers) Ltd, Imperial House,
21-25 North Street, Bromley, Kent BR1 1SD

British Library Cataloguing in Publication Data

Du Puy, David
The genus Cymbidium.
1. Orchids
I. Title II. Cribb, Phillip
584'.15 QK495.064

ISBN 0-7470-0607-5

Published in the USA in 1988 by
Timber Press,
9999 SW Wilshire,
Portland,
Oregon 97225,
USA

Library of Congress Cataloging-in-Publication Data

Du Puy, David.
The genus Cymbidium.

Bibliography: p.
Includes index.
1. Cymbidium. I. Cribb, Phillip. II. Title.
SB409.8.C95D8 1988 625.9'3415 88-8581
ISBN 0-88192-119-X

Typeset by Leaper & Gard, Bristol
Printed and bound in Great Britain by Butler and Tanner, Frome, Somerset

Contents

Colour plates by Claire Smith

Colour Photographs

List of Figures

List of Maps

List of Tables

Acknowledgements

We would like to thank the Orchid Digest Foundation, inspired by the late Walter Bertsch, who have so generously supported the last year's research and preparation of this book and also provided funding for preparation of the text and illustrations. The Orchid Digest were supported in this venture by the following sponsors: the San Diego Orchid Society, the Cymbidium Society of America, the South Bay Orchid Society and the San Gabriel Orchid Society. The following have also generously subscribed to the Orchid Digest's support: Helen Congleton, the Orchid House, Mr & Mrs Don Bradish and Mr Bill Bailey.

The authors are particularly grateful for the support of Ernest Hetherington, Harold Koopowitz, Trudi and Fordyce Marsh, Lance Birk, Don Herman and other members of the Foundation.

Many friends have helped us with advice, material and inspiration over the past six years, notably Kit Seth, Brian Ford-Lloyd, Finn Rasmussen, Gosta Kjellsson, Gunnar Seidenfaden, Tony Lamb, Jim Comber, Jeffrey Wood, Peter Taylor, Andrew Bacon, C.L. Chan, P.S. Shim, Gloria Barretto, Tom Reeve, G. Hermon Slade, Gwen Lee, Niklaus Trudel, Arnold Klehm, Lady Lisa Sainsbury, Keith Andrew, Chris Bailes, David Menzies, Alan Moon, Andrew Easton, Mark Clements, Meta and Fritz Held, the late Eric Young, P.S. Lavarack, Tem Smitinand, David Clulow, Ray Bilton and Brian Rittershausen.

Claire Smith, now of Cape Town, devoted much of her time and skill during a three year visit to Kew to painting the exquisite watercolour plates that grace this book. They were justly awarded a Gold Medal at the Royal Horticultural Society in spring 1985. The large composite figures are also her work. We would like to record our thanks to her for the inspiration and standards that her work has given us.

Much of the research recorded in these pages was undertaken by one of the authors (DD) while preparing a doctoral thesis for the University of Birmingham, funded by a three year grant from the Science and Engineering Council. Facilities and material were also kindly provided by the Keeper of the Herbarium of the Royal Botanic Gardens at Kew, where much of the living material has been grown under the supervision of John Woodhams. Visits have been made to several herbaria including Edinburgh, Paris, Leiden, Oakes Ames, Singapore, Bogor, Melbourne and Sydney. We would like to thank the Directors and staff of these institutions for their assistance. Field work has been made possible through funding generously donated by the Bentham-Moxon Trust, the Studley College Trust, D. Clulow, R. Bilton, the Orchid Review and B. Rittershausen.

The chapter on the influence of the species in modern hybridising has been kindly contributed by Ernest Hetherington and Andrew Easton. They are leading exponents of *Cymbidium* breeding at the present day and we gratefully acknowledge the value of their contribution. Comments on this aspect have also been gratefully received from Alan Moon and Chris Bailes of the Eric Young Orchid Foundation. Many thanks also to Blaise Du Puy, for her valuable input to the chapter concerning cultivation.

The diagram indicating the pollination strategy of *C. insigne* and *Dendrobium infundibulum* was kindly lent by Ben Johnson (Denmark).

Caron Mitchell has patiently and skilfully deciphered our manuscript into an immaculate typescript. Finally, we would like to thank our respective spouses, Blaise and Marianne, for their encouragement during the prolonged genesis of this work.

Preface

> It is strange that one of the World's most popular orchid groups would not have a technical description (monograph) ... A good monograph on any genus of plants serves as a foundation for knowledgeable work both from a taxonomic and hybridizing standpoint.
>
> I can only hope too, someone will eventually wish to compile a monograph on the genus. What value this would be to the *Cymbidium* world.
>
> Ernest Hetherington in *Orchid Advocate* 3(4): 126 (1977)

For much of the world, the words 'orchid' and '*Cymbidium*' are synonymous. Cymbidiums have been, for over one hundred years, and remain at the present day, the most important orchids in commerce, being primarily grown for their flowers which are attractive, long lasting and large. To a lesser but increasing extent they are gaining popularity as decorative pot-plants with smaller, but equally high quality flowers. In contrast, in the East they are prized for their attractive habit, and flowers which are strongly scented. The commercially important, large-flowered hybrids have been derived from only a few of the 44 species currently recognised in the genus. However, the search for novelty has led to the use of some of the smaller-flowered species, such as *C. floribundum* and *C. devonianum*, with conspicuous success, as parents in hybridising programmes. No doubt others will be tried in an attempt to satisfy the insatiable appetite of the orchid grower for novelty.

Ernest Hetherington's plea for a monograph of the genus comes from a grower in the vanguard of modern commercial orchid hybridising. His recognition of the need for sound taxonomic information, as the basis for the activities of orchid growers interested in both species and hybrids, comes from years of frustration at the lack of such information.

He would, no doubt, have been heartened to know that, even as he wrote, C.J. (Kit) Seth, in collaboration with the staff of the Orchid Herbarium at Kew, was engaged on a study of the sectional limits of the genus. The results were not published until 1984 (Seth & Cribb, 1984) and, by then, the research which has led to this publication was well advanced.

The aims of this study have been several, reflecting the needs of both botanist and horticulturist. First of all, we hope to have produced a classification that reflects the biological reality by examining the evidence from all available sources. Consistency of treatment between different parts of the genus has been an important objective. It has also been considered a priority to produce a usable classification and nomenclature. Finally, in achieving the last we have endeavoured to preserve the accepted nomenclature as far as possible.

In the course of this study, we have examined many hundreds of specimens, both preserved and living. Over three-quarters of the species of *Cymbidium* are now in cultivation at the Royal Botanic Gardens, Kew and we would like to thank many friends for their generosity in building this collection. Many species have also been seen in the wild. Field work in several countries has enabled us to gain a better understanding of natural variation and also to study the pollination syndrome of C. insigne for the first time.

Several species have proved to be much more widespread than had generally been appreciated. Such wide distributions are linked with variation and, where possible, this has been recognised by the use of infraspecific categories. Inevitably, some well established names have been sunk into synonymy as a result. This may be resisted by many horticulturists, but it is important to appreciate that such changes reflect our increasing knowledge of the genus, and are the result of examination of many specimens. We would particularly emphasise that morphological evidence, previously the only information available, has been supplemented by new information from extensive anatomical, micromorphological, cytological and biochemical studies.

Two species new to science have been discovered during the research. Although such novelties appear less frequently nowadays, we anticipate that more might yet be discovered as the last wild places in the world are opened up by the destruction of the tropical forests. It would be sad indeed if Cymbidiums could no longer be seen in their natural habitats. For several species that time may be close at hand!

INTRODUCTION

Cymbidium species have been cultivated in China for at least 2500 years, the earliest written records dating from the time of Confucius (about 500 BC). They continue to be of importance in cultivation in China to the present day. The first species now included in *Cymbidium* were introduced to Europe around the beginning of the eighteenth century, and Linnaeus described two species, *C. aloifolium* and *C. ensifolium* in the *Species Plantarum* in 1753, although they were then placed by him in the genus *Epidendrum*. Relatively few species were seen in cultivation in Britain until the time of the Industrial Revolution, which provided both the leisure time and the money for a great increase of interest in orchid cultivation. From about the mid-nineteenth century, extensive searching for and collection of new species took place, and many were introduced into cultivation. The genus *Cymbidium* is now considered to comprise 44 species, and is distributed from N W India to China and Japan, south through the Malay Archipelago to N and E Australia.

The advent of orchid hybridisation in the latter half of the nineteenth century led to the development of a novel line of *Cymbidium* cultivation, based on the large-flowered Himalayan species. For over a hundred years the species have been hybridised to produce plants with flowers of rich texture and colouring, and large size, that have formed the basis of a cut-flower and more recently a pot-plant industry which is now world-wide. *Cymbidium* hybrids are probably the most commercially important group of orchids in cultivation at the present time.

These modern hybrids are often complex, involving several species in their ancestry, but the species themselves have continued to be grown for their own merits and to play a role in hybridisation. The chapter on the role of the species in hybridisation has been kindly provided by Messrs Andrew Easton and Ernest Hetherington, two of the world's leading authorities and practitioners of *Cymbidium* hybridising at the present day. Species such as *C. devonianum*, *C. floribundum*, *C. tigrinum* and *C. ensifolium* are currently being crossed with large-flowered hybrids to produce smaller plants which are suitable for the pot-plant trade. This resurgence of interest in *Cymbidium* species has highlighted the taxonomic confusion that is to be found in the genus.

It is surprising, in view of their popularity, that no generally accepted, full taxonomic account of the genus exists. Indeed, until recently, the generic delimitation of *Cymbidium* itself was disputed. This study has attempted to produce a comprehensive account of the species of *Cymbidium* in which many of the taxonomic confusions are explained and resolved. Those which still defy clarification are pinpointed, and as complete a discussion of the problems as possible is given.

In fact, horticultural involvement in the genus, in both the Western world and in China and Japan, has itself created confusion. Minor differences have been given taxonomic recognition, particularly in species which are naturally rather variable. Further confusion has arisen in widely distributed species, through the naming of separate taxa in geographically restricted regional floras. In some cases, the complete range of variation within a species has not been examined by authors describing new taxa, and extremes of the variation have been recognised as distinct. The loss of the Berlin Herbarium during the last war has further hindered studies since it contained critical type material. These confusions are resolved in this work; synonymies are identified, and the literature concerning individual species is listed. Full accounts of all species, sections and subgenera are given in chapter 10.

Undoubtedly, one of the major causes of taxonomic confusion in the genus is the reliance placed on gross morphological characters, especially of the flowers. In addition, several species have been described from inadequate material, occasionally from a single flower or from a cultivated specimen of obscure provenance (eg *C. parishii*, *C. tracyanum* and *C. ensifolium*). In this study the morphological information has been re-examined and critically assessed. The binocular dissecting microscope has allowed the precise dissection and scale drawing of numerous specimens to produce comprehensive accounts of the floral variation. These drawings are available for examination at the Royal Botanic Gardens, Kew. In addition to the extensive herbarium collections principally from the Royal Botanic Gardens of Edinburgh and Kew, the Rijksherbarium at Leiden, and the herbaria at Paris, Bogor and Singapore, the large collection of spirit-preserved material and the extensive living collections at Kew, where over three-quarters of the species are in cultivation, have been examined and have provided further invaluable data. Additional and sometimes critical information has been gathered from leaf, seed and pollen micro-morphology, leaf anatomy, and other experimental studies. Numerical methods of taxonomic analysis have been used to clarify certain taxonomic problems, and a cladistic analysis has been used to indicate possible evolutionary relationships within the genus.

Field work in India, China, Thailand, Malaya, Java, Sabah, New Guinea and Australia has allowed the examination of many species in their wild habitats, contributing valuable information concerning the ecology and natural variation of wild populations.

CHAPTER 1

The Cymbidium Plant

Traditionally, the genus has been classified on the basis of gross morphological characters. These characters are described in detail for each species in chapter 10. The terminology applied to the various organs is summarised in figure 1.

Vegetative Morphology

The species of *Cymbidium* are usually autotrophic, terrestrial, epiphytic or lithophytic **herbs**. In many instances, normally epiphytic species are encountered as lithophytes. The species in subgenus *Jensoa** are entirely terrestrial, and include *C. macrorhizon* which is a saprophyte and grows entirely below the soil surface, except when the flower spike is produced. Plant size is variable, but some species form large clumps with the leaves reaching over a metre (39 in) in length.

Most species have thick **roots** up to about 8 mm (0.3 in) in diameter which are covered in thick, spongy, white velamen, and have a core of vascular tissue.

The erect stems are usually swollen to form prominent **pseudobulbs**, which are often slightly flattened, but which may be reduced to a slight swelling of the base of the stem. The usual growth habit is sympodial, with new pseudobulbs being produced annually on a short rhizome, resulting in the pseudobulbs being tightly clustered in the mature plant. The pseudobulbs of species in section *Eburnea* grow and flower continuously for 2–3 years before a new shoot is produced. In *C. mastersii*, *C. elongatum* and *C. suave*, each shoot grows continuously for many years, forming a long stem rather than a pseudobulb and giving the plant an apparently monopodial growth habit. *C. macrorhizon* lacks pseudobulbs and has a subterranean rhizome covered with small tubercles.

Cataphylls enclose the young shoot before the true leaves emerge, and can be distinguished from the true leaves by their lack of a distinct lamina and abscission zone. The cataphylls quickly become scarious and eventually disintegrate, leaving numerous fibres around the base of the pseudobulb.

Each growth has from 3–12 distichous **leaves**. In the saprophyte *C. macrorhizon* these are reduced to scales. The lamina of the leaf is articulated to the leaf base by an abscission zone, and although the lamina is eventually deciduous it usually persists on the pseudobulb for 2–4 years. The broad, sheathing leaf bases usually completely enclose the pseudobulb, and persist on the pseudobulb after the lamina has been shed. In *C. tigrinum* the pseudobulbs have 2–4 apical leaves, and are almost completely exposed.

The leaf texture may be almost rigid and very coriaceous, as in *C. aloifolium* and its allies and in *C. canaliculatum*, but it is usually thinner and flexible, and is often ribbed. *Cymbidium* leaves are always duplicate (folded along the mid-vein) in development, but may be conduplicate (V-shaped in transverse section) or plicate (ribbed) in appearance when mature. No species has leaves which can be categorised as truly plicate in the sense of Withner *et al.* (1975), as plicate leaves have convolute (folded like a fan) development. Other differences from the true plicate leaf category are evident in the leaf anatomy of *Cymbidium* (see chapter 3).

Most of the species in subgenus *Jensoa* and sub-

*For subgenera and sections see p. 56.

Figure 1: The morphology of *Cymbidium*

A Plant of *C. dayanum* (× 0.35)
B Flower: side view (× 0.4)
C Flower: front view (× 0.4)
D Lip: flattened to allow the measurement of constituent parts (× 1)
E Lip: front view (× 1). The mid-lobe is recurved
F Column and lip: side view with petals and sepals removed (× 1)
G Pollinarium (× 3). Enclosed by the anther-cap

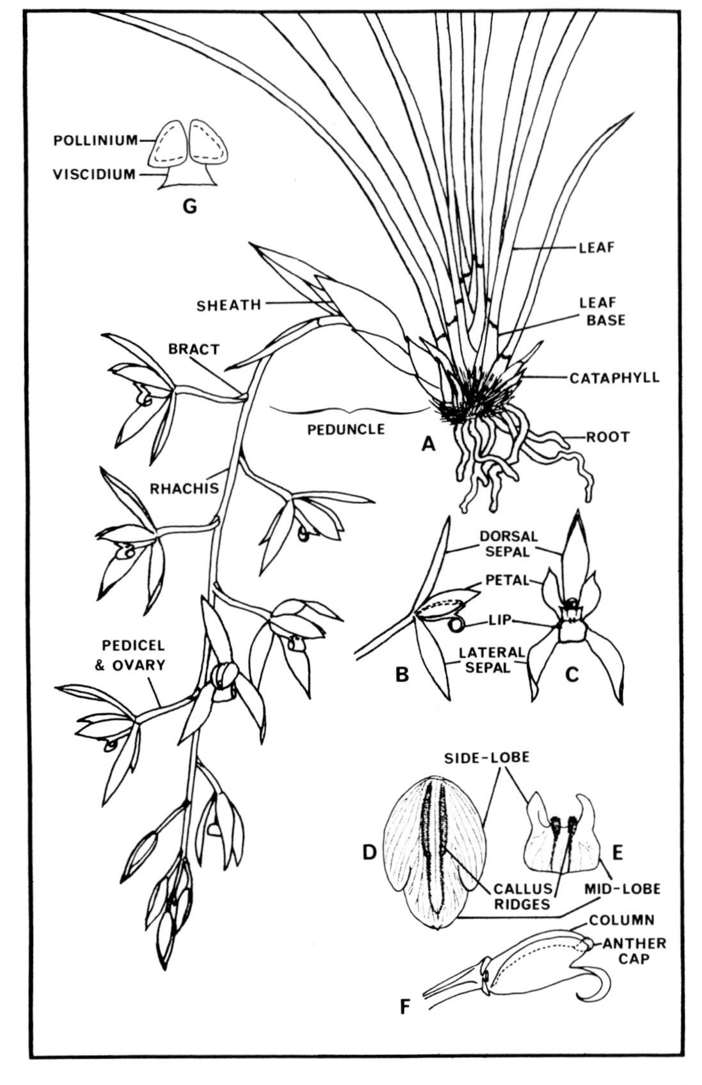

genus *Cyperorchis* have leaves of a similar ribbed appearance. However, subgenus *Cymbidium* contains many species with leaves which are duplicate in development, and conduplicate when mature that are consistent with the leathery leaf category of Withner *et al.* (1975). The leaves of the species in *C. aloifolium* and its allies and of *C. canaliculatum*, epiphytic or lithophytic species which are often exposed to arid conditions, have a hard, leathery leaf, with no protruding veins apart from the mid-vein, and many sclerenchyma fibres near the leaf surface making the leaf rigid. The leaves are also very thick (about 3 mm (0.12 in) or more), have a thick cuticle, and have a large volume of spongy mesophyll. The evidence therefore supports the suggestion of Withner *et al.* that leathery leaves are a xerophytic modification of the more primitive, thin leaves of plicate appearance. The leaves of these species appear to assume the function of water storage organs instead of the normal swollen pseudobulbs. It is possible that these adaptations may be linked to Crassulacean Acid Metabolism (Winter *et al.*, 1983). Several other anatomical and micromorphological characters can also be interpreted as xerophytic modifications in these species, and are discussed in chapter 3.

Other leaf shape modifications which can be linked to the ecology of the species include the broad, elliptic leaves of the shade-loving species *C. lancifolium* and *C. devonianum*, and the long, thin, relatively narrow leaves of most of the forest epiphytes which grow in shaded, humid conditions, including many of the species in subgenus *Cyperorchis*.

The leaf apex is usually acute and slightly unequal, but in section *Cymbidium* it is obtuse and often unequally bilobed. In *C. mastersii* and *C. eburneum* it is conspicuously forked, with a small mucro in the sinus formed as an extension of the mid-vein.

Floral Morphology

In *Cymbidium* the **inflorescence** is unbranched and may be erect, arching or pendulous. The peduncle is usually covered by about eight sheaths. One or two inflorescences are usually produced from the base of the mature pseudobulb. The species in section *Eburnea* produce inflorescences in the axils of the leaves towards the apex of the pseudobulb, and *C. elongatum* and *C. suave* also produce them in the axils of the upper leaves. In *C. lancifolium* they arise from one of the central nodes of the pseudobulb.

The **flowers** are usually about 3–5 cm (1.2–2 in) in diameter, but those in subgenus *Cyperorchis* are often 10 cm (4 in) or more across. Flower number varies from one in *C. goeringii* and *C. eburneum*, to about 50 in *C. canaliculatum*. Each flower has an inferior **ovary** and short **pedicel** which are difficult to delimit from each other, and is subtended by a small scarious **bract**. The flower comprises a dorsal and two lateral **sepals**, two free **petals** which may be spreading or covering the column, a lip, and a central column. The **lip** is hinged at the base of the column, or at the base of a very short column-foot in *C. devonianum* and some species in section *Cymbidium*. The base of the lip is fused with the base of the column for 3–6 mm (⅛–¼ in) in all species in subgenus *Cyperorchis*. The 3-lobed lip may be weakly saccate at the base in some species of section *Cymbidium*, but it is never spurred. It is usually 3-lobed, the two side-lobes being erect and loosely clasping the column, and the mid-lobe is usually somewhat recurved. Two parallel callus ridges are commonly borne on the upper surface of the lip. However, in *C. eburneum* and *C. erythrostylum* a wedge-shaped callus is found, while in *C. suave* and *C. madidum* the callus ridges are replaced by a glistening depression. Subgenus *Jensoa* is characterised by a callus of two ridges which converge towards their apices, forming a short tube at the base of the mid-lobe. In *C. devonianum* and *C. borneense*, the callus is reduced to two small swellings at the base of the mid-lobe, while in *C. aloifolium* and *C. bicolor* the callus ridges are often strongly sigmoid and may be broken in the middle.

The **column** is weakly winged, more strongly so towards the apex. The anther usually contains two **pollinia**, each of which is deeply cleft behind. Each pollinium may be visualised as a pair of unequal, flattened pollinia fused along the inner margin. In subgenus *Jensoa*, and in *C. borneense*, this fusion is absent, and there are two pairs (four) of unequal pollinia in the anther. The pollinia are usually triangular, but in section *Eburnea* they are usually quadrangular and in section *Cyperorchis* they are club-shaped. The pollinia are placed almost directly onto a sticky **viscidium**, and are held there by extremely elastic but very short caudicles. The caudicles are derived from the pollinia, while the viscidium is derived from the rostellum which is part of one of the stigma lobes. The viscidium, caudicles and pollinia constitute the **pollinarium**. The **stigma** is a sticky concavity located on the underside of the column, directly behind the anther.

CHAPTER 2

The Cymbidium Seed

Introduction

A range of *Cymbidium* species have recently been surveyed by Du Puy (1986) using a scanning electron microscope to allow high magnification. Several taxonomically useful conclusions have been drawn from the results. The pollinium micro-structure was also examined, but the results were taxonomically inconclusive.

The seed of *Cymbidium*, quite typical of the majority of orchid seeds, is minute and dust-like, 520–1900 µm (c. 0.5–1.9 mm, 0.02–0.07 in) long, and fusiform or filiform in outline. A spherical embryo without endosperm is enclosed in a thin testa, one cell layer thick. The testa cells are largest towards the centre of the seed. The anticlinal walls of the testa cells are thick, appearing as a net on the surface of the testa (figure 2A). The junctions of this net may protrude somewhat, sometimes giving the seed a spiny appearance. The surface walls between this net have fine secondary thickening which appears as minute striations which are either longitudinal, running parallel with the long axis of the seed (figure 2C) or transverse, running at right angles to the long axis of the seed (figure 3C,F).

Variation in *Cymbidium* Seeds

Cymbidium species in subgenus *Cymbidium* and subgenus *Cyperorchis* (figure 2) have fusiform seeds between 520–1200(1400) µm long. Seeds of species in sections *Bigibbarium* and *Cyperorchis*, and of *C. canaliculatum* (section *Austrocymbidium*), were shortest in the genus, only 520–650 µm long, while those of species in sections *Eburnea* and *Iridorchis* were the largest with a fusiform shape, usually 800–1200 µm long. Seed shape and size varied somewhat, both within and between samples.

A second, distinct type of seed has been found only in *Cymbidium* subgenus *Jensoa* (figure 3), and in some species of *Eulophia*. The seeds of these species are filiform, 1400–1900 µm long and very slender.

The surface cell walls of the seed have a striated appearance. Longitudinal striations were found in seeds with a fusiform shape (figure 2), while lateral striations were restricted to the filiform seeds (figure 3).

In many of the species examined, the junctions of the anticlinal walls were marked by an extension of the walls into a small, protruding, pyramidal structure. In a few species this was very pronounced, giving the seeds a spiny appearance.

The Implications for the Classification of Cymbidium

The fusiform seed shape found in *Cymbidium* subgenus *Cymbidium* and subgenus *Cyperorchis* was similar to the related genera *Grammatophyllum*, *Ansellia* (figure 4), *Porphyroglottis*, *Eulophiella*, and *Cymbidiella*. These seeds also have in common the longitudinal striations of the walls of the testa cells, and the somewhat protruding junctions of the net of cell walls which are often capped with waxy deposits. This type of seed is therefore characteristic of these two subgenera of *Cymbidium* and the related genera in the Cyrtopodiinae.

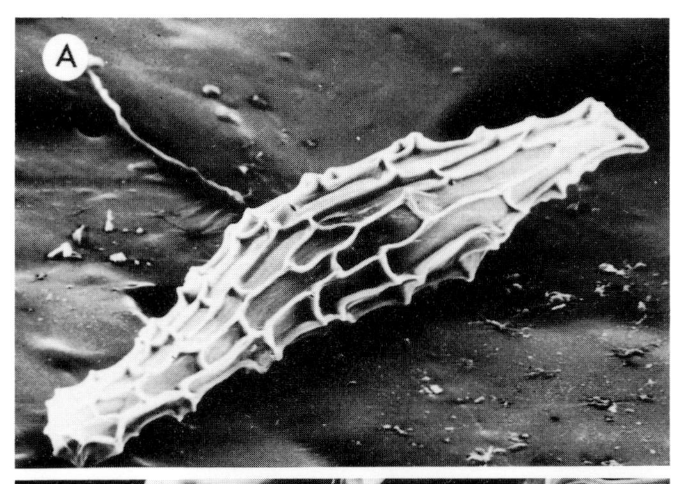

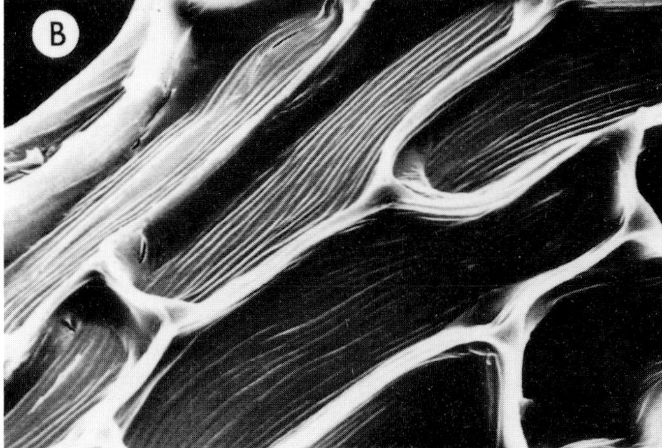

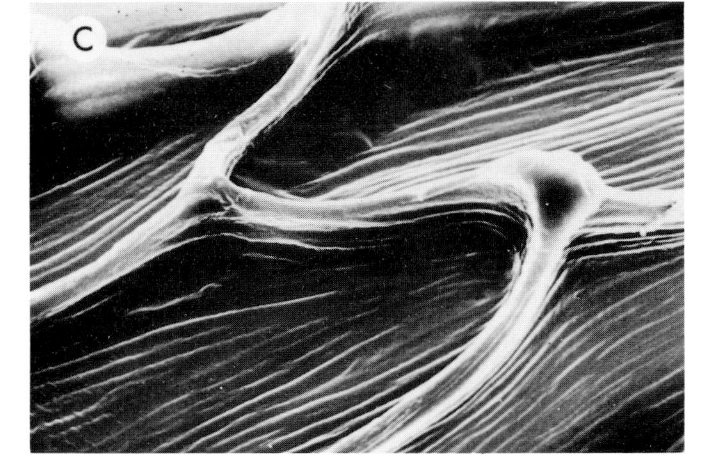

FIGURE 2: Subgenus *Cyperorchis*, section *Iridorchis* – *C. lowianum* (S. 879)

A Whole seed, showing the fusiform shape and network formed by the anticlinal walls of the testa cells (× 75)

B Testa cells, showing the slightly raised junction of the net of cell walls (× 380)

C Testa cells, showing the longitudinal secondary striations of the surface walls (× 765)

Within *Cymbidium* subgenus *Cymbidium* and subgenus *Cyperorchis*, the seeds vary in length, in their comparative breadth and in the degree to which the junctions of the anticlinal walls of the testa cells protrude. Seed length and comparative breadth vary both within and between species, so that only the most general taxonomic conclusions can be drawn. Species in sections *Cymbidium* and *Bigibbarium* usually have the most broadly fusiform seeds. The shortest seeds are found in sections *Cyperorchis* and *Bigibbarium* and in *C. canaliculatum*. Section *Austrocymbidium* shows great variation in seed morphology, in size, relative breadth and in the protrusions at the junctions of the net of cell walls. Seeds of *C. canaliculatum* are easily distinguished by the presence of long, distinct protrusions with a distinct waxy cap, giving the seeds a spiny appearance. This type of seed is also found in the related genus *Ansellia* and, to a lesser extent, in *Grammatophyllum* (figure 4). This is probably an adaptation to facilitate seed dispersal, by increasing the surface area to volume ratio of the seed, making it more buoyant in the air. This may be important in increasing the chance of seed finding a suitable niche in which to germinate, as for example in *C. canaliculatum* which grows in rotten wood in trees in the arid zones of northern Australia. The trees are often well-spaced, and the probability of the seed being deposited on a suitable site might be greater if the buoyancy of the seed were increased. It is unlikely that this type of seed indicates close taxonomic affinity between these taxa.

The narrow, filiform seeds with transverse secondary striations on the walls of the testa cells, and the junctions of the net of cell walls raised only slightly if at all, which are found in the species of subgenus *Jensoa*, are very distinct from those in the other two subgenera of *Cymbidium*. This combination of characters is also found in some species of the genus *Eulophia*, the Asiatic species *E. burkei* and *E. keithii* both having seeds with strikingly similar morphology to those of species in subgenus *Jensoa*.

Other workers (Clifford & Smith, 1969; Arditti *et al.*, 1979, 1980; Barthlott & Zeigler, 1981) have found very little or no variation within genera, but note occasional differences between related genera, and express the opinion that seed morphology should be most useful above the generic level. The similarity between seeds of *Cymbidium* subgenus *Cymbidium* and subgenus *Cyperorchis*, and the related genera such as *Grammatophyllum*, *Ansellia*, and *Porphyroglottis*, would seem to confirm this view. It is therefore highly unusual to find that the seeds of subgenus *Jensoa* are so distinct. Clifford & Smith (1969) found that variation in seed shape, seed length and testa wall secondary thickening were important at the tribal level. The latter character was also shown by Barthlott & Zeigler (1981) to be variable at the tribal level. They also found, however, that the genus *Habenaria* included species with both inflated and thread-like seeds. *Cymbidium* is therefore unusual, but not unique, in this variation. However, the variation in secondary thickening of the surface walls is highly unusual within a genus, and usually occurs at the tribal level.

FIGURE 3: **Subgenus *Jensoa*, section *Maxillarianthe*** – *C. cyperifolium* (S. 1051)

A Whole seed, showing the long, filiform shape (× 45)
B Testa cells, showing the weakly raised junctions of the net of cell walls (× 450)
C Testa cells, showing the transverse secondary striations of the surface walls (× 900)

Section *Jensoa* – *C. ensifolium* (*Streimann and Kairo* in NGF 39369)

D Whole seed (× 50)
E Testa cells (× 900)
F Testa cells (× 1010)

The species in subgenus *Jensoa* all exhibit a strong similarity to each other in seed shape and direction of the surface wall striations.

FIGURE 4: **Related genera**
Ansellia africana (S. 1041)

A Whole seed (× 85)
B Testa cells (× 400)

***Grammatophyllum* sp.** (S. 1018)
C Whole seed (× 65)
D Testa cells (× 650)

Both genera show strong similarity to the subgenera *Cymbidium* and *Cyperorchis*, having fusiform seeds and longitudinal striations of the periclinal walls. *Ansellia* particularly resembles *C. canaliculatum* (section *Austrocymbidium*) with strongly raised, wax-capped junctions of the anticlinal walls. This may create a greater buoyancy in the air, increasing the chances of the seed lodging in a suitable branch. This could be especially important in the more open, arid environments, with a sparse tree cover, in which these species grow.

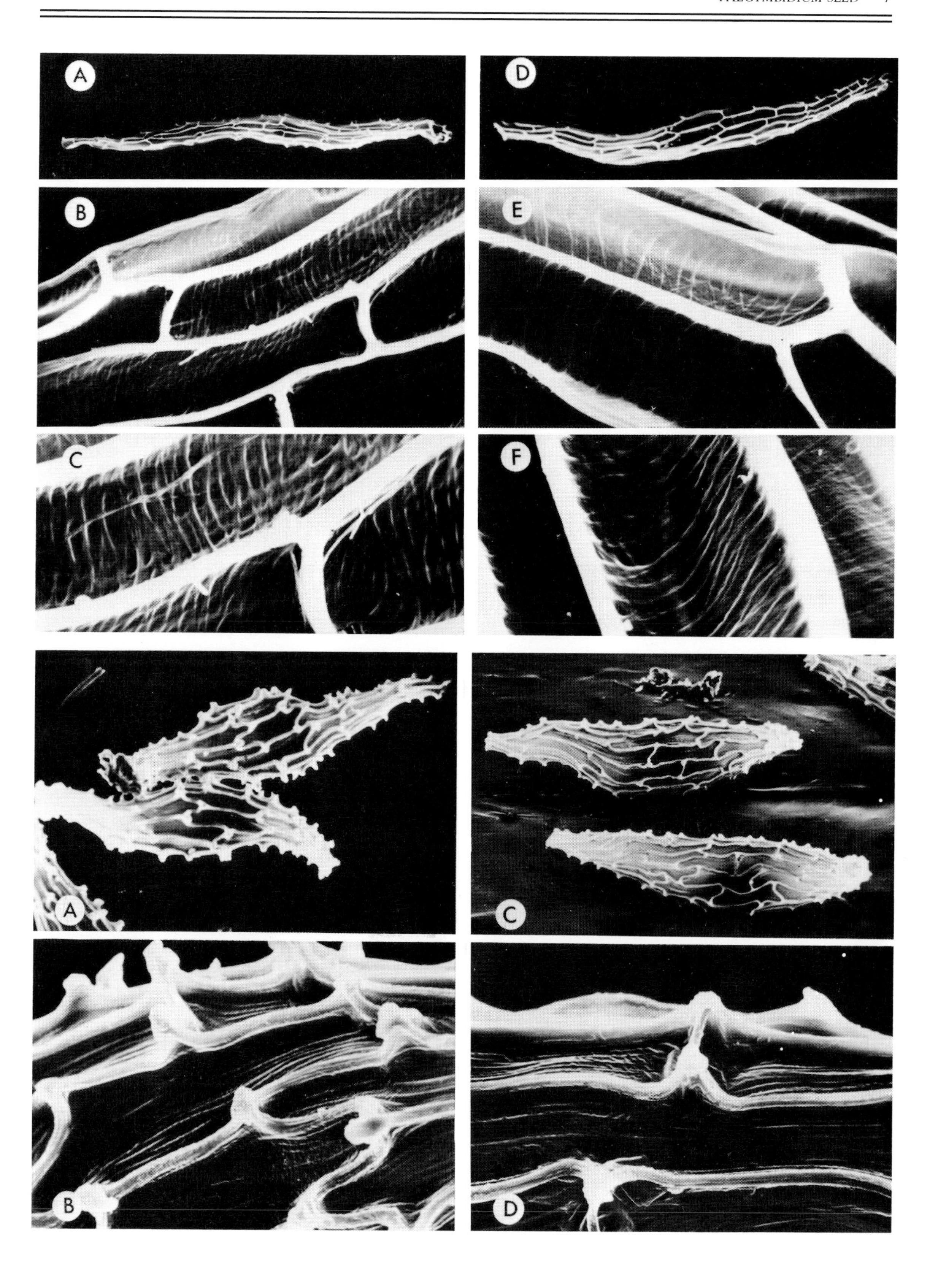
A
D
B
E
C
F
A
C
B
D

The large differences between subgenus *Jensoa* and the rest of *Cymbidium* are reflected in their leaf micromorphology and anatomy (see chapter 3). However, the relationship betweeen the three subgenera cannot be disputed as hybrids have been made, and flowered, between species in subgenus *Jensoa* and hybrids derived from species in subgenus *Cyperorchis*.

Furthermore, the genus most closely related to subgenus *Jensoa*, in its seed morphology, is *Eulophia*, which is usually regarded as being rather distantly related to *Cymbidium* (Dressler, 1981). Within *Eulophia* itself, seed morphology varies, some species being distinct with a clavate shape and a different distribution of cell sizes in the testa (small cells only towards one end of the seed). This type of seed is not found in *Cymbidium*.

It is possible that certain species of *Eulophia*, such as *E. keithii* and *E. burkei*, are more closely related to *Cymbidium* subgenus *Jensoa* than is the rest of *Eulophia*. Macromorphological studies suggest that *Eulophia* should not be split in this way, and it is perhaps more likely that this is a case of parallel evolution of similar seed types in the two genera. It is suggested, however, that further studies of seed morphology in the Cyrtopodiinae could provide valuable data which may elucidate the relationships within this subtribe. The combined data from studies of seed morphology and leaf morphology and anatomy provide evidence that subgenus *Jensoa* is the most distinct, and distantly related, of the three subgenera in *Cymbidium*.

CHAPTER 3

The Anatomy of Cymbidium

The *Cymbidium* leaf comprises an outer cell layer, the epidermis, and an inner, spongy mesophyll, through which run the vascular bundles, or veins.

The bulk of a *Cymbidium* leaf is made up of a mass of loosely packed, mesophyll cells whose chloroplasts provide sugars for the plant through photosynthesis. Running along the leaf, through this mesophyll tissue, are several vascular bundles, which transport water, sugars and micronutrients around the plant. These bundles have caps of thick-walled cells (sclerenchyma) which provide strengthening, preventing damage to the veins. The sugar transport system (the phloem) is completely enclosed in a cylinder of strong sclerenchyma. Further strengthening is provided by a series of fibre bundles (again of thick-walled sclerenchyma cells), which run along the length of the leaf, just below the epidermis. The epidermis itself is coated in a thick, waterproof cuticle, which covers the entire leaf and protects it from desiccation. A thin cross-section of a typical *Cymbidium* leaf is shown in photograph 8. The red dye indicates where the cell walls have been strengthened with lignin, while the blue dye has stained the thin cellulose walls of the other cells.

The stomata are mainly confined to the undersurface of the leaf. They are openings in the epidermis and cuticle, through which gases and water exchange with the surrounding air can be controlled. Each stoma is covered by a hemispherical dome of waterproof cuticle, with a small pore in the top. These can be seen in figure 6, a high magnification photograph of a typical leaf surface.

Variation in the Anatomy and Surface Morphology of *Cymbidium* Leaves

The taxonomically most informative character series concerns the subepidermal fibre bundles, composed of lignified sclerenchyma cells, which are a common feature in this genus and are characteristic of members of the Cymbidiinae (Kaushik, 1983). In most *Cymbidium* species fibre bundles are located in a row below the epidermis, on both sides of the leaf (photograph 8).

The species in section *Cymbidium** are characterised by the presence of a complete layer of sclerenchyma cells directly subjacent to the epidermis, which links together the usually distant strands, and is only broken below the stomata (figure 5A, photograph 7). This has a strengthening function, surrounding the soft mesophyll cells in a lignified protective layer. It may also be a xerophytic adaptation, helping to control water loss from the leaf.

A second modification of the basic distribution of fibre strands is found in section *Maxillarianthe* (subgenus *Jensoa*), where the subepidermal fibre bundles are entirely absent in *C. goeringii* and *C. faberi* (photograph 9), and only occur below the upper surface in *C. cyperifolium*. The absence of fibre bundles below the lower surface is therefore characteristic of this section. The two species *C. cyperifolium* and *C. faberi* are widespread in China and northern India, and can usually be easily distinguished from each other.

*For a list of subgenera and sections see p. 55.

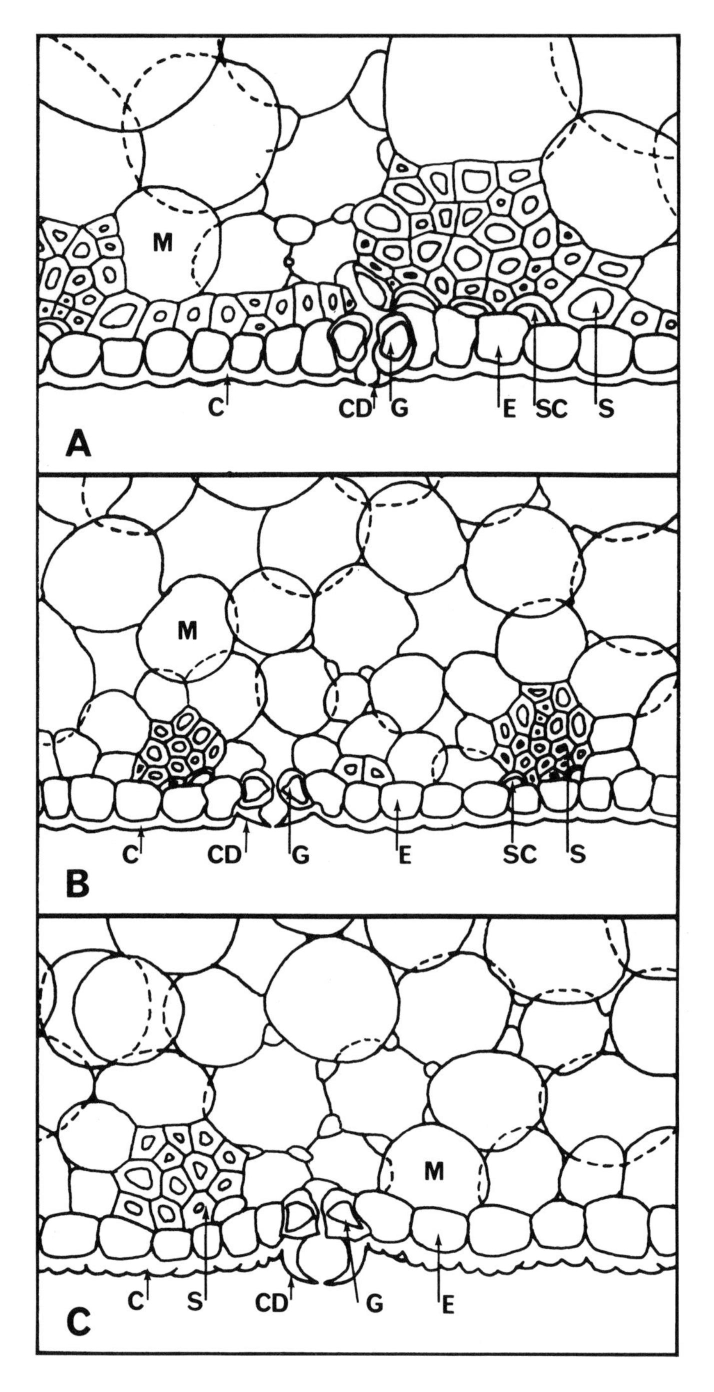

Figure 5: Transverse sections of the abaxial (lower) epidermis of the leaf, including a stoma (× 450)

A subgenus *Cymbidium* (section *Cymbidium*), *C. aloifolium* (Kew no. 180-34.11201). The cuticular dome covering the stoma is not raised above the surrounding cuticle; the epidermal cells are not papillose; there is a subepidermal layer of sclerenchyma fibres linking the fibre bundles and broken only below the stoma.

B subgenus *Cymbidium*, *C. dayanum* (Kew no. 120-83.01291). The cuticular dome is not raised; the epidermal cells are not papillose; the subepidermal fibre bundles are isolated from each other. This arrangement also occurs throughout subgenus *Cyperorchis*

C subgenus *Jensoa*, *C. ensifolium* (Kew no. 040-83.00337). The cuticular dome covering the stoma is raised well above the surrounding cuticle; the epidermal cells are strongly papillose; the subepidermal fibre bundles are isolated from each other.

Key:

C – cuticle
CD – cuticular dome covering the stoma
G – guard cell
E – epidermal cell
S – sclerenchyma cell, usually confined to subepidermal fibre bundles
SC – silica cell
M – mesophyll cell

FIGURE 6: Subgenus *Cymbidium*, section *Bigibbarium* – Scanning electron micrograph of the leaf of *C. devonianum* (Kew no. 102-83.01272)

A Abaxial epidermis (× 75)
B Stomata (× 375)
C Stomatal cover (× 750)

The stomatal covers are circular, not raised above the surrounding epidermis, and the pores are fusiform, both features typical of this subgenus and subgenus *Cyperorchis*. The absence of troughs marking the anticlinal walls of the epidermal cells is unusual and somewhat characteristic.

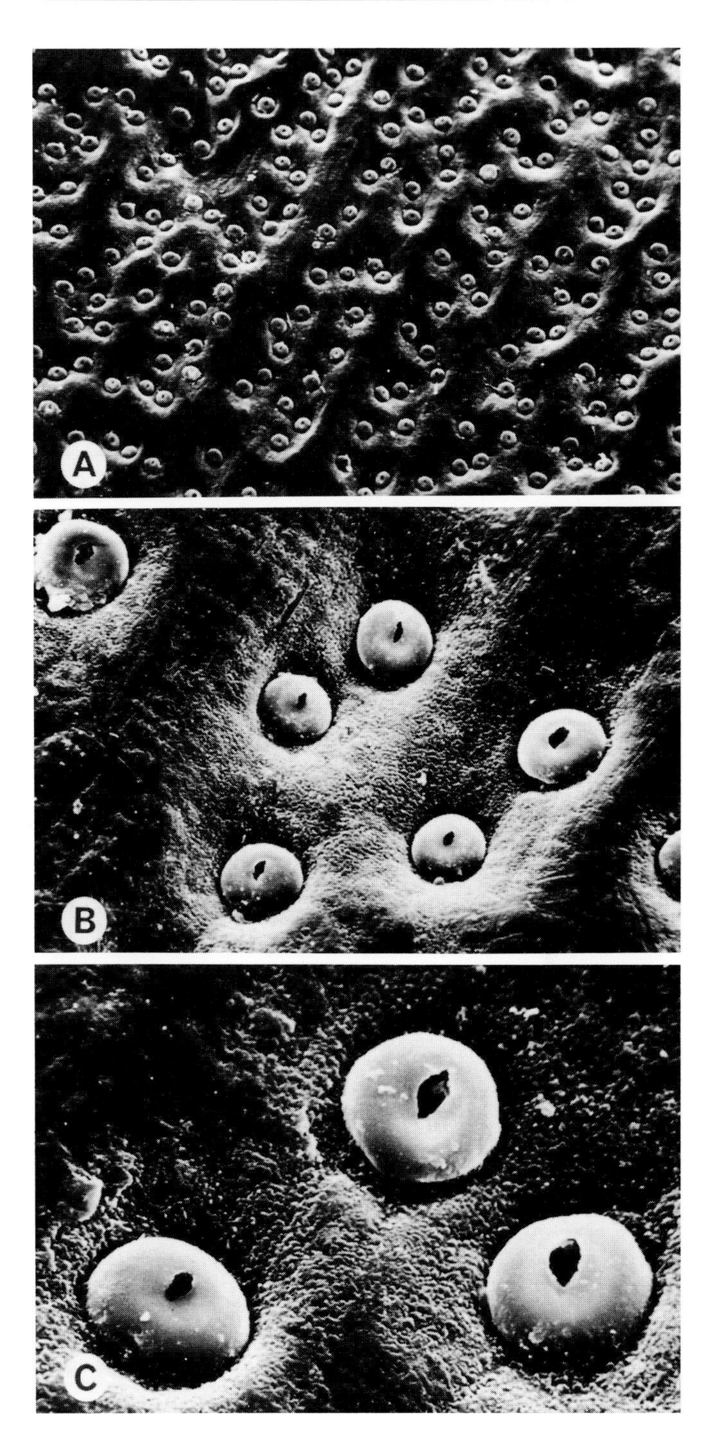

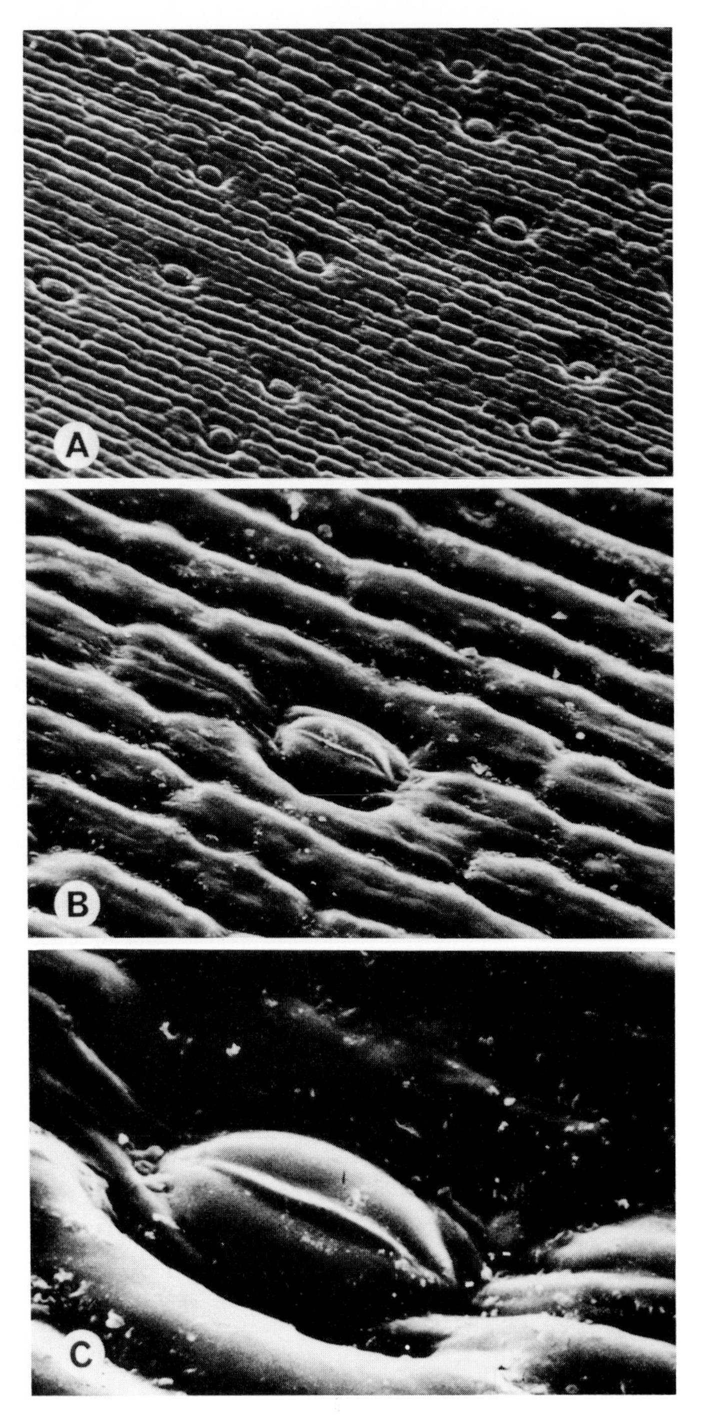

FIGURE 7: Subgenus *Cymbidium*, section *Cymbidium* – Scanning electron micrograph of the leaf of *C. atropurpureum* (Kew no. 417-63.41701)

A Lower epidermis (× 165), showing the scattered stomata
B Single stoma with the surrounding epidermal cells (× 750), showing the general shape of the cuticular stomatal cover and the related epidermal cells
C Single stoma (× 2065), showing the shape of the pore in the stomatal cover

Scanning electron micrographs show that the species in this section have characteristically elliptic stomatal covers, with a slit-shaped pore which extends almost the full length of the cover. The covers are not raised above the level of the surrounding epidermal cells.

1. *Cymbidium aloifolium. Seth* 154, cult. Kew

2. *Cymbidium bicolor* (Right) subsp. *bicolor*, Sri Lanka, *Wilhelm* 133, cult. Kew; (left) subsp. *pubescens*, Borneo, *Giles* 771, cult. Kew

Figure 8: Subgenus *Cymbidium*, section *Borneense* — Scanning electron micrographs of the leaf of *C. borneense* (Kew no. 248-82.02403)

A Abaxial epidermis (× 280)
B Single stoma and surrounding epidermal cells (× 560)
C Stomatal cover (× 1120)

The scattered stomata with elliptical covers which are not raised above the surrounding epidermis and with long, slit-shaped pores are very similar to those of species in section *Cymbidium*, to which this section is therefore closely related.

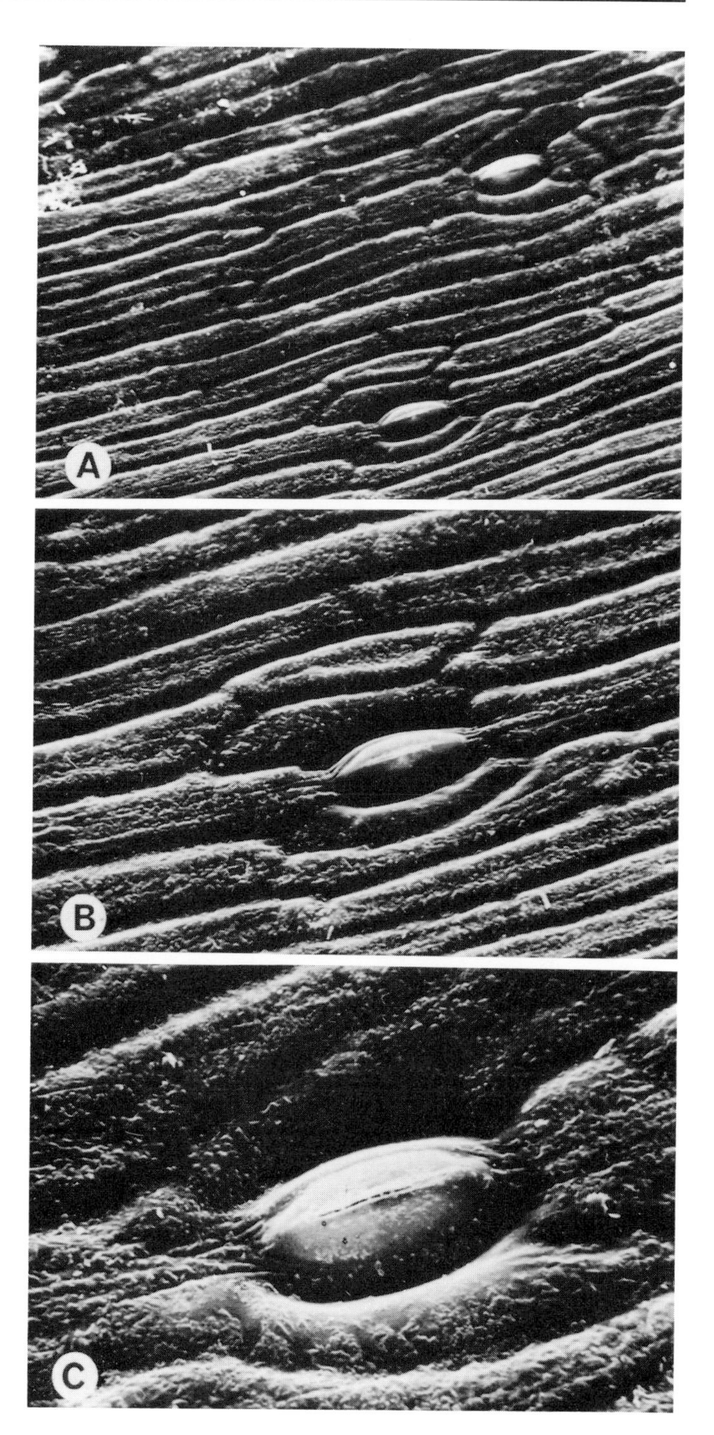

However, certain variants of *C. faberi* from northern India and Nepal closely resemble *C. cyperifolium* from the same region, and they can be morphologically difficult to distinguish. The presence of fibre strands below the upper epidermis in *C. cyperifolium* can be used to distinguish this species from *C. faberi*, in which they are totally absent. In this section the strengthening function of these fibre bundles is taken over by strong sclerenchymatous caps on the vascular bundles.

Section *Cymbidium* is characterised by several layers of elongated mesophyll cells which resemble loosely packed palisade tissue (photograph 7).

Subgenus *Cyperorchis* has a characteristic leaf margin in transverse section. While the species in subgenus *Cymbidium* and subgenus *Jensoa* have obtuse to rounded margins (photograph 11), those in subgenus *Cyperorchis* have slender, acuminate, decurved margins, appearing as narrow, recurved, hyaline margins on living leaves (photograph 10). One exception to this is *C. devonianum* (subgenus *Cymbidium*, section *Bigibbarium*) which also has this characteristic margin type, perhaps suggesting a closer relationship between this species and subgenus *Cyperorchis* than any other species in subgenus *Cymbidium*. However, *C. devonianum* lacks the fusion of the bases of the lip and column which characterises subgenus *Cyperorchis*, and there do not appear to be any other morphological characters which reiterate this connection. It is therefore maintained in subgenus *Cymbidium*.

The scanning electron microscope (S.E.M.) enables a picture of the micromorphology of the leaf surface to be produced, and does not destroy the cuticular cover over the leaf surface. The lower surface was examined because in most species the stomata are restricted to that surface. The only exceptions appear to be the distantly related species *C. canaliculatum* (section *Austrocymbidium*) and *C. borneense* (section *Borneense*).

Several useful characters have been obtained from the morphology of the dome of cuticular material formed by the outer ledges of the guard cells, and the shape of the pore in this dome. In the majority of species in subgenus *Cymbidium* and subgenus *Cyperorchis* these stomatal covers are not raised above the level of the rest of the epidermis, are round in outline, and have a central, spindle-shaped pore (figure 6). The strongly elliptic stomatal covers with narrow, slit-shaped pores found in section *Cymbidium* are highly distinctive (figure 7). This may be another xerophytic adaptation in this section.

The stomatal morphology of *C. borneense* (figure 8) is almost identical to that of the species in section *Cymbidium*, subgenus *Cymbidium*. Although the leaves lack the other anatomical characters of that section (complete subepidermal sclerenchymatous layer and palisade-like mesophyll cells), the similarity of the distinctive stomata suggests that they must be closely related.

The presence of about 6–16 rounded papillae on each of the epidermal cells in subgenus *Jensoa*, giving them a raspberry-like appearance (figure 9), is unique in the genus and highly characteristic.

These anatomical and leaf surface characters are summarised in tables 3.1 and 3.2.

FIGURE 9: Subgenus *Jensoa*, section *Maxillarianthe* – Scanning electron micrographs of the leaves of *C. goeringii* (Kew no. 351-81.03828)

A Abaxial epidermis (× 140)
B Stomata (× 350)
C Stomatal cover (× 700)

section *Geocymbidium*, *C. lancifolium* (Kew no. 102-83.01227)

D Abaxial epidermis (× 125)
E Stomata (× 325)
F Stomatal cover (× 625)

These species show the characteristic features of subgenus *Jensoa*, including the papillose epidermal cells, the raised stomatal covers and the rather circular pores. The epidermal cells of *C. lancifolium* are broader, with correspondingly more numerous papillae, than the other species in this subgenus.

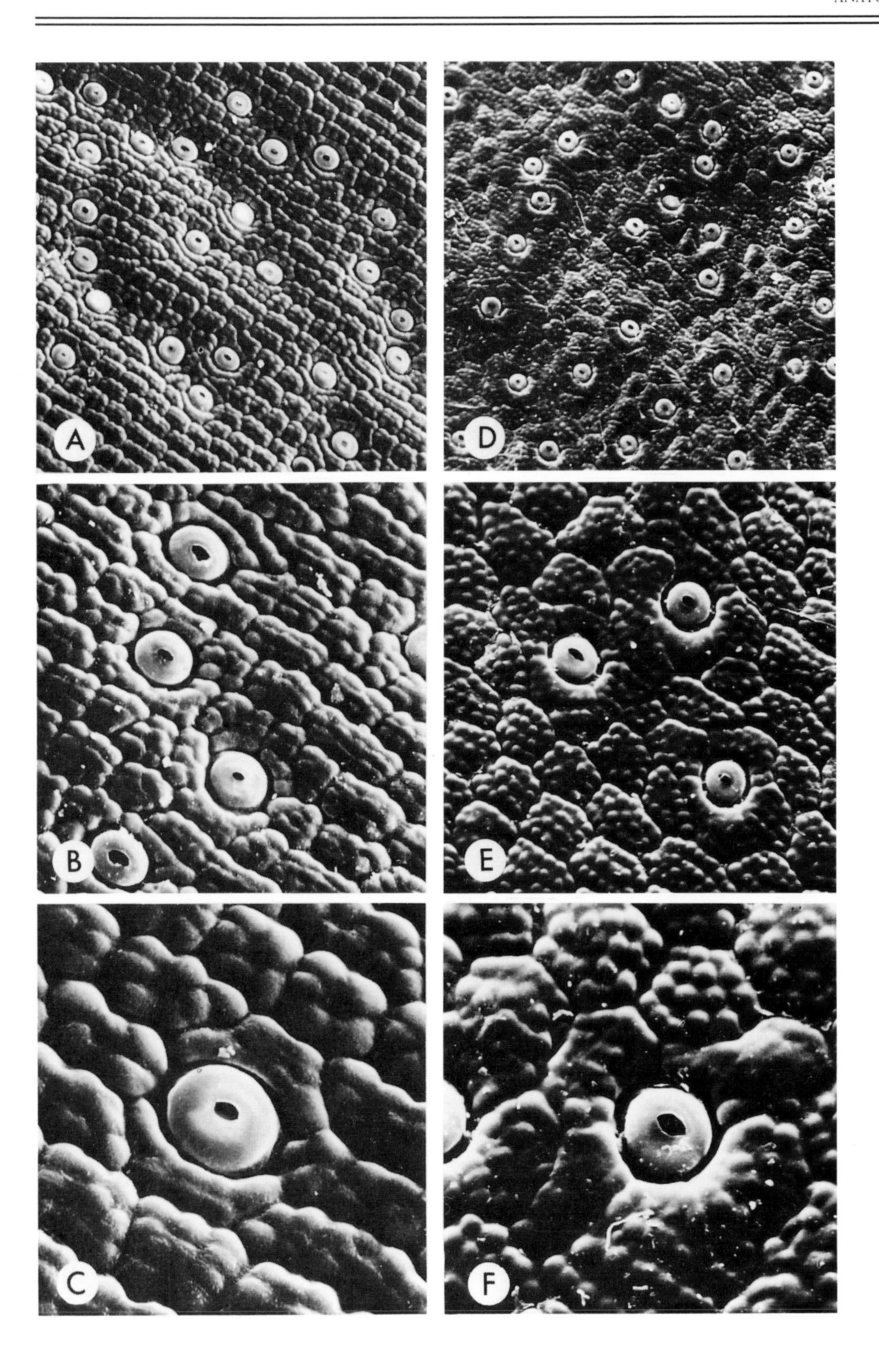
A
D
B
E
C
F

Table 3.1: Taxonomically useful characters of the leaf anatomy: table of results

Section	Species	Kew Accession No.	Presence of the sub-epidermal fibre strands
Subgenus Cymbidium			
Cymbidium	*C. aloifolium*	120-34.11201	+
	C. bicolor	558-65.55825	+
	C. atropurpureum	417-63.41701	+
	C. rectum	214-83.02467	+
Borneense	*C. borneense*	248-82.02403	+
Himantophyllum	*C. dayanum*	324-78.03481	+
	C. dayanum	120-81.01695	+
	C. dayanum	102-83.00863	+
	C. dayanum	151-83.01893	+
	C. dayanum	102-83.01291	+
Austrocymbidium	*C. chloranthum*	214-83.02460	+
	C. canaliculatum	040-83.00318	+
	C. canaliculatum	481-81.06520	+
	C. madidum	189-78.01942	+
	C. elongatum	229-83.02923	+
Floribundum	*C. floribundum*	340-82.03508	+
	C. floribundum	120-81.01688	+
	C. floribundum	120-81.01664	+
	C. suavissimum	120-81.01681	+
Bigibbarium	*C. devonianum*	102-83.01272	+
	C. devonianum	120-81.01670	+
Subgenus Cyperorchis			
Iridorchis	*C. tracyanum*	120-81.01676	+
	C. longifolium	120-81.01684	+
	C. hookerianum	323-81.03514	+
	C. lowianum	269-80.02506	+
	C. insigne	120-81.01683	+
	C. sanderae	120-81.01678	+
Eburnea	*C. eburnea*	120-81.01673	+
	C. mastersii	102-83.01182	+
	C. mastersii	120-81.01662	+
Annamaea	*C. erythrostylum*	430-81.04913	+
Cyperorchis	*C. elegans*	120-81.01685	+
	C. whiteae	551-82.05677	+
Parishiella	*C. tigrinum*	323-81.03531	+
Subgenus Jensoa			
Jensoa	*C. ensifolium*	102-83.01053	+
	C. ensifolium	481-81.06516	+
	C. ensifolium	474-82.05087	+
	C. ensifolium	040-83.00324	+
	C. ensifolium	040-83.00336	+
	C. ensifolium	040-83.00338	+
	C. ensifolium	040-83.00337	+
	C. ensifolium	102-83.01111	+
	C. sinense	325-81.03538	+
	C. sinense	077-83.00594	+
	C. sinense	424-82.05088	+
Maxillarianthe	*C. cyperifolium*	439-84.04707	Below adaxial surface only
	C. faberi	439-84.04748	–
	C. faberi	077-83.00593	–
	C. goeringii	351-81.03828	–
	C. goeringii	077-83.00592	–
Geocymbidium	*C. lancifolium*	340-82.03514	+

Note: + = present.

Complete layer of sub-epidermal sclerenchyma cells	Palisade-like cells in mesophyll	Leaf margin acuminate in T.S.
+	+	−
+	+	−
+	+	−
+	+	−
−	−	−
−	−	−
−	−	−
−	−	−
−	−	−
−	−	−
−	+	−
−	−	−
−	−	−
−	−	−
−	−	−
−	−	−
−	−	−
−	−	−
−	−	−
−	−	+
−	−	+
−	−	+
−	−	+
−	−	+
−	−	+
−	−	+
−	−	+
−	−	+
−	−	+
−	−	+
−	−	+
−	−	+
−	−	+
−	−	+
−	−	−
−	−	−
−	−	−
−	−	−
−	−	−
−	−	−
−	−	−
−	−	−
−	−	−
−	−	−
−	−	−
−	−	−
−	−	−
−	−	−
−	−	−
−	−	−
−	−	−

Table 3.2: Taxonomically useful characters of epidermal and stomatal morphology: table of results

Section	Species	Kew Accession No.	Stomatal cover shape
Subgenus Cymbidium			
Cymbidium	*C. bicolor*	032.75.00438	elliptical
	C. rectum	214-83.02467	elliptical
	C. atropurpureum	417-63.41701	elliptical
Borneense	*C. borneense*	214-83.02513	elliptical
	C. borneense	248-82.02403	elliptical
Himantophyllum	*C. dayanum*	102-83.01291	circular
	C. dayanum	151-83.01893	circular
Austrocymbidium	*C. chloranthum*	731-60.73101	circular
	C. chloranthum	214-83.02460	circular
	C. canaliculatum	040-83.00318	circular
	C. madidum	336-82.03312	circular
	C. suave	340-82.03504	circular
Floribundum	*C. floribundum*	120-81.01688	circular
	C. suavissimum	120-81.01681	circular
Bigibbarium	*C. devonianum*	102-83.01272	circular
Subgenus Cyperorchis			
Iridorchis	*C. lowianum*	120-81.01685	circular
	C. sanderae	120-81.01678	circular
	C. insigne	102-83.01280	circular
Eburnea	*C. eburneum*	120-81.01673	circular
	C. mastersii	102-83.01182	circular
Annamaea	*C. erythrostylum*	040-83.00316	circular
	C. erythrostylum	551-82.05676	circular
Cyperorchis	*C. elegans*	120-81.01685	circular
	C. cochleare	167-84.01164	circular
	C. whiteae	551-82.05677	circular
Parishiella	*C. tigrinum*	323-81.03513	circular
Subgenus Jensoa			
Jensoa	*C. ensifolium*	214-83.02491	circular
	C. ensifolium	474-82.05087	circular
	C. ensifolium	340-82.03501	circular
	C. ensifolium	102-83.01053	circular
	C. kanran	429-83.05475	circular
	C. sinense	325-81.03538	circular
	C. sinense	474-82.05088	circular
Maxillarianthe	*C. faberi*	077-83.00593	circular
	C. goeringii	195-79.01884	circular
	C. goeringii	351-81.03828	circular
	C. goeringii	077-83.00592	circular
Geocymbidium	*C. lancifolium*	340-82.03514	circular
	C. lancifolium	102-83.01277	circular
Other genera			
Grammatophyllum speciosum		032-75.00440	circular
Ansellia africana		052-77.00323	circular
Cymbidiella humblottii		340-82.03517	circular
Eulophia keithii		400-65.40007	circular
Eulophia stenophylla		040-81.01284	circular

Stomatal cover raised above the surrounding epidermis or not	Pore shape	Epidermal cell surface
level	slit-shaped	smooth
level	slit-shaped	smooth
level	slit-shaped	smooth
level	slit-shaped	smooth
level	slit-shaped	smooth
level	fusiform	smooth
level	fusiform	smooth
level	fusiform	smooth
level	fusiform	smooth
level	fusiform	smooth
level	fusiform	smooth
level	fusiform	smooth
level	fusiform	smooth
level	fusiform	smooth
level	fusiform	smooth
level	fusiform	smooth
level	fusiform	smooth
level	fusiform	smooth
level	fusiform	smooth
level	fusiform	smooth
level	fusiform	smooth
level	fusiform	smooth
level	fusiform	smooth
level	fusiform	smooth
level	fusiform	smooth
level	fusiform	smooth
raised	circular	papillose
raised	circular	papillose
raised	circular	papillose
raised	circular	papillose
raised	circular	papillose
raised	circular	papillose
raised	circular	papillose
raised	circular	papillose
raised	circular	papillose
raised	circular	papillose
raised	circular	papillose
raised	circular	papillose
raised	circular	papillose
level	slit-shaped	smooth
level	fusiform	smooth
level	slit-shaped	papillose
level	fusiform	smooth
level	fusiform	smooth

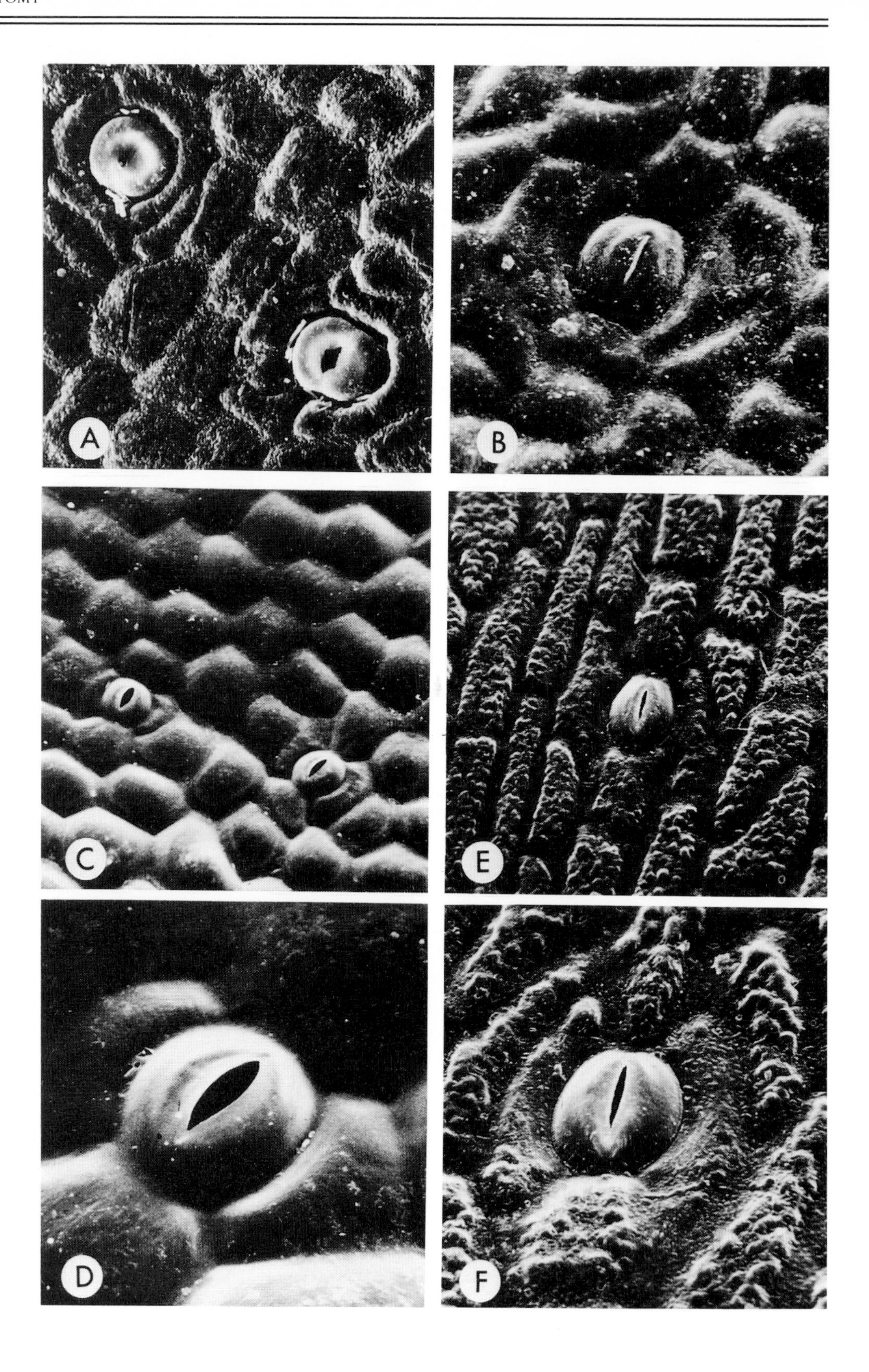
A
B
C
E
D
F

Figure 10: Related genera. Scanning electron micrographs of the abaxial leaf surfaces of:

Ansellia africana (Kew no. 052-77.00323)
A Stomata (× 325)

Grammatophyllum speciosum (Kew no. 032-75.00440)
B Stomata (× 825)

Eulophia stenophylla (Kew no. 040-81.01284)
C Epidermis with stomata (× 230)
D Stomatal cover (× 800)

Cymbidiella humblottii (Kew no. 340-82.03517)
E Epidermis (× 420)
F Stomatal cover (× 850)

These related species all differ from the *Cymbidium* species examined in their epidermal morphology. The most similar are the leaves of *Ansellia* and species in subgenera *Cymbidium* and *Cyperorchis*, and *Grammatophyllum* resembles section *Cymbidium* with its slit-shaped pores in the stomatal covers. A more detailed survey may help to indicate the relationships between the genera in the subtribe Cyrtopodiinae.

Related Genera

The related genera have a very variable epidermal and stomatal morphology (figure 10; table 3.2). Of the more closely related genera (Dressler, 1981), *Ansellia* appears to be the most similar to *Cymbidium* species in the subgenera *Cymbidium* and *Cyperorchis*. The epidermal cells are smooth, and the stomatal covers circular with a central fusiform pore (figure 10A), suggesting a close relationship between these taxa. *Grammatophyllum* has a slit-shaped pore somewhat similar to section *Cymbidium*, but the cover is circular, not elongated, and the arrangement of epidermal cells differs (figure 10B). *Cymbidiella* has papillose epidermal cells similar to those characteristic of subgenus *Jensoa*, but the stomatal cover is not raised, is somewhat elongated, and has a slit-shaped rather than a circular pore, suggesting that these two taxa are not closely related, and that the papillae on the epidermal cells have evolved on two separate occasions (figure 10E,F). *Eulophia* is very distinct (figure 10C,D) and the epidermal morphology does not echo the affinity to subgenus *Jensoa* suggested by seed micromorphology. Characters of the epidermis and stomata, on the above evidence, would appear to be useful in the elucidation of the relationships between genera in the Cyrtopodiinae.

CHAPTER 4

The Cytology of Cymbidium

Chromosome Numbers

The diploid chromosome number in all species of *Cymbidium* is 40 (Wimber, 1957a; Tanaka & Kamemoto, 1974; Leonhardt, 1979; Du Puy, 1986). The only exceptions are a few named triploid or tetraploid cultivars of some species; *C. floribundum* 'Geshohen' (2n = 80) in Tanaka & Kamemoto (1974), *C. insigne* 'Bieri' (2n = 60) and *C. floribundum* 'Yoshina' (2n = 60) in Leonhardt (1979). There have been occasional counts other than 2n = 40, but these are normally contradicted by other counts of the same species, and are either miscounts or variants of little importance to the taxonomy of the genus.

Meiosis and Sporad Development

Meiosis and pollen formation in the genus *Cymbidium* has been studied by Wimber (1957b, 1957c). Several species from subgenus *Cyperorchis*, and *C. floribundum* (subgenus *Cymbidium*), were investigated. The normal cycle of meiotic and mitotic divisions produces a tetrad which does not separate into individual pollen grains even when mature. This was found in most cases, but a small percentage of abnormalities in individual sporads was noted in many of the species. These abnormalities included the formation of univalents and fragments which later formed micronuclei, creating polyads instead of tetrads. Bridges were also noted at telophase 1. These abnormalities were common in two species only. In *C. lowianum* 'Concolor' the higher percentage of abnormal sporad development was explained by Wimber as resulting from this cultivar being an extreme variant of the species. In *C. insigne* the high percentage of abnormality could only be a result of some nonhomologies inherent in the genomes of the specimens examined, perhaps due to some past hybridisation of *C. insigne* (Wimber, 1957b). It was also shown that the percentages of abnormal sporad development varied within a single plant, probably due to minor environmental variations.

When primary hybrids were examined, abnormal meiosis and sporad develoment was found to be more common. The percentage varied, but as all crosses were between species within the subgenus *Cyperorchis*, the parental species were closely related, and the genomes although divergent to some extent were still close enough to allow production of some normal, fertile tetrads. In general, the irregularities were not common enough to reduce noticeably the fertility of the pollinium. For this reason there has been little problem in interbreeding to form the modern complex hybrids (Wimber, 1957c).

Several *C. floribundum* hybrids were shown to have a meiotic cycle which was almost entirely disrupted. These hybrids are almost completely sterile and very few crosses with them have been successful. This is explained by the large difference between the genome of *C. floribundum* and the genomes of the hybrids involving the species in subgenus *Cyperorchis* which were used as the second parent (Wimber, 1957c).

In recent studies by Du Puy (1986), very occasional disruption of the meiotic cycle was noted in *C. lowianum* and *C. tracyanum*. This showed as the presence of univalents or fragments at telophase 1.

The meiotic cycle of C. Pumilow (*C. floribundum* × *C. lowianum*) was very badly disrupted. Many abnormalities of the reduction division cycle and sporad development were observed. Univalents, fragments, bridges and unequal divisions were common. This disruption resulted in the production of many polyads instead of tetrads. There were also some normal looking tetrads produced but in very reduced numbers, resulting in high sterility. Examination of primary hybrids of species in different sections may indicate the degree of divergence between the taxa involved.

CHAPTER 5

Pollination and Floral Fragrances in Cymbidium

Pollination

Few observations of the natural pollination of *Cymbidium* species have been made. Two reports on the pollination of *C. madidum* suggest that the native Australian honey bee (*Trigona* sp.) is the pollinating insect. Macpherson & Rupp (1935) studied a population of *C. madidum* near Prosperine, Queensland, in September 1934 and suggested that the bees were initially attracted by the sweet perfume of the flowers, and later appeared to collect the viscid secretion from the disc of the lip. The bees were observed to gnaw at this exudation, which was transferred from the front legs to the second pair of legs, and finally it was fastened to the pollen sacs on the rear legs. It was suggested that this substance is used as 'bee glue', to repair small cracks and gaps in the nest. The lip of the flower was observed to be 'irritable', in that it closed up against the column in 'several jerky movements' as the bee moved into the flower, imprisoning the bee and bringing its back into contact with the sticky stigmatic surface, on which pollinia were deposited. As the bee struggled backwards out of the flower, the anther was ruptured and new pollinia were deposited on the hairs on the back of the bee. The bee was later (1936) identified as *Trigona kockingsi.* Smythe (1970) reported that another species, *Trigona caponaria,* was collected carrying pollinia of *C. madidum.*

A. Lamb (pers. comm.) has observed the pollination of *C. finlaysonianum* by an unidentified bee species in Sabah.

On a recent expedition to northern Thailand during April 1983, one of us (DD) was able to conduct field observations on the pollination of *C. insigne.* The expedition also included two Danish botanists from the University of Copenhagen; Dr Finn N. Rasmussen and G. Kjellsson, with whom these observations have been presented for publication (Kjellsson *et al.*, 1985).

During this stay, in Phu Luang Wildlife Sanctuary, it was observed that *Cymbidium insigne* Rolfe and *Dendrobium infundibulum* Lindley were pollinated by a bumble bee, *Bombus eximius* Smith, which is a regular pollinator of *Rhododendron lyi* Leveille in the area. Observations suggested that flowers of the two orchid species mimic flowers of *R. lyi*, and that these three species constitute a three-way floral mimicry system operated by *B. eximius.*

The flowers of the three species appear quite similar in shape and size (photographs 14, 15). *R. lyi* has large, white flowers with a pale yellow patch at the base of one petal. The petals are fused basally into a slightly zygomorphic, conical corolla about 6 cm (2.4 in) across. The flowers of *D. infundibulum* are white with a light orange patch at the base of the lip. *C. insigne* occurs in two colour variants in the area. Both have creamy-white flowers with a pale yellow patch on the mid-lobe of the lip. However, one form has red stripes on the lip while the other is unstriped (photographs 12, 13). The unstriped form has so far only been recorded from this locality. Flowers of both colour forms turn dull red when pollinated or when the pollinarium has been removed.

D. infundibulum flowers from the beginning of March until about the middle of April. *C. insigne* usually flowers from early February until the middle of April. *R. lyi* flowers mainly in the months of March and April. The flowering season of *R. lyi* thus coincides to a great extent with that of the two orchids. Observations indicate that flowers of the two orchid species can last for about a month if the pollinia are not removed. *B. eximius* also shows seasonality. The worker females are found only during early spring (Frison, 1934). The condition of the specimens caught suggests that they had been flying for some time (O. Lomholdt & P. Williams, pers. comm.). Both orchid species were found growing in the immediate vicinity of *Rhododendron* scrub. *D. infundibulum* usually grows on rocks and boulders, and occasionally epiphytically. *C. insigne* is always terrestrial, and appears to require at least a thin layer of soil. Often the flowering spikes are seen protruding through the foliage of low-growing bushes.

The bumblebee *Bombus eximius* was found to be the main pollinator of the two orchids (photographs 16, 17). Most of the bees were worker females and, less frequently, queen bees. A second species, *B. rotundiceps* Friese, was seen only a few times carrying *D. infundibulum* pollinaria. Some other insects (*Anthophora himalayensis* Rad. and *Apis dorsata* Fabr. (bees), *Clinteria* sp. (a Cetoninae beetle) and two species of hawkmoth) sometimes visited the flowers, but were never seen carrying pollinaria.

The flowers were visited most regularly during the morning between 5 and 10 am, except for the period from 6 to 7 am, when visits to flowers nearly doubled. It may be that the early foraging bees are most likely to make mistakes in their search for a food source. In the hot hours, from about 10 am to 3 pm, bumblebee activity was much reduced, but it increased somewhat in the late afternoon. However, afternoon visits to orchid flowers were only observed twice, to *C. insigne* in both cases. The visits that resulted in removal and/or deposition of pollinaria lasted 4 to 12 seconds in *C. insigne*, but only 2 to 3 seconds in *D. infundibulum*.

Examination of the flowers of *D. infundibulum* and *C. insigne* revealed neither nectar, nor any other reward for the pollinators, and no odour could be perceived.

It was possible for both orchid species to use the same bee as pollinator because they utilised different parts of the bee (figure 11, photograph 17). *C. insigne* has a long column, and its pollinia were caught on the back of the main body of the bee (figure 11B,D), whereas *D. infundibulum* had a shorter column, and placed its pollinia nearer to the head of the bee (figure 11A,C), precisely on a small bald, shiny patch on the bee's otherwise hairy back. The two orchids in this way ensured that neither collected the wrong pollen, their respective columns being the correct length to collect the appropriate pollinia when the bee entered the next flower.

This specialised morphology and the precise positioning of the pollinaria on the bees indicate a high degree of adaptation to the pollinator. The spatial separation of the different pollinaria on the bee ensures that there is no interference between the reproductive systems of the two orchids.

About 6 per cent of all individuals of *B. eximius* seen were carrying pollinaria of *D. infundibulum*, while only a few carried pollinaria or remnants of pollinaria from *C. insigne*. However, the frequency of flower visits was much lower in *D. infundibulum* than in *C. insigne*. A possible explanation could be that the total number of flowers of *D. infundibulum* was much higher than that of *C. insigne*, so that a high number of *D. infundibulum* pollinaria were accumulated. Furthermore, the *Dendrobium* pollinaria are perhaps affixed more firmly to the bee than the *Cymbidium* pollinaria. The flowering season of these orchid species is extended over a period of nearly two months, thereby increasing the chances of pollination.

The observations and the experiments indicate pollination by floral deceit, which has previously been reported in several orchids (Ackerman, 1983a; Bierzychudek, 1981; Boyden, 1980; Dafni, 1983; Dafni & Ivri, 1981a, 1981b; Nilsson, 1980, 1983). The flowers are similar enough in colour, size and shape to trick the bees into visiting the orchids, even though they offer no food reward, by mimicking the *Rhododendron* flowers with their abundant nectar and pollen. The similarity between the *Rhododendron* and *C. insigne* appears to be increased in a large number of individuals in this population, by the loss of the red markings on the lip.

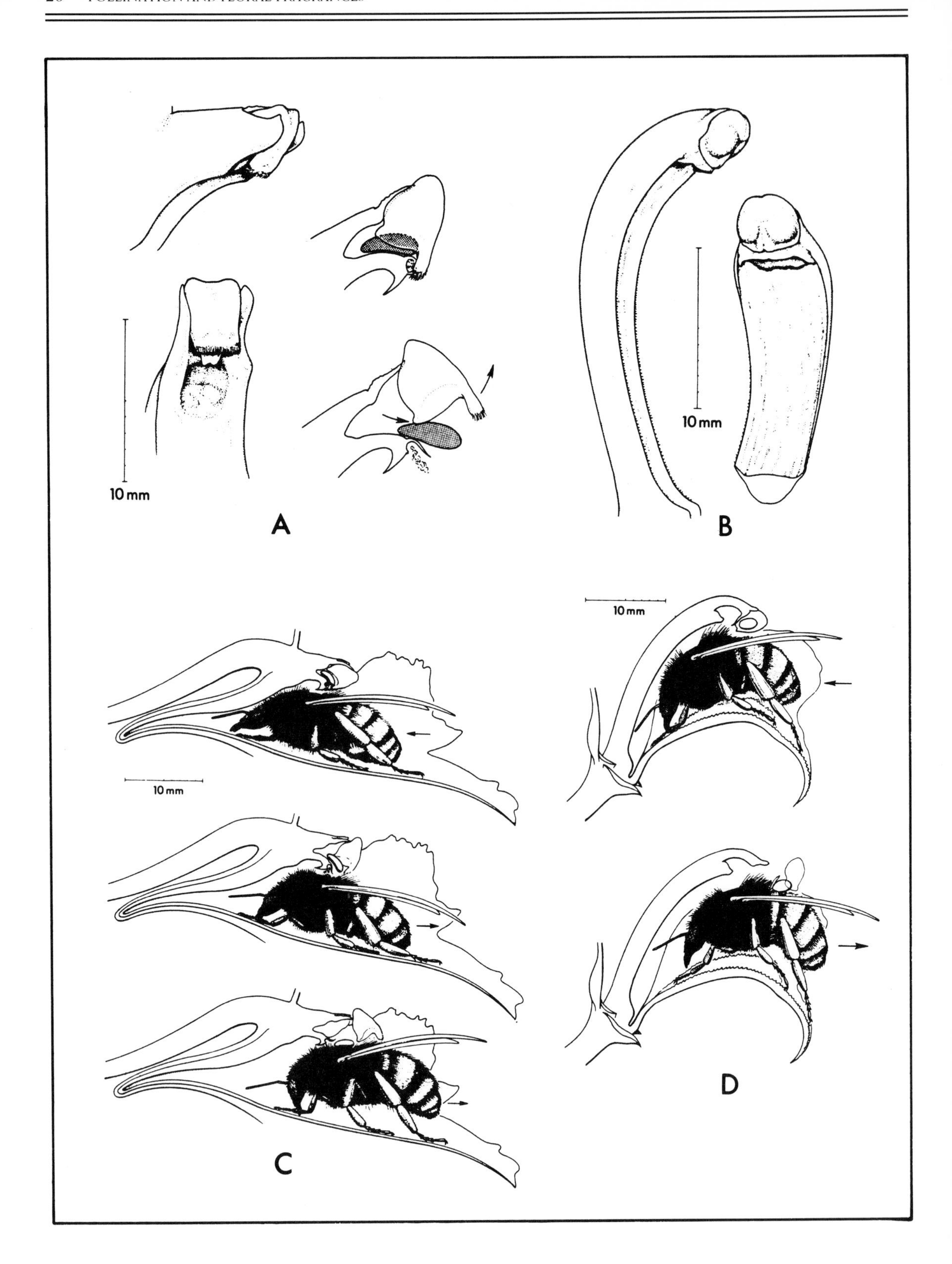
10 mm
A
10 mm
B
10 mm
10 mm
C
D

Figure 11:

A – The column of *Dendrobium infundibulum*
B – The column of *Cymbidium insigne*
C – Pollination of *D. infundibulum*
D – Pollination of *C. insigne*

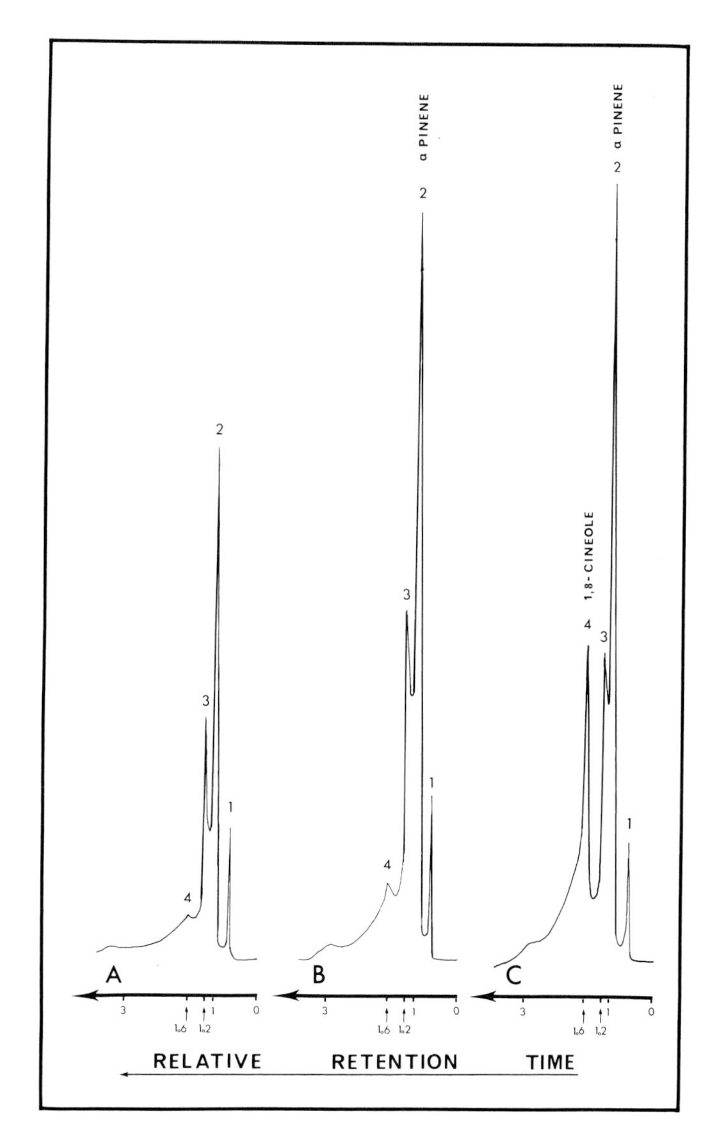

Figure 12: Analysis of the fragrance of *C. tracyanum*

A Pure sample of *C. tracyanum*
B *C. tracyanum* enriched with alpha-pinene
C *C. tracyanum* enriched with alpha-pinene and 1,8-cineole

Floral Fragrances

Analyses of floral fragrances have provided important biological and taxonomic evidence in the study of several genera of tropical American orchids from the subtribes Catasetinae and Stanhopeinae (Dodson & Hills, 1966; Hills, Williams & Dodson, 1968, 1972; Dodson *et al.*, 1969; Ackerman, 1983b) and also in several European species (Nilsson 1978a, 1978b, 1979a, 1979b, 1981). In some of these cases the composition of floral fragrances has been shown to be species specific and has also been demonstrated to attract a particular pollinating insect.

The fragrant species of *Cymbidium* include the perfumed *C. eburneum* and the almond-scented *C. mastersii* (section *Eburnea*), the coconut-scented *C. atropurpureum* and *C. borneense* (subgenus *Cymbidium*), the sweet-scented *C. tracyanum*, *C. iridioides* and *C. hookerianum* (section *Iridorchis*) and the sweetly fruit-scented *C. sinense*, *C. ensifolium*, *C. kanran*, *C. faberi* and *C. cyperfolium* (Subgenus *Jensoa*). Others, including *C. madidum* and *C. suave* (section *Austrocymbidium*), are less strongly fragrant.

Two of these species have been analysed by Du Puy (1986) for their floral fragrances. The fragrance of *C. tracyanum*, when analysed in a gas chromatograph, showed four peaks of which peak 2 was identified, by enrichment with pure chemicals, as alpha-pinene, and peak 4 as 1,8-cineole (figure 12). Only two major peaks were observed in the analysis of *C. hookerianum*, one of which was identified as alpha-pinene (figure 13).

These preliminary results indicate that floral fragrance analysis could provide useful taxonomic evidence when more species are surveyed.

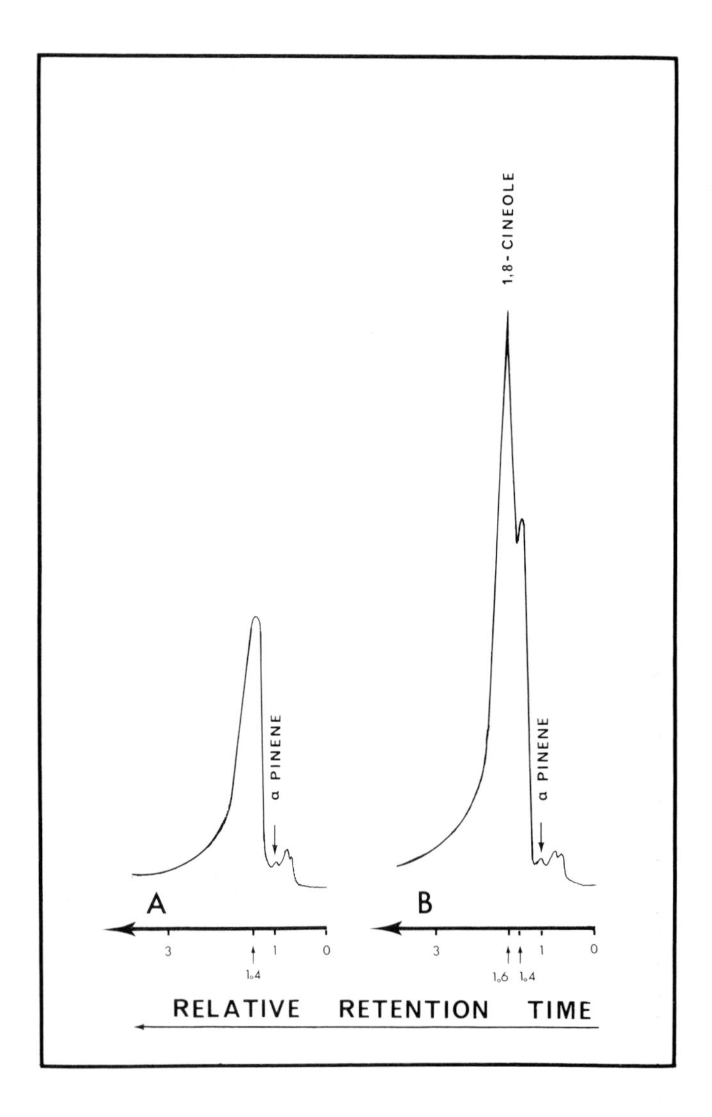

FIGURE 13: Analysis of the fragrance of *C. hookerianum*

A Pure sample of *C. hookerianum*

B *C. hookerianum* enriched with 1,8-cineole

. *Cymbidium rectum.* Sabah, *Lamb* K12, cult. Kew

4. *Cymbidium finlaysonianum*. Sarawak, *Giles* 600, cult. Kew

5. *Cymbidium atropurpureum.* Sarawak, *Giles & Woolliams* s.n., cult. Kew

6. *Cymbidium borneense*. Sabah, *Lamb* K7, cult. Kew

CHAPTER 6

Distribution and Biogeography of Cymbidium

Cymbidium Sw. is widely distributed in S E Asia, from N W India to Japan, and south to Australia (map 1A). The greatest concentration of species occurs in N E India, S W China, Indo-China and Malesia (table 6.1). Individual species' distributions are illustrated in maps 3–11, and are discussed in the accounts of the species (see chapter 10). The distributions of the three subgenera are also illustrated (maps 1B, 2A, 2B).

Centres of Diversity

The three subgenera have distinct centres of diversity, and table 6.1 lists the number of species which occur in eleven geographical regions. Subgenus *Cymbidium* has a centre of diversity in W Malaysia (with seven species) and the Malay Islands (ten species). Subgenus *Cyperorchis* has its greatest diversity of species in the Himalaya of N E India and Nepal (nine species) and in S W China (nine species), with southerly extensions into Indo-China (nine species). Subgenus *Jensoa* is best represented in S W and S E China (six and eight species), and in N E India (seven species). Taiwan contains a diversity of species, and some unusual variants have been recorded there (see *C. goeringii* and *C. ensifolium*), suggesting that subgenus *Jensoa* may be rapidly evolving in that region at present.

Table 6.1: Numbers of species in the subgenera of *Cymbidium* Sw. which occur in various regions of the total distribution

Regions	Subgenus *Cymbidium*	Subgenus *Cyperorchis*	Subgenus *Jensoa*	Total
Japan	1	0	4	5
S E China	3	1	6	10
S W China	5	9	8	22
N India	3	9	7	19
S India	2	0	1	3
Indo-China	4	9	5	18
W Malaysia	7	1	2	10
Malay Islands	10	2	2	14
New Guinea	0	0	2	2
Philippines	4	0	2	6
Australia	3	0	0	3

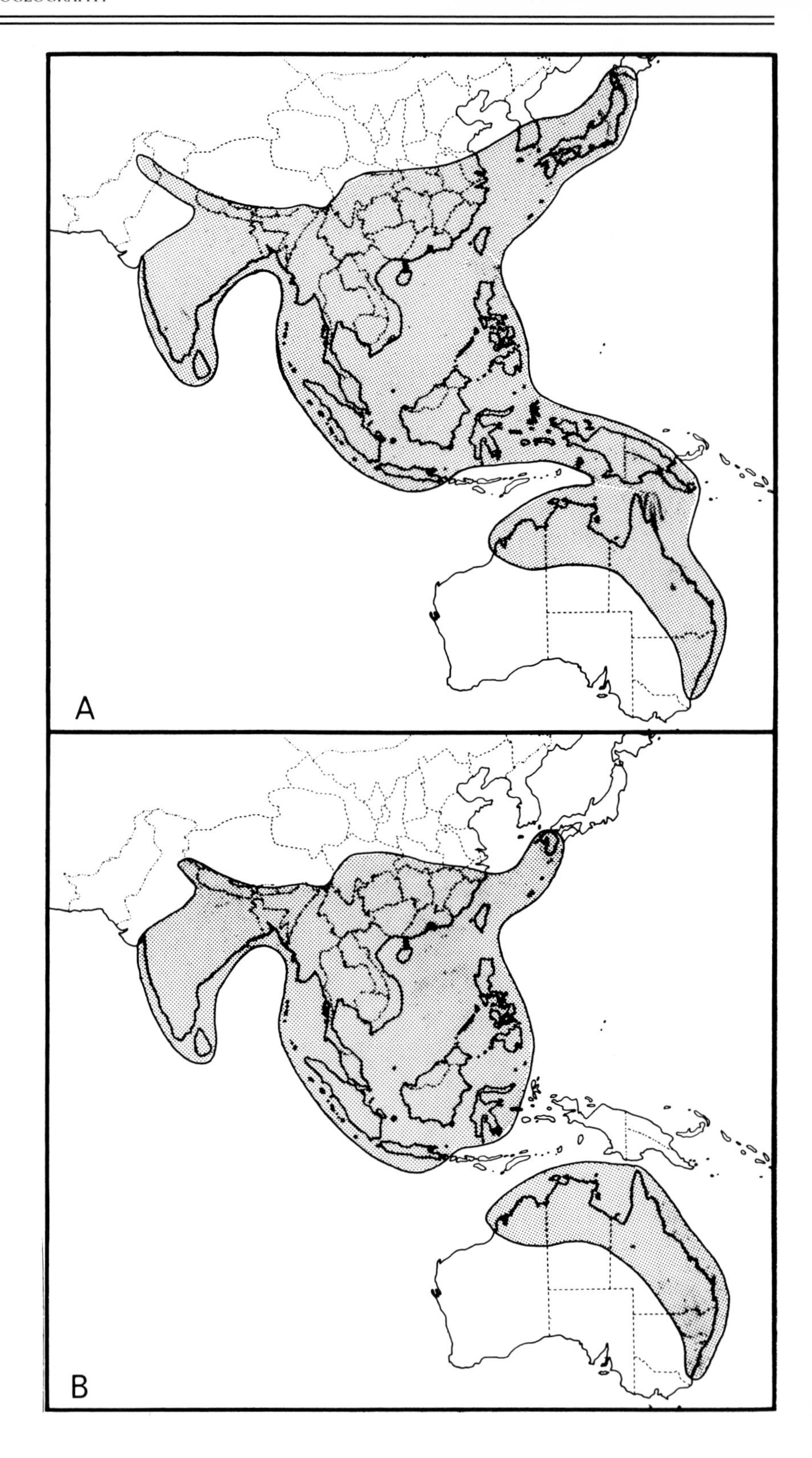

Map 1

A The distribution of the genus *Cymbidium* in E and S E Asia, Malesia and Australia

B The distribution of subgenus *Cymbidium*

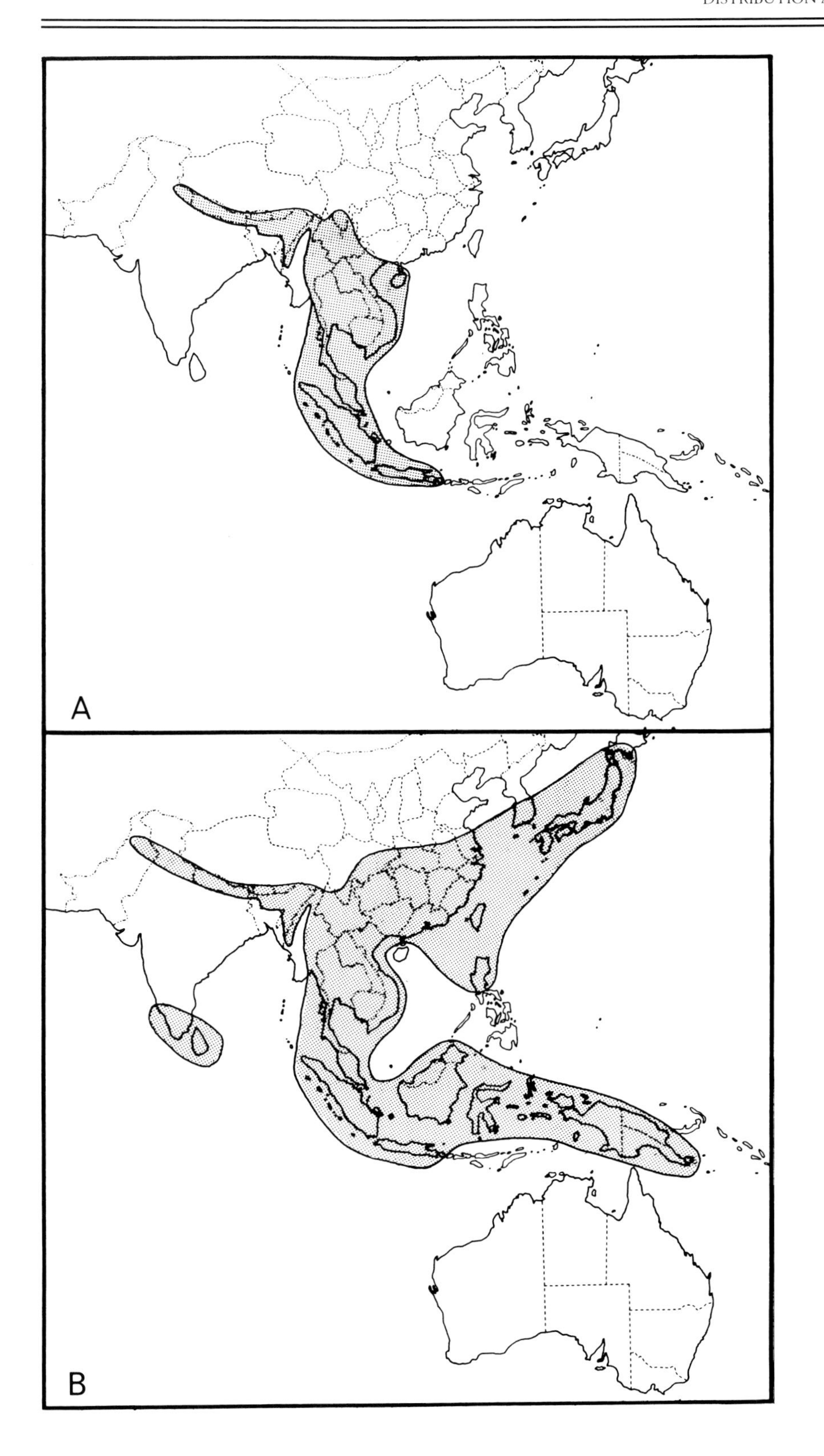

Map 2

A The distribution of subgenus *Cyperorchis*
B The distribution of subgenus *Jensoa*

Distribution Patterns

Cymbidium species usually occur in upland areas, often above about 1000 m (3280 ft) above sea level. Notable exceptions to this are the species in subgenus *Cymbidium* which often occur near sea level. *C. aloifolium*, for example, is the only species which occurs throughout much of lowland India, and *C. finlaysonianum* has often been reported from coastal areas in Malesia. Towards the northern extreme of the distribution in Japan, and the southern extreme in Australia, *Cymbidium* species often occur at correspondingly lower altitudes.

In northern India, Nepal, and Bhutan, most *Cymbidium* species occur in the lower elevations (700–2500 m, 2295–8200 ft) of the Himalaya, and often also in Meghalaya (the Khasia Hills). These two mountain ranges are separated by the Brahmaputra plains, and some species show disjunct distributions in this area (*C. devonianum*, map 6D), although more widespread species are also distributed around the Himalaya of N E Assam and in the hills of Nagaland, linking these two mountainous regions (*C. iridioides*, map 7B). The range of *C. macrorhizon* extends further westwards in the Himalaya than any other *Cymbidium* species (map 11E).

Many of the Himalayan species have distributions which extend eastwards into Burma and south-western China. In China itself, Hu (1971) suggested that *Cymbidium* species are generally distributed in a U-shaped pattern in the mountains of the south of China, with the western arm extending northwards into the mountains of Sichuan, and the eastern arm following the coastal hills into Zhejiang. However, the distributions of several Chinese species are now known to be more extensive, notably *C. goeringii*, *C. faberi* and *C. floribundum* (maps 11C, 11B, 6A), being recorded from the provinces of central southern China, including Anhui, Hubei, Henan, Hunan and Jiangsi (Wu & Chen, 1980). There is a relatively flat region of lowland plains along the Yangtze River in northern Hunan, central Hubei and northern Kiangsi in which *Cymbidium* is absent. However, these plains are bordered by a chain of hills to the north where several species are found, linking the western and eastern arms of the U-shaped distribution pattern suggested by Hu, and forming a circular pattern instead. The northern limit of *Cymbidium* species in China is about 32°N. The eastward distribution of the genus continues through Taiwan, and the Ryukyu Islands, to Japan and South Korea. The north-eastern limit of *Cymbidium* is about 42°N in Japan (map 1A).

The southward distribution of *Cymbidium* extends through Indo-China into W Malaysia and south-east into the Malay Islands. Subgenus *Cyperorchis* does not occur in Borneo, and only subgenus *Jensoa* is represented in New Guinea. The Philippines can be divided into two regions: the northern island, Luzon, has affinities with southern China in that it contains variants of *C. ensifolium* and *C. dayanum* which are typical of the Asiatic mainland, and a variant now included in *C. cyperifolium* (syn. *C. aliciae* Quis.) which occurs in India, China and Indo-China. This affinity has also been noted by Hu (1971) in other orchid genera. The southern islands contain *C. atropurpureum* and *C. bicolor* subsp. *pubescens*, species typical of the Malay Islands. One species, *C. finlaysonianum*, is distributed throughout the whole island group (Quisumbing, 1940).

Few species occur south or east of Wallace's Line, and those in Sulawesi and New Guinea have probably spread there from the Malay Islands, using the general west-east wind currents which would carry the dust-like seed. The three *Cymbidium* species which occur in Australia are endemic, and although their distributions extend well into northern Queensland and Cape York, they do not occur in New Guinea (maps 5C, 5D, 5E). They are included in section *Austrocymbidium* of subgenus *Cymbidium*, along with *C. chloranthum*, *C. hartinahianum* and *C. elongatum* from W Malaysia, Sumatra and Java. Subgenus *Cymbidium* has a centre of diversity in Malesia, and it is interesting to speculate on the biogeographical history of the Australian species.

Cymbidium in Australia

Hooker (1860) was the first to recognise an Indo-Malesian element in the Australian tropical and subtropical rainforest flora. Athough many genera and species are distinct from those found in Malesia, the families represented are similar. Barlow (1981) used plate tectonic theory to show that the Australian plate collided with the Sunda Island arc system to the north during the Miocene epoch, about 15 million years BP, forming New Guinea and a chain of islands linking Australia with Malesia. During the following millennia, intrusive elements of the Malesian flora used this chain of islands to colonise the lowland tropical regions of northern Australia, which at that time were extensively covered in tropical forest (Burbridge, 1960). The temperature was probably 5°C (9°F) higher than at present (Specht, 1981). The Orchidaceae would have been able to spread rapidly by their wind-dispersed seed, and this route through New Guinea and Cape York has been suggested as the main route for colonisation by the epiphytic orchids (Lavarack, 1981). This route was taken by such genera as *Dendrobium* (Lavarack, 1981), and subsequent evolution has produced new taxa which are very distinct, certainly at the sectional level. The presence of the three *Cymbidium* species in Queensland and on the east coast appears to support this pattern, and the distribution of the complete genus does not refute it (map 1A). However, an examination of the distribution of subgenus *Cymbidium* alone shows that no species related to section *Austrocymbidium* occur in New Guinea, and the closest locality of the related *C. chloranthum* is in Java (map 5A). It seems more likely, therefore, that *Cymbidium* entered north-western Australia from Java, perhaps using the Sunda Islands and Timor in the process, rather than New Guinea. This probably occurred during a wetter period when tropical forest extended throughout northern Australia. The efficiency of *Cymbidium* seed dispersal suggests that Java could have been the source despite its present distance from north-western Australia. Those two land masses were probably closer than at present, but continued northern movement of the Australian plate has subsequently recreated a larger ocean barrier between them (Barlow, 1981). It is probable that *Cymbidium* colonised Australia at a later date than *Dendrobium*, allowing the evolution of distinct endemic species which differ substantially from related species, but not distinct sections and subgenera.

Increasingly frequent periods of aridity are thought to have started about 15 million years ago, causing great periodic reductions of the rainforest area, and consequently natural selection has produced evolution of xerophytic characteristics in the plant communities of northern Australia (Specht, 1981). *C. madidum* and *C. suave* are confined to the damper forest refugias of eastern Australia (maps 5C, 5E). *C. canaliculatum* is widespread over northern Australia (map 5D), undoubtedly because it has evolved such xerophytic characteristics as very thick, leathery leaves which resist desiccation, the channelled leaf shape which directs any available water to the base of the plant, a rough epidermal surface which helps to maintain a boundary layer around the leaf, CAM photosynthetic pathways, and the somewhat spiny seeds which have a large surface/volume ratio making them more buoyant in the air and increasing the probability of a seed encountering a suitable tree on which to grow. Its roots are usually well protected from desiccation as they grow into the moist rotten heartwood of *Eucalyptus*. This species is perhaps the most successful of the Australian Orchidaceae in its adaptation to arid environments.

CHAPTER 7

The History of the Cultivation of Cymbidium

PART A:

The Far East

Orchids have probably been cultivated in China since before the time of Confucius (551–479 BC). *Cymbidium* species were amongst the earliest to be cultivated (Chen & Tang, 1982), their attraction being the simple but elegant form of the plant, and the shape and delicate fragrance of the flowers (photographs 1, 2, 4–6). Since then, selection of desirable and unusual variants has taken place, and some clones in cultivation are now far removed from their more typical wild ancestors. Historically, hybridisation has not been used to produce new variation, but naturally occurring hybrids have been selected and brought into cultivation (Chow, 1979). These two factors have made the classification of cultivated specimens very difficult, especially in the complexes related to *C. ensifolium*, *C. kanran* and *C. goeringii*, and have led to the publication of many specific and varietal names for variants which have little or no relevance to the biology of the wild populations. These might be better recognised by cultivar names. Examples of such names include the albino variants of *C. ensifolium* which have been named *C. gyokuchin*, Mak., or *C. ensifolium* var. *susin* Yen, and the various colour variants of *C. kanran* which have been named as distinct varieties.

For 2000 or more years, *Cymbidium* species have been significant in the culture and ethnic history of China. Their cultivation has been linked with the elite, and the plants have come to epitomise such human qualities as elegance, refinement and nobility.

The Chinese word *lan* means 'orchid', but in writings before the time of Confucius this word referred to several aromatic plants which were used in religious ceremonies and were said to ward off evil spirits. Indeed the Chinese verb *lan* means to ward off, or to check. Perhaps the first use of *lan* in connection with orchids in Chinese literature is in the *Shih Chi* (The Classic Songs), a collection of poetry and songs which predates Confucius. In this, the plant was used to signify love and the courtship of a young couple (Hu, 1971). This theme is continued in a story reputed to be from the Shin (Ch'in) Dynasty (249–207 BC). Yohki-hi, a woman of legendary beauty, and the wife of the Emperor Shi-kotei, was unable to have any children. However, the Emperor acquired a plant of *Cymbidium* for his wife who placed it in her room where it flowered in the following autumn, emitting a sweet fragrance which perhaps had an intoxicating effect on the lady. In due course she bore a son, and this was repeated each year until they had thirteen wise and brave sons. This orchid is reputed to have been the albino variant of *C. ensifolium* which is said to have about thirteen flowers on each scape (Nagano, 1960).

Confucius was influential in attaching the name *lan* solely to orchids. He knew of the wild habitat of orchids as is evident in the line:

> The *chih-lan* that grows in the deep gorges does not withhold its fragrance because of lack of appreciation.

This saying continues:

> The superb person strives for self-discipline, maintenance of principle and establishment of virtue. He does not alter his integrity because of poverty and distress.

That he was familiar with cultivated *Cymbidium* species is also evident through such sayings as:

> The association of a superior person is like entering a hall of *chih-lan* [*Cymbidium*]. In the course of time one becomes accustomed to the superior ways of life, and gets used to the fragrance.

Considering the importance of Confucius' teaching to Chinese philosophy, it is easy to see how such comparisons could lead to the high esteem in which *Cymbidium* species were, and still are, held. For example, Confucius commented that *lan* produced the fragrance of a King. Later scholars developed this theme, making *lan* a symbol of royalty and the loyalty of the King's subjects (see Hu, 1971).

During the Eastern Chin Dynasty, the ruling Wang family built an orchid pavilion (AD 354), to serve as a gathering place for the leading scholars of the time. It is still on the original site at Nanking (Nanjing).

The earliest book on orchids was published in AD 1233, by Chao Shih-keng, of Fukien Province, the area which was at that time the centre of orchid cultivation. It was called *Chin Chang Lan Pu*, and includes descriptions of 22 orchids, mostly *Cymbidium* species. In AD 1247, Wang Kuei-Lsueh, also from Fukien, published a second treatise, *Lan Pu*, in which 37 orchids are described and full cultural details are given. He wrote:

> It is a symbol of perfect personality, the quality of a superb person.

These two works illustrate the popularity of orchid cultivation, and especially of cymbidiums, at that time (Nagano, 1960; Hu, 1971). Their popularity and, to a large extent, their mystique remain today. Two ink paintings of *C. goeringii*, reproduced by Hu (1971), make the same point. One was painted by Chao Meng-chien (1199–1264) during the Southern Sung Dynasty, the second by Cheng Sze-shiao in 1306. Chinese orchid paintings are generally in black and white, reflecting the Chinese appreciation of the simplicity and refinement of the *Cymbidium* orchids, which they express most effectively using this technique (Hu, 1971). Two examples of ink paintings from the Ming Dynasty are given in photographs 1 and 2.

In the literature of the Ming and Ch'ing dynasties (ca. AD 1400–1800), *lan* was used in front of a noun to modify it in such a way that it took on the sentiment of good, fine, elegant or refined. Examples of this are *lan-chang* (= orchid writing) which meant 'fine manuscript', *lan-i* (= orchid manner) which meant 'handsome person', *lan-hui* (= orchid instruction) which meant 'good teaching' and *lan-hsin* (= orchid heart) which meant 'refined lady' (Hu, 1971).

Orchid cultivation in Japan does not appear to be so ancient. The first Japanese orchid book was published in the early eighteenth century, written by Matsuoka. In it, several orchid genera were described, and two cymbidiums were illustrated. Orchid cultivation was very popular at that time in Japan, with different sectors of the class system preferring to grow different orchids. Thus, the Imperial family and its circle grew *Dendrobium monile* Sw., and the Samurai and even some Shoguns grew *Neofinetia falcata* (Thunb.)H.H. Hu. The merchants and other wealthy people, including immigrants from China and other parts of Asia (Nagano, 1953, 1960), grew *Cymbidium*.

Orchids are grown in Japan primarily for their scent, and for the beauty of their leaves, and although the form and beauty of their flowers is also appreciated, they are of less importance. Specimens with variegated leaves are very highly prized (photographs 5, 6). The scents of orchids such as *C. kanran* were said to be transmitted to the clothes of lovers (Nagano, 1952, 1953, 1955).

PART B: The Influence of Cymbidium Species in Modern Hybrids*

The Formative Years

The history of the development of modern-day cymbidiums is quite distinct from the history of other generic hybrid groups. Changes in *Cymbidium* hybrid quality have often occurred suddenly through the appearance of a certain species or, in many cases, of an individual hybrid or clone of a species. We shall examine this more closely. Until the appearance of *Cymbidium insigne, C. sanderae* and perhaps *C. lowianum*, most *Cymbidium* hybrids were ugly ducklings. Furthermore, they did not flower well in the glasshouses of England and Europe, and their basic culture was not understood. Their growth habit was rank compared with the other genera that were coming in from all over the world. Other orchids were so much more colourful, easy to grow, varied in type and flower, and were of more all-round appeal.

The appearance of *C. insigne* from French Indo-China (Vietnam) in the early years of this century was the most important event in the history of *Cymbidium* hybridising. Until then, no *Cymbidium* species was known that even resembled it and, more importantly, here at last was a *Cymbidium* that had erect spikes, well spaced, white or light pink flowers, a compact growth habit and thin erect leaves: what potential! We shall look at *C. insigne* further. However, let us see which other species have been our building blocks.

The genus *Cymbidium* is one of a moderate number of species. Of the 44 species, not more than seven or eight have been responsible for 90 per cent of the hybrids we have today. A count of *Cymbidium* species which have been used gives us a total of 18 or 19, and it is debatable whether some others should be included despite records which show that hybrids have been made from them. This group can be divided into two categories; those of major and those of minor influence. Let us look at the major group first: this includes *C. eburneum, C. hookerianum, C. insigne, C. lowianum, C. sanderae, C. floribundum, C. tracyanum* and *C. erythrostylum*. In our second group we must include *C. aloifolium, C. atropurpureum, C. dayanum, C. suavissimum, C. madidum, C. devonianum, C. ensifolium, C. sinense, C. kanran, C. tigrinum* and *C. mastersii.*

We have not listed *C. i'ansonii*, now considered to be a variety of *C. lowianum*, which is reported to have been a major influence in breeding red-flowered hybrids. Another point of interest is that some species, such as *C. eburneum*, while seldom used, have made a notable contribution through only one or a few hybrids. The famous tetraploid *C.* Alexanderi 'Westonbirt' (photograph 18) is probably the most influential single orchid cultivar from a parental standpoint in any genus. The influences in its ancestry are *C.* Eburneo-lowianum 'Concolor' and *C. insigne* 'Sanderi'. Here we have a gene pool that has exerted its influence for over 50 years. Let us go further and, first of all, examine some of the other major species to see what their influence has been. A note of interest at this time is that *C. floribundum* has been included as a major influence, because it has given rise to an entire race of magnificent miniatures.

If taken in alphabetical order, we shall not be faced with the problem of which might be the most important. We should make a slight exception by saying that *C. insigne* stands pre-eminent.

C. erythrostylum is truly outstanding in breeding the white, pinkish or light-coloured, early-flowering cymbidiums of compact growth habit and great freedom of bloom. The white petals are carried somewhat forward, while the lip is rather small. However, when used with tetraploids which do not have these characteristics, the results have been superb, especially in creating early-blooming cymbidiums which flower months before mid-season. As mentioned, 'great events' have often been associated with the appearance of a single parent in orchid breeding. So it was with *C.* Early Bird 'Pacific' AM/RHS (Edward Marshall × *erythrostylum*) (photograph 19). This chance near-tetraploid was the magic key to the production of high-quality, early-flowering clones. When crossed with the tetaploid *C.* Balkis 'Silver Orb', the patriarch hybrid, *C.* Fred Stewart resulted (photograph 21). Here was a tetraploid for further advancement. Cymbidiums Stanley Fouraker (photograph 22), Earlyana, White Christmas and others followed.

*by E. Hetherington and A. Easton

For many years, the magnificent green or yellow-green, early-blooming *C. hookerianum* was used to breed large-flowered earlies. It has imparted large size, good open carriage and good spacing on the spike to its progeny. Regrettably, there is no perfect package in hybridising. *C. hookerianum* is heat sensitive and, if subjected to too high a temperature when the flower spikes are developing, the buds cease to grow, set for some weeks or months and fall off. Additionally, many of the *C. hookerianum* hybrids are difficult to flower, unlike the *C. erythrostylum* hybrids which are free-blooming. As we check the early hybrid lists, however, we find several of its hybrids such as *C.* Lowio-grandiflorum and *C.* Coningsbyanum. The cultivar 'Brockhurst' of *C.* Coningsbyanum (photograph 25) has been a valuable parent. *C.* Grand Monarch, a *C. hookerianum* hybrid, when crossed with *C.* Baldur, made the famous and noteworthy *C.* Sicily. Another hybrid from this species was *C.* San Miguel (photograph 24).

C. insigne has been of extraordinary value, both in the diversity of its hybrids and in the range of its influence. There have been a number of superior cultivars of this species. The key to all hybridising is in the selection of precise cultivars for their gene-pool influence. Through breeding with *C. insigne* influence, hybrids have been obtained which have ranged from clear vivid pinks through to dark rose-violet. Some of the best pre-1939 hybrids from *C. insigne* have been Cymbidiums Lyoth, Lillian Sander and Susette. A hybrid of quality in the dark pink to rose-red range was *C.* Ceres (*lowianum* var. *i'ansonii* × *insigne*), bred by Hamilton Smith in 1919.

C. lowianum, first described in 1879, marvellously widened the range of *Cymbidium* hybrid quality. It differed so very much from the later *C. insigne* in having a more rank growth habit, with wider leaves and arching spikes. It has green flowers with a pale yellow lip, distinctively marked on its apical portion with a maroon, V-shaped bar. It should also be recognised that *C. lowianum* and its hybrids are 'late bloomers', flowering at the end of the season. With all the major species, one very important fact must be borne in mind; fine selected cultivars are the key to developing successful breeding lines. With *C. lowianum*, a typical form is rather a nondescript green or greenish-yellow, perhaps mustard-coloured, with small flowers of poor shape and a rather poor lip marking. Which have been the fine cultivars? *C. lowianum* 'Pitts' was crossed with *C.* Mirabel, FCC/RHS, to make the splendid diploid *C.* Blue Smoke (photograph 28). *C. lowianum* 'Pitts' is a very fine vivid green with a yellow lip and a dark red-maroon bar. *C. lowianum* 'St Albans' was another of similar type. *C. lowianum* 'Concolor', a greenish-yellow with an ochre bar on the lip, is dominant and has been used to breed some beautiful chartreuse or greenish pure colours; hybrids such as *C.* Esmeralda (*lowianum* × Venus) made by McBeans in 1937. *C. lowianum* was one parent of *C.* Pauwelsii 'Compte de Hemptine', the second of the naturally occurring tetraploids. The dominance of the 'lowianum' lip, which shows as a bar on the apical portions of hybrid lips, remarkably carries through for many generations, as does the late-blooming habit. A few of the notable early hybrids from this species are Cymbidiums President Wilson, Eburneo-lowianum 'Concolor' (one of the parents of *C.* Alexanderi 'Westonbirt') and Pauwelsii.

C. sanderae has been considered by some authorities to be a notable major species. However, it is difficult to trace its influence as easily as can be done for other species. This attractive species has white flowers with a dark maroon-spotted lip which is very attractive and can be considered to have hybridising potential especially in breeding some of the 'decorator' types, as well as proven warmth-tolerance.

The last of our major species is the dwarf *C. floribundum*, still better known as *C. pumilum*. We can wonder how long it will take the orchid world to change over to the new name. Prior to 1939, this species was largely unknown from a hybridising standpoint having produced only two hybrids: *C.* Minuet (photograph 36) in the cross with *C. insigne*, and *C.* Pumander from *C.* Louis Sander. *C.* Minuet was bred by Sidney Alexander, W.H. Alexander's son, who was killed in World War II. Both hybrids showed the way for post-1945 breeding. *C. floribundum*, when used for breeding, transmits its dwarf growth habit, freedom of bloom, good clear colouring, and a tendency for its hybrids to bloom considerably before the conventional ones. This species came dramatically into its own in miniature-breeding during the 1960s and 1970s. Only a representative group of its hybrids can be listed. Cymbidiums Flirtation and Fairy Wand were perhaps the first to help establish 'miniatures' in horticulture. Interestingly, prior to the introduction of these two, 'miniatures' as a horticultural type did not even exist. The term was coined by the late Emma Menninger and was popularised by Ernest Hetherington of Stewart Orchids. A few of the earlier *C. floribundum* hybrids are Cymbidiums Evening Star, Tom Thumb, Mimi (photograph 37), Dag, Starbright, Jill, Dr Baker, Orchid Conference (heavily influenced by *C. devonianum*) and Pipeta.

The large-flowered *C. tracyanum* has received mixed attention from hybridisers. *C. tracyanum* and its hybrids flower very early. However, perhaps its main fault has been its flowers which do not have a clear colour and do not last very well. Still, this species received much attention prior to 1945 in such hybrids as Cymbidiums Verulam, Hanburyanum, Wiganianum, Louisiana (photograph 26) and Moira.

Let us now look at the second grouping of *Cymbidium* species which have been used for hybridising. The first of these is *C. aloifolium* which is also wrongly known as *C. simulans* by breeders. This is a tropical species with coriaceous leaves and long, pendulous spikes of rather smallish flowers. Its most important characteristic is that it flowers very well in the true tropics. *C. atropurpureum* also has strap-like, coriaceous leaves and is tropical-growing. *C. dayanum* is of miniature growth-habit, with rather pretty flowers of some hybridising promise, though these are short-lived. *C. devonianum* is a charming miniature species with long, arching, pendulous sprays of miniature flowers in shades of greenish-brown. It has been used with success by Keith Andrew, of Keith Andrew Orchids in England, to create a distinct horticultural type with pendulous spikes (photographs 42, 43). Some of the most notable have been Cymbidiums Bulbarrow, Tiny Tiger, Corfu, Brook Street, Devon Park, Devon Parish and Devon Wood.

The fragrant miniature species *C. ensifolium*, with its many and varied cultivars, has just begun to be used for hybridising. It has been in cultivation in the Orient for well over 2000 years! Happily, its fragrance can be inherited, as well as its compact growth-habit and free-blooming, erect spikes, and this species holds great potential. Another characteristic not to be overlooked is that *C. ensifolium* blooms in the late summer to early autumn as opposed to spring for most cymbidiums. *C.* Peter Pan (photograph 40) and Korintji point the way for further use of this species. We must place *C. kanran*, first identified in 1902, together with *C. ensifolium* and *C. sinense*, all having similar, erect spikes, fragrant flowers and a miniature habit.

The large-growing but small, green-flowered species *C. madidum* from Australia was not used much, if at all, until Mary Bea Ireland of Santa Barbara, California used it extensively in quite a large group of successful hybrids in the 1960s. What characteristics do *C. madidum* hybrids have? With few exceptions, the growth habit is as large as a conventional hybrid, which is a fault. However, the flower sprays are pendent and often 1.5 or 1.8 m (5 or 6 ft) long while the colours are generally in shades of yellow. When used with diploids or tetraploids, the flowers are enlarged and resemble conventional hybrids in shape, but still retain a miniature size. An outstanding *C. madidum* hybrid has been *C.* Sunshine Falls (King Arthur × *madidum*) made by Stewarts. The *C.* King Arthur used here was a tetraploid which dominated for small bulb size and kept the plant size down without sacrificing flower quality. Other *C. madidum* hybrids of note have been Cymbidiums Gladys Whitesell (photograph 45), Nonna, Pat Ann, Cricket and Sweet Lime.

The true miniature species *C. tigrinum* has given some charming hybrids (see photograph 47). Why it has not been used more by other hybridisers is difficult to say. Clearly the most notable was bred by Stewart Orchids from *C.* Alexanderi 'Westonbirt' to make *C.* Tiger Tail. Several of the 'Tiger' series such as *C.* Tiger Cub (photograph 46) and *C.* Tiger Hunt are distinctive.

Let us look at some of the men and women who have featured in *Cymbidium* hybridising. Prior to the rediscovery of cymbidiums by the orchid world after 1945, there were very few growers in England or Europe who paid much, if any, attention to cymbidiums. The leaders were H.G. Alexander, grower to Sir George Holford and later owner of his own nursery, A.A. McBean of McBeans Orchids, and the firm of Frederick Sander & Sons at St Albans in England and Bruges in Belgium. A few of the wealthy in England, such as the Rothschilds, and Baron Schroeder of Dell Park, also recognised the potential of hybrid cymbidiums. We have spoken of the influence of the species in hybridisation but we can gain a better perspective by realising that all plant exploration, hybridising, growing and indeed merchandising is the work of people. People who have dedicated their lives to these efforts. Some were commercial growers and others enthusiastic amateurs. We shall make acquaintance here with a few. For some years prior to 1950, Mr Barnard-Hankey bred cymbidiums in the south of England at Plush in Dorset. Sir William Cooke of Wyld Court Orchids also had a keen interest in cymbidiums. He also raised race horses and his hybrid *C.* Blue Smoke (photograph 28), a memorable post-World War II green, was named after a race horse. Mr Arno Bowers of Duarte, California did splendid pioneer work in breeding 'miniature' cymbidiums such as *C.* Sweetheart and many of the 'King Arthur' series. Mrs Emma Menninger of Arcadia, California was an all-round *Cymbidium* hybridiser of note. Comte de Hemptine of Belgium must be remembered for the splendid tetraploid, *C.* Pauwelsii 'Comte de Hemptine' FCC/RHS. The name Rothschild of Exbury, England tells of the history of cymbidiums, especially *C.* Rosanna 'Pinkie' FCC/RHS (photograph 23) and others. Four Rothschilds have raised cymbidiums. Lord Hothfield of Surrey, England will always be remembered for *C.* Babylon 'Castle Hill' FCC/RHS. It is difficult to mention the name McBeans of Cooksbridge in England and still be brief. In the years preceding World War II and continuing to the present time, few *Cymbidium* nurseries in the world have made as substantial a contribution both to quality and colour in *Cymbidium* hybrids. Founded originally by A. McBean and now owned by Ray Bilton, the nursery's contribution to the *Cymbidium* world continues. Surely, the name of Sander & Sons of St Albans, England, will be recorded in history for their notable contribution, not only to the improvement of high colour cymbidiums but also for the introduction of new species such as *C. insigne.* Sir George Holford and his grower, H.G. Alexander, who in time ran the business under his own name, established some of the very foundations of orchid breeding with *C.* Alexanderi 'Westonbirt'. Truly, Alexander was a giant in his field. Hamilton Smith, the Rothschilds, Brummitt, Colman, Veitch, Cowan, Schroeder, Wrigley and the Edinburgh Botanical Garden in Scotland are a few gleaned from the list of registrants before 1939. After 1945 the interest and activity in cymbidiums exploded throughout the world in areas where they could be grown.

The Modern Era

It is useful to take the 1960s as a watershed period in the transition from the dominance of British-influenced *Cymbidium* hybrids, largely rooted in the 1930s and 1940s, to that of the American hybrids, often from quite distinct or new breeding lines. The World Orchid Conference (WOC) held in Long Beach in 1966 gave over 40 medal awards to cymbidiums, of which more than 90 per cent were American-bred. Additionally, the species *C. floribundum* featured for the first time as an immediate parent or grandparent of more than a quarter of all the awarded cymbidiums.

Even a cursory analysis of the various species in the background of modern hybrids will reconfirm the integral role of *C. insigne* and its various cultivars as by far the most significant parental influence. Closer analysis suggests that there are three approximate percentages of species' involvement which reflect in distinct and identifiably different progeny. If we allow for differences in chromosome number (ploidy), which computerised parental analysis cannot easily take into account, we have first a group of modern hybrids with 50 per cent or more of *C. insigne* in their family tree. In the 1940s and 1950s these were the Cymbidiums Alexanderi 'Westonbirt' (photograph 18), Rosanna 'Pinkie' (photograph 23), Balkis (photograph 20) and Joan of Arc hybrids. Since then we have seen a decline in the high percentage *C. insigne* hybrids, largely resulting from the shift in emphasis away from mid-season breeding. An Armacost & Royston cross of the early 1950s, *C.* Dainty, was used by Stewarts to produce *C.* Dainetta. At 65 per cent *C. insigne* parentage, it is a spectacular diploid with a tall arching spike often over 1.5 m (5 ft) in length. Another Dainty hybrid, although with less *C. insigne* influence, is *C.* Victoria Arvanitis. One of its most outstanding characteristics is its ramrod straight spikes with the lowest flowers on the spike at least 46 cm (18 in) above the tips of the foliage.

The second easily discernible level of *C. insigne* influence is between 30 per cent and 50 per cent. It encompasses a much wider range of colours.

Examples of the older parents in this range are Cymbidiums Apollo 'Exbury', Carisona 'Glendessary', Pearl Easter 'McBeans', Snowsprite 'Jean', Stanley Fouraker (photograph 22) and Flavian. The dominance of *C. insigne* is reduced but still makes a vital contribution. In *C.* Apollo it reduces leaf width, in *C.* Carisona and *C.* Flavian it is the lilac-pink of *C. insigne* in association with strong *C. lowianum* var. *i'ansonii* influence which gives the remarkable colour intensity.

Two hybrids from the third group where *C. insigne* has very low influence are *C.* Blue Smoke (photograph 28) and *C.* Early Bird (photograph 19). Both are very important progenitors of present-day hybrid lines — Blue Smoke for its lateness and Early Bird for imparting shape and earliness.

Does *C. insigne* have much to offer present-day hybridisers? It should probably not be reintroduced to modern breeding lines. *C.* Lyoth, which is 75 per cent *C. insigne,* typifies many of the shy-blooming traits attributable to *C. insigne.* When crossed with the floriferous *C. floribundum,* the resultant hybrid *C.* Starbright is one of the less free-spiking primary hybrids from *C. floribundum.*

Complex hybrids with a high *C. insigne* influence are not often seen in registrations today. Certainly, they would be of little use in a breeding programme where compact-spiked potted plants were the desired result. But for free-standing upright cut-flower types, the *C. insigne* dominated parents will still continue to be used. Diploid mid-season pinks like *C.* Finetta and its recent offspring *C.* Candy Floss or the tetraploid intermediate *C.* Alison Shaw 'Valentine' AM/AOS are all present-day stars producing quality offspring (see photograph 38).

The rising influence of *C. erythrostylum* has probably yet to peak. Sander's hybrid, *C.* Early Bird 'Pacific' (photograph 19), gained an Award of Merit in England in 1950 yet has never featured in an English hybrid because it was astutely bought the same day by Emma Menninger. She saw its potential and, in doing so, established *C. erythrostylum* as a significant parental influence of the post-1945 period. Both Stewarts and Emma Menninger saw a potential in *C.* Early Bird 'Pacific' far overriding any incidence of 'rabbit-eared' petals in some of its offspring. Certainly most of the diploid *C. erythrostylum* hybrids were disappointing, with indifferent colours, a high incidence of forward-held petals and thin substance. But some were further hybridised and, for example, *C.* Redwood, which was the parent of *C.* Mighty Mouse, passed on an invaluable *C. erythrostylum* contribution to the superbly shaped, chocolate brown *C.* Kiri Te Kanawa cultivars (photograph 31).

Three mighty building blocks of earliness and shape in present-day cymbidiums are strongly influenced by *C. erythrostylum.* They are Cymbidiums Stanley Fouraker (photograph 22), Fred Stewart (photograph 21) and Rincon 'Clarisse': the first two are progeny of *C.* Early Bird 'Pacific' (photograph 19) and tetraploids, while Rincon 'Clarisse', a diploid, has more recently been colchicine-converted to a tetraploid form. In percentages of major species Cymbidiums Stanley Fouraker and Fred Stewart are identical, with exactly the same levels of *C. insigne, C. erythrostylum* and *C. eburneum* in their parentage. *C.* Fred Stewart has only 30 per cent *C. sanderae* influence, which nevertheless has probably contributed significantly. *C.* Rincon has exactly 25 per cent *C. erythrostylum* and 25 per cent *C. insigne* but with an equivalent 25 per cent *C. hookerianum. C. hookerianum* apparently stimulates the appearance of soft, shell pinks. *C.* Rincon 'Clarisse' is well known for breeding pinks of this type, and in earlier days *C.* Coningsbyanum 'Brockhurst' (photograph 25), with *C. hookerianum* and *C. insigne* in exactly the same ratio as *C.* Rincon, was similarly useful. There have been problems with many of the *C. erythrostylum* offspring in that not all will draw water and keep when cut. The problem seems to be accentuated by relatively high levels of *C. tracyanum* in conjunction with *C. erythrostylum.*

As modern tetraploids from the better *C. erythrostylum* lines appear, it is obvious they will force us to revise earlier concepts of perfection in flower shape. The truly circular *Cymbidium* flower with 6 cm (2.4 in) petals and overlapping lateral sepals has been achieved and in the 'early' or 'early mid-season' types rather than in later-flowering hybrid lines where the high award winners were previously concentrated. Tetraploid hybrids like Cymbidiums Winter Fair, Winter Wonder, Fancy Free (photograph 32), Solana Beach and Via del Playa (photograph 33) have advanced flower shape close to perfection while achieving substance almost unbelievably improved from the initial diploid *C. erythrostylum* hybrids. *C.* Fancy Free 'Mont Millais' (photograph 32) is one of the best, derived from the mid-season *C.* Snowsprite crossed with the diploid *C.* Rincon 'Clarisse'.

Another example of species' characteristics in a modern hybrid is seen in *C.* (Liliana × Angelica) (photograph 27), created in the search for colourful varieties for early season flowering. Here, an early season pastel pink has resulted from repeated crossing of *C. tracyanum*-based hybrids, into mid-season pinks derived from *C. insigne* and *C. eburneum*. The resultant hybrid (which has *C. tracyanum* four times on one side of its parentage, and twice on the other), shows the influence of *C. tracyanum* in its overall shape and lip. However, it represents a considerable advance on the pale pastels, particularly yellows, which otherwise dominate *C. tracyanum* breeding. *C. Louisiana* (photograph 26), another *C. tracyanum* hybrid, also shows how strongly the species can express itself.

C. Pontiac 'Grouville' (photograph 34), which received a Silver Medal at the 12th WOC, and *C.* (Lady McAlpine × Bay Sun) (photograph 35) show how diverse can be the results obtained by careful selection in breeding on from the major species building blocks *C. insigne*, *C. eburneum* and *C. lowianum* var. *i'ansonii*. Both plants have in their background that crucially important parent of standard cymbidiums, *C.* Rio Rita 'Radiant' FCC/RHS (photograph 29). Selection over numerous generations, in the one case for colour, and in the other for pastel flowers of high quality, has created hybrids of a standard which breeders of only 30 years ago could hardly have thought possible.

The other species which has exploded into prominence since 1945 is the miniature *C. floribundum*. Credit must again be given to the British for making the first two *C. floribundum* hybrids, but even more credit to the Americans for visualising the potential of 'miniature' cymbidiums and then expanding and developing the type. As with earlies from *C. erythrostylum*, it was the late Emma Menninger and Ernest Hetherington of Stewarts, plus Henry Tanaka and Mary Bea Ireland of Dos Pueblos Orchids, who pioneered development of 'miniature' and later 'polymin' cymbidiums, as intermediates were then known. By today's standards the early crosses were woeful. Many had dull flowers which were often short-lived, and some seedlings or even whole crosses were notoriously shy-blooming. Gradually, selected clones of *C. floribundum* from warmer habitats and 'alba' forms were brought into cultivation. An improved range of primary *C. floribundum* hybrids appeared with a fairly complete colour spread. Hybridisers were plagued by infertility problems; few pods contained viable seed and many of the seedlings were aneuploids (with odd chromosome numbers — see chapter 4). The occasional plants which produced reasonable quantities of seed or normal seedlings were highly prized and closely guarded. Unbloomed second generation seedlings in the *C. floribundum* line were rarely offered for sale, and even then almost invariably were from crossings back to 'standard' hybrids. It was not until the 1970s that the first hybrids between 'miniatures' or 'intermediates' appeared; most of these were crosses with *C.* Showgirl. Invariably when the odd fertile *C. floribundum* hybrid gave seed it was as the result of eggs or pollen which had not proceeded normally through meiosis. There are few, if any, second or third generation *C. floribundum* hybrids which are diploid (see photograph 38).

As the *C. floribundum* influence for miniature flower and plant size becomes diluted, we are seeing that the species can still make an important contribution. Among the second and third generation hybrids are some, which although essentially standard in growth and flower size, have retained the finest *C. floribundum* attributes. Culturally, these valuable influences are: improved spike production, quicker re-establishment from division and some resistance to bud drop. Although primary *C. floribundum* hybrids now have little to offer *per se*, intercrossing of the various descendants is proving to be rewarding for hobbyists as well as for commercial growers.

C. ensifolium, through its most notable offspring the tetraploid *C.* Peter Pan 'Greensleeves' (photograph 40), and to a much lesser extent *C.* Korintji, is proving to be a very popular inclusion in many breeding programmes (see photographs 39, 41). Many diploid *C. ensifolium* hybrids are quite infertile which was the case with *C.* Peter Pan 'Greensleeves' until Dr Donald Wimber converted it to a tetraploid form. The drawback with *C. ensifolium* hybrids is their limited colour range and short flower life, especially when cut. More than balancing this, however, are qualities like scent, summer blooming, erect spikes and miniature habit. The second generation C. Peter Pan hybrids, particularly those with another tetraploid, are remarkably improved, and when these are intercrossed in combinations like C. Australia Fair (Peter Pilot × Pink Peach), the flowers are essentially a miniature C. Fred Stewart, while the plant habit, blooming time and warmth-tolerance are just like tetraploid C. Peter Pan 'Greensleeves', which is the grandparent on both sides of the cross.

We do not think it is too early to start recognising the tetraploid C. Peter Pan 'Greensleeves' as the potential C. Alexanderi 'Westonbirt' of 'miniature' and 'intermediate' *Cymbidium* breeding. Even in the first generation 'Peter Pan' hybrids, a full colour range has appeared: rich yellows with C. Coraki 'Margaret' and clear shell pinks with the C. Prettipink grex. These all have the advantage of being tetraploids, so the second generation prospects are truly exciting.

Possiby the species that offers most to present-day hybrids by direct involvement is *C. sanderae* and *C. sanderae* 'Emma Menninger', its tetraploid form. It has been found to be a vital contributor to warmth-tolerance, and the hybrid C. Everglades, between *C. sanderae* and tetraploid C. Peter Pan 'Greensleeves', is quite outstanding in this respect. We also commented earlier that *C. sanderae* seemed to make a significant contribution to the generic background of C. Fred Stewart (photograph 21). Certainly, it has been our experience that the *C. sanderae* characteristic of blooming for two or three years from the same mature bulb is more pronounced in the C. Fred Stewart hybrids, and their flowers are on the average more shapely than those from C. Stanley Fouraker (photograph 22). The other major species contributing to reblooming on the mature bulb is *C. eburneum*; both C. Stanley Fouraker and C. Fred Stewart have the same percentage of it in their parentage.

When grown under intermediate conditions, particularly night temperatures in the 13–17°C (56–62°F) range, it is normal for *C. sanderae* to have two distinct flowerings, one in the autumn and a heavier one in the spring. This quality is also of great interest and should encourage hybridisers, aiming to satisfy both hobbyist and cut-flower demands, to use the species and its immediate offspring quite widely.

C. lowianum still exerts considerable influence through hybrids like C. Blue Smoke (photograph 28), although largely at the late end of the *Cymbidium* season. The big arching sprays so characteristic of *C. lowianum* and its progeny are largely out of fashion with commercial growers, and interest in greens is also at a low ebb. However, it does appear that *C. lowianum*, both directly and through *C. lowianum* var. *i'ansonii*, has had an important role in colour intensification. Consider an orchid like C. Chironla 'Tabasco', burnt orange in colour, which is predominantly *C. lowianum*, and breeds intense-coloured fiery oranges with other cymbidiums that are predominantly *C. lowianum* with a good splash of *C. insigne* as the second most significant background species.

Cut-flower growers recognise the toughness that *C. lowianum* passes on to its progeny. In their day, the C. California hybrids were highly prized for quality flowers late in the season, and C. Dorama and its tetraploid descendants are still widely sought for intercrossing to inbred *C. erythrostylum* lines, so as to strengthen the flowers and improve their keeping qualities.

Of the remaining major species, *C. tracyanum* has little to offer directly, other than its novelty value, best seen in hybrids like C. Dr Pepper and C. Gladrags. *C. eburneum* is little grown, yet its older hybrids like C. Caroll have recently been rediscovered in Japan. The attractively compact foliage of *C. eburneum* makes a pot-plant which is particularly appealing to the discerning Japanese eye. The newest hybrid of this type, C. Tussock (*eburneum* × *sanderae*), was made to satisfy these requirements.

A flurry of hybridising with *C. devonianum* by Keith Andrew, Featherhill and others in the 1960s and 1970s is continuing (see photograph 42). C. Bulbarrow, a triploid bred from C. Western Rose, established the standard of excellence for pendulous types. Although they are extremely difficult to transport without damage, the graceful and richly coloured sprays of *C. devonianum* hybrids are popular with hobbyist growers, and provide a colourful spectacle at most spring shows. While there was for some time an attitude that C. Bulbarrow and more recently C. Jack Hudlow (photograph 43) cultivars (also triploids) could not be improved upon, recent hybrids with new diploid parents like C. Tapestry (photograph 44), C. Kalinka and C. Caligold plus the inspired cross *C.* Plush Canyon utilising the talent of *C.* Rio Rita 'Radiant', nearly 50 years old, are proving that *C. devonianum* has breeding avenues which are far from exhausted.

Among the minor contributors, the three Australian species offer much to interest 1980s hybridisers. *C. madidum*, *C. suave* and *C. canaliculatum* are being re-examined in the light of the success of some of their most recent hybrids. *C.* Fifi 'Harry', a *C. madidum* hybrid, was crossed with tetraploid *C. sanderae* 'Emma Menninger' to produce *C.* Gladys Whitesell (photograph 45), a superb white 'intermediate' with both scent and warmth-tolerance. In the 'novelty' types, *C.* Sweet Devon (*suave* × *devonianum*) and *C.* Kuranda (*suave* × *madidum*) are extremely floriferous miniatures, with the latter regularly in full flower as late as the week of the summer solstice.

The newest hybrid in *C. madidum* breeding is *C.* Phar Lap (Flame Hawk × *madidum*), another named after a racehorse. Of particular appeal has been the potent miniaturising influence of a *C. devonianum* grandparent, and the appearance of dark clones more intensely coloured than even the *C.* Flame Hawk parent. Although only about 20 per cent of the seedlings have produced deep red shades, they show a remarkable improvement for colourful late-blooming intermediates and, in the southern hemisphere at least, are very useful around the Christmas season.

In California recently, we noted several plants of *C.* Phar Lap in full bloom while nearby were 20 or more *C.* Peter Pan hybrids with spikes opening. The date was 1 September and the *C.* Phar Lap cultivars had been awarded in early June 1987. For the first time, we realised that a long-time cymbidium hybridising dream had been achieved. Judicious use of the minor species *C. madidum* and *C. ensifolium* has been the key to this breakthrough. Now a full 12-month flowering of *Cymbidium* hybrids is a reality, and new generations should see the colour range and flower quality strengthened.

Careful study of the nutritional requirements and flower initiation prerequisites of the major *Cymbidium* species would be a most rewarding future project. Already, several of the leading Dutch cut-flower growers have acquired a comprehensive collection of *Cymbidium* species, so that by better understanding them, they will be able further to refine cultural techniques applied to their hybrids. It is astonishing and disappointing that so many people who claim to be *Cymbidium* hybridisers have such a limited knowledge of the species building blocks.

CHAPTER 8

The Cultivation of Cymbidium

In general, the popularity of cymbidiums is a reflection of their ease of cultivation. Contrary to popular belief, most orchids are relatively easy to grow, provided that care is taken to maintain an open, free-draining compost, and to prevent damage from unsuitably low temperatures or over-watering. In warmer regions, cymbidiums may be grown out of doors, but in northern USA and northern Europe they require protection from frosts in winter. Cultivation may be improved through the provision of appropriate shading and fertilisers, and the control of pests. Cymbidiums in cultivation can be divided into three groups.

1. Large-flowered Himalayan, Indo-Chinese and S W Chinese Species, and the Majority of Hybrids

This group includes the following species: *C. tracyanum*, *C. iridioides*, *C. erythraeum*, *C. hookerianum*, *C. wilsonii*, *C. lowianum*, *C. schroederi*, *C. insigne*, *C. sanderae*, *C. eburneum**, *C. parishii*, *C. roseum**, *C. mastersii**, *C. erythrostylum**, *C. elegans*, *C. cochleare*, *C. whiteae*, *C. sigmoideum**, *C. tigrinum**.

(*These species require somewhat warmer conditions than the others.)

These species are distributed from the Himalaya of northern India and Nepal, to the mountains of S W China, and south into Indo-China, at altitudes of about 1200–2800 m (3935–9185 ft). They mostly grow on the larger branches of trees in damp, montane forest, amongst epiphytic ferns and mosses, forming the largest specimens in hollows with rotting wood, and in the forks of branches where leaf litter accumulates, or less frequently as lithophytes. The climate in these areas is seasonal, with cool, dry winters and warm, wet summers.

In their natural environment, they flower during the relatively dry, sunny period between December and April. Although brief frosts may occasionally occur, the plants are usually well-protected by the trees on which they grow, and the roots are dry enough to prevent rotting. Shortly afterwards the summer rains break, and it is during this warm, damp weather that rapid growth takes place.

These conditions should be mirrored in cultivation. During the summer, high light levels are required to maintain growth, but growth is inhibited at temperatures above 27°C (81°F). Cymbidiums must therefore be grown in shaded glasshouses, where a warm but humid environment is maintained. Frequent misting, and daily spraying of the paths and staging, will help to provide a suitably humid environment. Flowering is initiated during the summer, and is dependent upon a maximum temperature of 12°C (54°F) during the night. Higher night temperatures will promote rapid growth, but will retard flowering.

This group of cymbidiums can be cultivated outside in mediterranean climates where night temperatures never fall below 5°C (41°F), such as in

7. *Cymbidium dayanum*. Thailand, *Menzies & Du Puy* 72, cult. Kew

8. *Cymbidium canaliculatum.* E Australia, *Easton* s.n., cult. Kew

9. *Cymbidium madidum.* (Left) E Australia, ex Canberra B.G. 722657, cult. Kew; (right) E Australia, *Easton* s.n., cult. Kew

10. *Cymbidium floribundum*. China, *Andrew* s.n., cult. Kew

California and southern Australia. Even in colder climates, the plants can be placed outside during the summer months. They will benefit from the high light levels, and will even enjoy full sun for the part of the day. However, it is preferable if broken shade is provided, such as beneath an open-crowned tree or under a lattice of wooden laths, especially in regions of strong, hot sunshine. This helps to maintain moisture and humidity and, in conjunction with regular spraying, will prevent the foliage from becoming burnt at the tips. Plants growing in the wild often flower very profusely in exposed situations, but the foliage is usually badly damaged, and the plants are unsightly. Protection from drying winds is also important if the plants are to be successfully grown in the open air.

The onset of autumn and winter produces lower temperatures and light levels. Shading should be removed from the glasshouse, and both water and fertiliser should be reduced. These measures copy as closely as possible the conditions of the cool, dry season which occur in the wild. Cymbidiums may also be overwintered indoors, on a bright windowsill. Most of the large-flowered species flower during late winter and early spring. Only small amounts of water are required at this stage, as the new roots will not yet have developed.

2. Small-flowered Tropical and Subtropical Species (including the Australian Species)

The species included here are: *C. aloifolium*, *C. bicolor*, *C. rectum*, *C. finlaysonianum*, *C. atropurpureum*, *C. borneense*, *C. dayanum*, *C. canaliculatum**, *C. hartinhianum*, *C. chloranthum*, *C. madidum*, *C. suave**, *C. elongatum*, *C. suavissimum*.

(*These species occur in drier areas, and require higher light levels.)

The distribution of this group of species extends from India and southern China, through Indo-China and Malesia, to Australia. They mainly occur at low altitudes, and often near sea level, ascending to 2500 m (8200 ft) in the tropics of Malesia. The climate varies from continuously hot and humid in Malesia, through seasonally wet in India, Indo-China and N E Australia, to the arid conditions of N W Australia. These species are usually epiphytes in tropical forests, but are often able to tolerate extended periods of exposure to full sunshine and low rainfall. Some of these species have thick, leathery leaves, and can be amongst the few orchids which thrive on isolated trees, on trees in plantations and on exposed rocks.

Outside the tropics, these species require glasshouse protection, in order to maintain a consistently humid, warm atmosphere. They grow and flower continuously throughout the summer months, and should be watered copiously and fed regularly. Shading is required to maintain an even temperature below 30°C (86°F) during the day and 18–20°C (64–68°F) at night. Those species which occur in exposed situations appreciate high light levels, and do best near to the roof of the glasshouse. Insufficient light during winter reduces the growth rate, and prevents flowering. The plants consequently require less water, and can rot if watering is not reduced. The Australian species often develop a long root system, and can benefit from longer containers such as terracotta pipes or bark tubes.

3. Small-flowered Japanese, Chinese and Himalayan Species

The following species are included: *C. floribundum*, *C. devonianum*, *C. ensifolium**, *C. munronianum*, *C. sinense*, *C. kanran*, *C. cyperifolium*, *C. faberi*, *C. goeringii*, *C. lancifolium**, *C. macrorhizon*.

(*These species also occur in the tropics.)

The majority of these species are distributed from Japan, through southern China, to the Himalaya of Nepal and northern India, from sea level in temperate areas of Japan, up to 2000–3000 m (6560–9840 ft) in southern China and northern India. They are mostly terrestrials, growing in open to dense woodland, often in deep leaf litter, in damp situations. They have a distinct seasonality, but different species flower at different times of the year. Several of these species have exceptionally wide distributions, extending into the tropics. Care must be taken to check the provenance of your plants, and to grow them in an appropriate temperature regime.

These warm-temperate species mainly require cool, moist growing conditions. Some are sufficiently hardy to withstand winters outside in the south of England, and grow best in sheltered, well-drained positions. Although they are terrestrials, they have thick, fleshy roots, similar to other epiphytic species, and their requirements are similar. They grow in well-drained sites, in deep leaf litter, and require similar composts to other cymbidiums, perhaps more finely chopped and with added leaf mould, peat, or moss, and charcoal. They are best grown in cool, well-ventilated greenhouses which are kept frost-free in winter.

Composts

Cymbidiums are usually epiphytic in nature, and even the terrestrial species have similar requirements to the epiphytes. Although they grow best in moisture-retaining rotting wood, leaf litter or other accumulated plant debris, they dry out rapidly and their roots are never wet for long. The potting mix should be open enough to allow free drainage, but should also be able to absorb and hold moisture. The openness of the compost will to some extent depend upon the frequency with which the grower can water his plants. More frequent watering will allow a more open mix to be used. This has the advantage of allowing enough air to reach the roots, as in the wild epiphytic plants, and also allows easier control of soil moisture in the winter, when excess water can easily cause the roots to rot.

Ideally, a compost should maintain its openness for several years, so that repotting is only necessary once every three years or so. Large chips of pine bark form the basis of many potting mixtures, serving to keep the compost open. Perlite, and sometimes coarse peat, are also added to hold moisture, especially important for amateur growers who cannot ensure regular watering. Coarsely crushed charcoal helps to keep the compost fresh, by absorbing detrimental chemicals, such as salts from excess fertiliser or water and plant decomposition products. Small amounts of dolomitic limestone are sometimes added as well, to balance the acidity caused by phenolic acids released as the pine bark breaks down.

A general compost which could be varied to meet the growers' personal requirements would include approximately 3 parts of coarse bark flakes, 1 part of coarse peat, 1 part of perlag and 1 part of coarse charcoal chips.

There are other ingredients which may be used successfully. Orchid roots often cling to expanded polystyrene chips more than to any other compost component, and are therefore recommended, although they are rather unsightly. Furthermore, the chips are lightweight, keep the compost open and do not decay. Sphagnum moss is moisture-retentive and open, and is especially beneficial when chopped into seedling or terrestrial mixes. Coarse fern fibre, shredded bark, and coconut or synthetic fibres are slow to decompose and are useful alternatives to bark. Clay pot or brick shards assist drainage. All of these are slightly porous and hold various amounts of water.

Slow-release fertilisers in pellet form may be added to the compost, but should be used at half of the rate recommended for other horticultural crops.

Repotting

Repotting should be undertaken in spring, after flowering but before rapid new growth has started, minimising the set-back inevitably caused through root disturbance. It is preferable to repot only once every two or three years, if a large enough pot and a durable compost have been provided initially. Repotting will be required if the compost is old and decomposing, or the plant has grown too large for its pot. The plant should be knocked out of its pot, or the pot cut away, and the old compost removed from between the roots. Any dead roots can be trimmed. The plant can be divided, provided that each division has at least three fresh pseudobulbs. Old pseudobulbs ('backbulbs') which are still sound are usually left, as they are not harmful, and probably help to store water and nutrients. Alternatively, they can be removed and placed in damp compost, where they will often sprout, and eventually produce a new plant.

Repotting is carried out by placing the plant in the pot, and working fresh compost between the roots with a finger. Gentle rocking of the plant and tapping of the pot will settle the compost, and more is added and firmed until the bases of the pseudobulbs are just covered. After an initial soaking, watering should be decreased and misting increased for two or three weeks, until growth resumes. It is advisable to wet the compost for two or three days prior to use.

Care should be taken to ensure cleanliness of the pruning knife or secateurs when the roots are pruned or the plants are divided during repotting, or when cutting flower spikes. If a virus disease is present, it will be transferred from one plant to the next, through the infected sap on the cutting blade. A wash in dilute bleach (10 per cent), or flaming after dipping in methylated spirits, should ensure that no virus is transferred between plants. Large cuts should be treated with powdered sulphur or another fungicide to prevent rotting.

Watering

More plants are killed through poor drainage and over-watering than by other mistreatments. *Cymbidium* roots must not be allowed to stand in waterlogged composts, as this will surely cause them to rot and die. Rainwater is preferable for watering whenever possible; it is less likely to contain dissolved calcium or mineral salts. Frequency of watering will depend upon the amount of water retained by the particular compost used, and will vary for each grower, but the compost should never be allowed to dry out completely. Enough water should be given so that it runs freely through the drainage holes, every time the plant is watered. Water requirements are greatly reduced during the winter months.

Fertilisers

A dilute fertiliser should be added to the water once a week during the summer months, as cymbidiums are heavy feeders. Care should be taken not to overfeed the plants, as detrimental salts will build up. The compost should be 'flushed' regularly when watering, to wash away these excess salts. Never water in dribs and drabs; a less frequent but thorough soaking is preferable. Excess salts and fertiliser can accumulate in the leaf tips, causing them to become scorched.

Pests

Cymbidiums are not generally prone to fungal diseases, but some insects can seriously weaken them. Red spider mites and false spider mites are a frequent cause of yellowing of the leaves. These creatures are so small that they can be difficult to see, but a very fine web on the underside of the leaf is characteristic of a heavy infestation. They can be discouraged to some extent by maintaining a humid atmosphere, and spraying with water will wash some of them away. However, if heavy infestations persist, *Phytocelius persimilis*, a predatory mite, can be introduced to control the attack. It is sometimes necessary to resort to chemical pesticides, such as diazanone.

Aphids will sometimes attack young shoots and flower spikes, causing distortion and spotting of the flowers. They can often be treated by wiping them away with soapy water. Mealy bugs and scale insects can be gently scraped off or killed by painting them with methylated spirits. Mealy bugs may also be controlled using a ladybird *Cryptolaemus montrouzieri*, an Australian species which lays its eggs into the eggs of the pest. If there is no alternative, full strength pesticide sprays such as diazanone or malathion should be applied with due care. A minimum of two sprays should be given, seven to ten days apart, as the pest may survive the first application in the form of eggs. This will also help to prevent the pest becoming resistant to the chemical. The chemicals applied should be alternated for the same reason, and occasional 'preventive' spraying should be avoided.

Slugs and snails can cause damage, especially if the plants are grown outside, but they are usually easily removed or killed with slug pellets. Earthworms can reduce the porosity of the compost; a fine mesh over the drainage holes in the base of the pot will avoid this problem. Woodlice can also damage roots, and a serious infestation can be treated by repotting the plant.

Viruses may cause yellow or brown mottling and streaking of the leaves. They are impossible to cure, and badly affected plants should be destroyed. Viruses are transmitted in the sap, and all pruning blades should be sterilised between plants, to prevent the virus from spreading.

Ventilation

Ventilation is necessary all through the year, but especially during the summer season when the temperatures in greenhouses are likely to rise. Cymbidiums grow best in moving air, and a fan can be used to keep the air circulating when venting is reduced during winter. The leaves should move slightly in the air flow.

CHAPTER 9

Other Uses of Cymbidium

The ethnobotanical uses of orchids have been comprehensively surveyed by Lawler (1984) and, unless otherwise stated, the original sources of the information in this account are cited there. The uses can be categorised as follows.

Medicinal Uses

In China, the mucilaginous root (perhaps also referring to the pseudobulb) of a *Cymbidium* species has been used against rheumatism and neuralgia. An aqueous decoction of the thickened root and rhizome of cultivated specimens of *C. ensifolium* has been used to treat gonorrhoea and syphilis and, in a mixture with fermented glutinous rice, for stomach ache (Hu, 1971). This species, along with *C. floribundum*, *C. goeringii* and *C. pendulum* (probably *C. aloifolium*), has been prescribed as herbal medicine by local countryside doctors for the improvement and strengthening of weak lungs and for the relief of coughs (Chen & Tang, 1982).

In northern India, *C. macrorhizon* rhizomes have been used as a diaphoretic and febrifuge, and to treat boils. *C. ensifolium* has been used in ayurvedic medicine, and as a constituent of oils that were applied to tumours, both malignant and benign. In India this species, powdered with ginger and extracted with water to produce a liquid, has been used to excite vomiting and diarrhoea and cure chronic illnesses, weakness of the eyes, vertigo and paralysis. The Sinhalese name for *C. aloifolium*, literally translated, means 'poison dust', but this probably refers to irritations of the eye caused by the fine seed, rather than to any inherent chemical properties.

The flowers of *C. ensifolium* have been used in Indo-China to prepare an eyewash and a diuretic, and the roots form part of a pectoral medicine. *C. aloifolium* (probably *C. finlaysonianum*) has been used to treat sores and burns, and has also been used to bathe weak infants and to combat menstrual irregularity. The roots of *C. finlaysonianum* are an ingredient of a mixture given as medicine to sick elephants.

In Australia, all of the native *Cymbidium* species have been widely reported to be effective against diarrhoea and dysentery. The pseudobulbs are usually chewed, but they may be dried and powdered and other parts of the plant appear to have been used as well. The seed of *C. madidum* has been used by the Aborigines as an oral contraceptive, but no antifertility activity has been found.

These species should be examined chemically for their active constituents.

Food Uses

In Japan, the flowers of *C. virescens* (= *C. goeringii*) have been made into preserves with plum vinegar, or preserved in salt and made into a drink with hot water. In Bhutan, the flower buds of *C. hookerianum* are used in curries, but are soaked in water beforehand to remove the bitter flavour (C. Bailes, pers. comm.). The pseudobulbs of *C. madidum* and *C. canaliculatum* can be eaten as a rather mucilaginous food. They are chewed raw by the Australian Aborigines, or turned into a food resembling tapioca or sago. The grated pseudobulb may also be cooked, and is reputed to be indistinguishable from arrowroot. Starch can be prepared from the pseudobulbs of both of these species. The tender parts of the stem and the base of the leaves may be eaten, and *C. suave* has also been reported as edible.

Other Uses

The juice of the pseudobulbs of *C. canaliculatum* has been used in Australia to produce a glue which may be used to affix feathers to dancers' bodies or to totemic emblems, or as a fixative for stone and bark painting. The pseudobulb may be chewed to extract the juice, which is then applied directly to the rock or bark surfaces, or over the pigments, or it may be ground together with the colour before it is used. The roasted pseudobulbs of *C. lancifolium* also produce a sticky substance which has been used in western Java to fasten Sundanese knives to their handles.

Fibres from the crushed pseudobulbs of *C. canaliculatum* may be used by the Aborigines of Australia as packing for pipe bowls to retain the smoke, or may be made into brushes. In Malaya, brooms made from *C. finlaysonianum* have been used ritually to sprinkle water in the house of a recently deceased person to prevent his spirit from haunting the living.

A *Cymbidium* species has been used in China for colouring.

CHAPTER 10

The Classification of Cymbidium

Introduction

The taxonomic position of *Cymbidium* Sw. within the Orchidaceae

John Lindley, who was the first to produce a working classification of the Orchidaceae (1830–40), placed *Cymbidium* in the tribe Vandeae. The work of Rudolf Schlechter (1924) incorporated many of Lindley's and later botanists' ideas and has provided the basis for most modern classifications. He divided the Orchidaceae into two subfamilies; the Monandreae, with one anther, comprising the majority of orchid species, and the Diandreae with two, or rarely three, anthers. The Monandreae he further split into three tribes, placing *Cymbidium* in the tribe Kerosphaerae. He used the names given by Pfitzer (1887) rather than those published by earlier authors such as Lindley. Consequently, because of priority under the International Code of Botanical Nomenclature, many of these names were synonyms. Dressler & Dodson (1960) and Dressler (1974) published classifications based on Schlechter's work, but correcting the nomenclatural mistakes. These were updated by Dressler (1981), and this latter is the classification used in this study. He recognised six subfamilies in the Orchidaceae, with *Cymbidium* included in the Vandioideae Endlicher. Within this, there are four tribes, including the Cymbidieae Pfitzer, which contains all of the sympodial vandoid orchids with two pollinia. By this criterion, subgenus *Jensoa*, with four pollinia, should not be included here. Dressler noted, however, that four pollinia is probably the primitive condition, with the fusion along the inner margin of each pair producing the more advanced condition of two deeply cleft pollinia. He stated that this reduction has occurred at least twice in the Vandeae, and perhaps more often. Certainly, this appears to have happened within the genus *Cymbidium* itself, and such criteria as compatibility in hybridisation, morphological and cytological similarity, and leaf anatomy strongly suggest that this is the case. There is little justification, therefore, for removing *Jensoa* to the tribe Maxillarieae simply on the basis of its pollinium number.

Cymbidium is placed in the subtribe Cyrtopodiinae Bentham, characterised by having pseudobulbs of several internodes, articulated leaves, usually a lateral inflorescence, the column usually with a prominent column-foot and a pollinarium with two pollinia and a viscidium. Again the character of two pollinia would split *Cymbidium*, placing *Jensoa* in the subtribe Zygopetalinae. It is evident that pollinium number has limited usefulness in differentiating tribes and subtribes within the Vandoideae.

Summary of the classification of *Cymbidium* (Dressler, 1981)

Family **Orchidaceae**
Subfamily **Vandoideae** Endlicher
Tribe **Cymbidieae** Pfitzer
Subtribe **Cyrtopodiinae** Bentham
Alliance 4, including *Cymbidium*, *Ansellia* and *Grammatophyllum*

Figure 14: The pollinaria of *Cymbidium* and related genera in the Cyrtopodiinae (all × 5 approx.)

A *Cymbidium* subgen. *Cymbidium* sect. *Cymbidium* – *C. bicolor* (Cult. Kew no. 434-63.43407)
B *Cymbidium* subgen. *Cymbidium* sect. *Austrocymbidium* – *C. canaliculatum* (Cult. Kew no. 040-83.00318)
C *Cymbidium* subgen. *Cyperorchis* sect. *Cyperorchis* – *C. elegans* (Cult. Kew no. 120-81.01685)
D *Cymbidium* subgen. *Cyperorchis* sect. *Iridorchis* – *C. tracyanum* (Cult. Kew no. 269-80.02507)
E *Cymbidium* subgen. *Cyperorchis* sect. *Eburnea* – *C. eburneum* (Cult. Kew no. 120-81.01673)
F *Cymbidium* subgen. *Jensoa* sect. *Jensoa* – *C. ensifolium* (Cult. Kew no. 481-81.06516)
G *Ansellia gigantea* (*Philcox and Leppard* 8865, Kew spirit colln. no. 29047)
H *Grammatophyllum wallisii* (Cult. Burnham Nurseries, Kew spirit colln. no. 45825)
I *Dipodium scandens* (*Lamb* SAN 91508, Kew spirit colln. no. 45489)
J *Grammangis ellisii* (*Mason* 24, Kew spirit colln. no. 46442)
K *Eulophiella roempleriana* (*Pottinger* s.n., Cult. Kew no. 548-82.05641)
L *Eulophia keithii* (Cult. Kew no. 400-65, Kew spirit colln. no. 37630)
M *Cymbidiella flabellata* (*Andrew* s.n., Cult. Kew no. 340-82.03516)
N *Porphyroglottis maxwelliae* (*Schage* s.n., Kew spirit colln. no. 33762)

The genus *Ansellia* closely resembles *Cymbidium* in its pollinarium structure, with the pollinia placed directly on to the top of the viscidium. The remaining genera (excluding *Porphyroglottis*) differ in the presence of a distinct stipe. In *Grammatophyllum* and *Dipodium* the stipe is deeply bifid, and in *Dipodium*, *Grammangis*, *Eulophiella*, *Eulophia* and *Cymbidiella* the stipe arises in the centre of the viscidium.

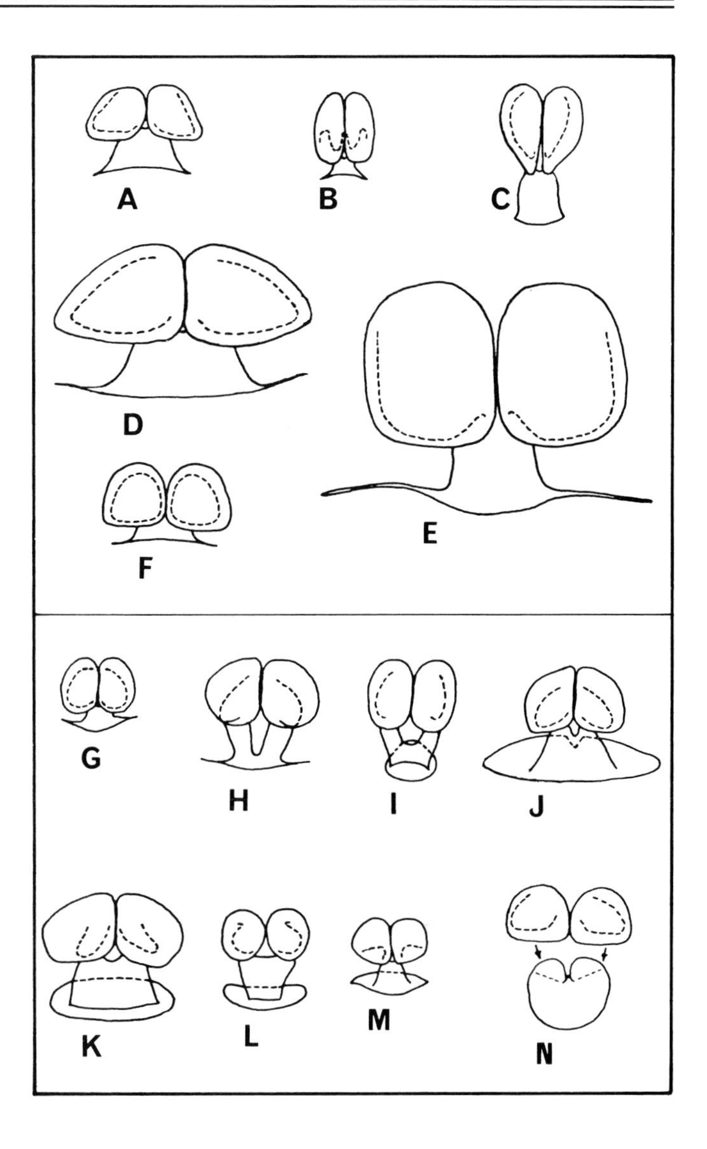

The genera in subtribe Cyrtopodiinae may be divided into five tentative alliances (Dressler, 1981): Alliance 1 – *Bromheadia*, *Claderia*; Alliance 2 – *Eulophia*, *Eulophiella*; Alliance 3 – *Cymbidiella*, *Grammangis*; Alliance 4 – *Ansellia*, *Grammatophyllum*, *Cymbidium* (and perhaps also *Porphyroglottis* and *Poicilanthe*); Alliance 5 – *Dipodium*.

All of these related genera have two cleft pollinia, but can usually be distinguished from *Cymbidium* by the structure of the pollinarium, especially by the presence of a stipe. *Ansellia* and *Grammatophyllum* are considered to be closely related to *Cymbidium*, and have similar chromosome numbers (*Ansellia*, 2n = 42; *Grammatophyllum*, 2n = 40; *Cymbidium*, 2n = 40 – Tanaka & Kamemoto, 1984). The pollinarium structure of *Ansellia* is very close to that of *Cymbidium*, with two cleft pollinia attached by elastic caudicles almost directly on to a narrow viscidium (figure 14H). It is an African genus, and can be distinguished by its paniculate rather than racemose scape, which is produced terminally rather than basally or laterally on the pseudobulb. *Grammatophyllum*, however, has two long strap-shaped stipes connecting the two cleft pollinia to the narrow viscidium (figure 14G). It is otherwise similar to *Cymbidium*, with a basal, racemose scape, and the short fusion of the base of the lip and the base of the column resemble especially subgenus *Cyperorchis*. A third genus with cane-like pseudobulbs is *Porphyroglottis*. It has a motile, hairy lip on a long column-foot. Its pollinarium is highly distinct, with the two cleft pollinia placed directly in the centre of a disc-shaped viscidium (figure 14N).

Dipodium also has a distinct pollinarium (figure 14I), with a broad, bilobulate stipe from the centre of a disc-shaped viscidium, somewhat resembling that of *Grammatophyllum*, from which it differs in its lateral scape. The structure of the lip is highly specialised and distinct from *Cymbidium*; it has a strongly hairy mid-lobe of the lip, and the side-lobes are very reduced. The base of the lip is again fused to the column.

Grammangis, *Cymbidiella*, *Eulophia* and *Eulophiella* all have a similar type of pollinarium which is distinct from that of *Cymbidium*. A broad, somewhat bilobed stipe arises from the centre of a disc-shaped viscidium (figure 14J-M). *Cymbidiella* is otherwise similar to *Cymbidium*, although it does have a column-foot, an unusual character in *Cymbidium*, and a distinct chromosome number of 54. *Eulophia* is easily distinguished by its spurred lip. *Eulophiella* has a short mentum formed by a fusion of the lateral sepals to the column-foot.

Taxonomic treatment within *Cymbidium* Sw.

The genus *Cymbidium* was established by Swartz in 1799, based on *Epidendrum aloifolium* L. (= *C. aloifolium* (L.)Sw.). The generic delimitation was broad, and it contained many species which are not now considered to belong to the genus. Only two of his species are still recognised within *Cymbidium*; *C. ensifolium* and *C. aloifolium*. Lindley (1833) also included many species which are now placed in other genera, and noted that the genus would probably be split when sufficient information became available,

Table 10.1: A comparison of the classifications of the genus *Cymbidium* Sw. (Roman type indicates sections)

Blume (1848, 1849, 1858)	Schlechter (1924)	P.F. Hunt (1970)	Seth & Cribb (1984)
Cymbidium Sw.	*Cymbidium* Sw.	*Cymbidium* Sw.	*Cymbidium* subgen. *Cymbidium*
	Eucymbidium	Cymbidium	Cymbidium
	Himantophyllum	Himantophyllum	Himantophyllum
	Austrocymbidium	Austrocymbidium	Austrocymbidium
			Floribundum
			Suavissimum
	Bigibbarium	Bigibbarium	Bigibbarium
	Cyperorchis Bl.		*Cymbidium* subgen. *Cyperorchis*
Iridorchis Bl.	Iridorchis	Iridorchis	Iridorchis
			Eburnea
	Annamaea	Annamaea	Annamaea
Cyperorchis Bl.	Eucyperorchis	Cyperorchis	Cyperorchis
	Parishiella	Parishiella	Parishiella
	Cymbidium Sw.		*Cymbidium* subgen. *Jensoa*
	Jensoa	Jensoa	Jensoa
	Maxillarianthe	Maxillarianthe	Maxillarianthe
	Geocymbidium	Geocymbidium	Geocymbidium
	Macrorhizon	Macrorhizon	Pachyrhizanthe

stating 'I presume that each section will be hereafter recognised as distinct, for which reason I have given names which may be retained either as generic or sectional'. He distinguished the S E Asian plants in subgenus *Eucymbidium*, giving the genus the form it presently resembles.

Blume (1848, 1849, 1858) removed *C. elegans* to form the genus *Cyperorchis* on the basis of its connivent perianth, its sessile lip which is parallel to the column and is shortly trilobed at the apex, and its elongated column with a beaked rostellum and a pair of pyriform pollinia on the middle of a flat, transversely ovoid viscidium. He emphasised the importance of the structure of the pollinarium, and the shape of the pollinium. Also in 1858, Blume removed *C. giganteum* (= *C. iridioides*) to the genus *Iridorchis*, this latter being distinguished from *Cymbidium* and *Cyperorchis* by its lip and column which are fused at the base for several millimetres, and the erroneous observation that the anther contained four pollinia. Blume saw the lip and column fusion as a link with *Grammatophyllum*, which has a similar structure, and distinguished these two genera on the mistaken basis of their differing pollinium number.

Reichenbach f. (1852) initially accepted the separation of *Cyperorchis* from *Cymbidium*. However, he later (1864) replaced *C. elegans* in *Cymbidium*, noting that although the pollinarium and pollinium shape are different, they were not sufficiently so to merit the separation of a distinct genus.

The discovery of more species blurred the distinction between *Cymbidium* and *Cyperorchis*. Hooker (1890) recognised *Cyperorchis*, and included in it *C. elegans*, *C. cochleare* and *C. mastersii* from N India. He stated, however, that 'except by the narrow lip, long hypochile and small usually orbicular epichile (or mid-lobe), it is not easy to separate this genus from *Cymbidium*, for the pollinia vary in both genera, and *Cyperorchis mastersii* resembles very much *Cymbidium*'. The first major revision of the complete genus was published in 1924, by Schlechter, following the discovery and description of many new species. He recognised *Cyperorchis* as a distinct genus, but emphasised the fusion of the base of the lip and the base of the column as the distinguishing feature, rather than the pollinium shape, thereby extending the limits of *Cyperorchis* to include all of the large-flowered species, including those in Blume's *Iridorchis*. Within the genus *Cyperorchis* he recognised four sections (table 10.1), including section *Eucyperorchis* in which he placed *C. elegans*, *C. cochleare*, *C. mastersii*, *C. rosea* and *C. whiteae*, and section *Iridorchis* in which he included *C. eburnea*, *C. iridioides*, and most of the other related species. *C. tigrinum* and *C. erythrostylum* were placed in separate, monotypic sections. He further distinguished *Cyperorchis* by its large pseudobulbs, its thinner and more pointed leaves, its larger flowers and the presence of a squarish stalk joining the pollinia to the viscidium. This latter character is not, in fact, evident in any of the species in *Cyperorchis*. In his revision he proposed a total of eight sections in *Cymbidium*, and four in *Cyperorchis*. This framework is the basis of the modern infrageneric classification of *Cymbidium*, most of these sections still being recognised more or less in their original form (table 10.1). His species lists included 79 species, but he does not appear to have critically examined them.

Proposed in this revision

Cymbidium subgen. *Cymbidium*

Cymbidium
Borneense
Himantophyllum
Austrocymbidium
Floribundum

Bigibbarium

Cymbidium subgen. *Cyperorchis*

Iridorchis
Eburnea
Annamaea
Cyperorchis
Parishiella

Cymbidium subgen. *Jensoa*

Jensoa
Maxillarianthe
Geocymbidium
Pachyrhizanthe

P.F. Hunt (1970) revised the sections in the genus, including *Cyperorchis* within *Cymbidium*, but maintaining Schlechter's sectional divisions (table 10.1). Although he listed the characters which have been used to split *Cyperorchis sensu* Blume *non* Schlechter from *Cymbidium* as 'the narrower labellum, relatively longer hypochile, smaller epichile, and broader stipes (viscidium) of the pollinia ... the very short column-foot adnate to the column, the fusion of the lip and column, the flowers not opening very wide ... and its much denser, pendulous racemes', he also noted that there are intermediate species, and that the characters mentioned are 'not of generic status'. This treatment is now generally accepted.

Seth & Cribb (1984) divided the genus into three subgenera. Subgenus *Cyperorchis* is distinguished from subgenus *Cymbidium* by the fusion of the lip and column-base, and corresponds to Schlechter's *Cyperorchis*. Blume's more restricted delimitation of *Cyperorchis*, and his *Iridorchis*, are maintained as sections within this subgenus. They emphasised that the four pollinia found in the species in Schlechter's sections *Jensoa*, *Maxillarianthe*, *Geocymbidium* and *Macrorhizon* (*Pachyrhizanthe*) could equally be regarded as evidence for the splitting of the genus, and they placed these in a third subgenus, *Jensoa*. They also created three new sections (table 10.1).

Although the sectional limits have been revised by Schlechter (1924), Hunt (1970) and Seth & Cribb (1984), there has been no adequate revision of *Cymbidium*. Several floras have dealt with the species of *Cymbidium* on a regional basis. These include the works by J.D. Hooker (1891), King & Pantling (1898) and Pradhan (1979) for the Himalayan and Indian species; S.Y. Hu (1973) and Wu & Chen (1980) for China; Lin (1977) for Taiwan; Maekawa (1971) for Japan; Seidenfaden (1983) for Thailand; Holttum (1964) for W Malaysia; J.J. Smith (1905), Backer & Bakhuizen (1968) and Comber (1980) for Java; Du Puy & Lamb (1984) for Sabah; and Dockrill (1969) for Australia.

Taxonomy

Cymbidium Sw. in Nov. Acta Soc. Sci. Upps. 6: 70 (1799); Lindley, Gen. Sp. Orchid. Pl.: 161 (1833); Schltr. in Fedde, Repert. 20: 96 (1924); P. Hunt in Kew Bull. 24: 93 (1970); Seth & Cribb in Arditti (ed.), Orchid Biology, Reviews and Perspectives 3: 283 (1984). Type: *C. aloifolium* (L.) Sw.

Jensoa Raf., Fl. Tellur, 4: 38 (1836). Type: *Jensoa ensata* (Thunb.) Raf. (= *C. ensifolium* (L.)Sw.).

Cyperorchis Bl., Rumphia 4: 47 (1848), Mus. Bot. 1: 48 (1849) and Orchid. Arch. Ind. 1: 92 (1858). Type: *Cymbidium elegans* Lindley.

Iridorchis Bl. Orchid. Arch. Ind. 1: 91, t.26 (1858). Type: *Cymbidium giganteum* Wall. ex Lindley (= *C. iridioides* D. Don).

Arethusantha Finet in Bull. Soc. Bot. France 44: 178–80, t.15 (1897). Type: *Arethusantha bletioides* Finet (= *C. elegans* Lindley).

Cyperocymbidium A. Hawkes in Orchid Rev. 72: 420 (1964).

Epiphytic, lithophytic or terrestrial *herbs*, with vegetative growth from the base or lower nodes of the persistent pseudobulb, which is usually produced annually, but may grow indeterminately for two or occasionally many more years; autotrophic or rarely saprophytic. *Pseudobulbs* ovoid to spindle-shaped, occasionally absent and replaced by a slender stem, often inconspicuous and concealed within the leaf bases. *Roots* thick, white, velamen-covered, branching, usually arising from the base of the new growth. *Cataphylls* several, surrounding the young growth, often becoming scarious and fibrous with age. *Leaves* up to 13, distichous, linear-elliptic or narrowly ligulate to elliptic, acuminate to strongly bilobed at the apex, articulated close to the pseudobulb to their persistent, broadly sheathing bases. *Inflorescences* racemose, densely to laxly flowered, usually arising from within the cataphylls, but occasionally from within the axils of the leaves; peduncle erect, arching or pendulous, usually covered with inflated, cymbiform sheaths. *Flowers* 1-many, up to 12 cm in diameter, often large, showy and sometimes fragrant, subtended by a small bract, pedicel and ovary. *Sepals* and *petals* free, subsimilar, often spreading or with the petals porrect and covering the column. *Lip* 3-lobed, free or fused at the base to the base of the column for 3–6 mm; side-lobes erect and weakly clasping the column; mid-lobe often recurved; callus usually two parallel ridges from near the base of the lip to the base of the mid-lobe, usually swollen towards the apex, sometimes broken in the middle, convergent towards the apices or reduced to a pair of small swellings at the base of the mid-lobe, rarely absent and replaced by a glistening shallow depression or fused into a single cuneate ridge. *Column* somewhat arcuate, weakly or strongly winged, semi-terete in cross-section, concave on the ventral surface. *Pollinia* usually 2, deeply cleft but sometimes 4 in two unequal pairs, triangular, quadrangular, ovoid or club-shaped, subsessile and attached by short, elastic caudicles to a usually triangular viscidium with the lower corners usually drawn out into short, thread-like appendages. *Capsule* fusiform to ellipsoidal or oblong-ellipsoidal, narrowing at the base to a short pedicel and at the apex to a short beak formed by the persistent column.

DISTRIBUTION. N W Himalaya to Japan, and south through Indo-China and Malesia to the Philippines, New Guinea and Australia (map 1A).

A list of the subgenera, sections and species in *Cymbidium:*

Subgenus **Cymbidium**
section **Cymbidium**
1. **C. aloifolium** (L.)Sw.
2. **C. bicolor** Lindley
subsp. **bicolor**
subsp. **obtusum** Du Puy & Cribb
subsp. **pubescens** (Lindley) Du Puy & Cribb
3. **C. rectum** Ridley
4. **C. finlaysonianum** Lindley
5. **C. atropurpureum** (Lindley)Rolfe
section **Borneense** Du Puy & Cribb
6. **C. borneense** J.J. Wood
section **Himantophyllum** Schltr.
7. **C. dayanum** Reichb.f.
section **Austrocymbidium** Schltr.
8. **C. canaliculatum** R.Br.
9. **C. hartinahianum** Comber & Nasution
10. **C. chloranthum** Lindley
11. **C. madidum** Lindley
12. **C. suave** R.Br.
13. **C. elongatum** J.J. Wood, Du Puy & P.S. Shim
section **Floribundum** Seth & Cribb
14. **C. floribundum** Lindley
15. **C. suavissimum** Sander ex C. Curtis
section **Bigibbarium** Schltr.
16. **C. devonianum** Paxton

Subgenus **Cyperorchis** (Bl.)Seth & Cribb
section **Iridorchis** (Bl.)P. Hunt
17. **C. tracyanum** L. Castle
18. **C. iridioides** D. Don
19. **C. erythraeum** Lindley
20. **C. hookerianum** Reichb.f.
21. **C. wilsonii** (Rolfe ex Cook)Rolfe
22. **C. lowianum** (Reichb.f.)Reichb.f.
var. **lowianum**
var. **i'ansonii** (Rolfe)Du Puy & Cribb
23. **C. schroederi** Rolfe
24. **C. insigne** Rolfe
25. **C. sanderae** (Rolfe)Du Puy & Cribb
section **Eburnea** Seth & Cribb
26. **C. eburneum** Lindley
27. **C. parishii** Reichb.f.
28. **C. roseum** J.J. Smith
29. **C. mastersii** Griffith ex Lindley
section **Annamaea** (Schltr.)P. Hunt
30. **C. erythrostylum** Rolfe
section **Cyperorchis** (Bl.)P. Hunt
31. **C. elegans** Lindley
32. **C. cochleare** Lindley
33. **C. whiteae** King & Pantling
34. **C. sigmoideum** J.J. Smith
section **Parishiella** (Schltr.)P. Hunt
35. **C. tigrinum** Parish ex Hook.

Subgenus **Jensoa** (Raf.)Seth & Cribb
section **Jensoa** (Raf.)Schltr.
36. **C. ensifolium** (L.)Sw.
subsp. **ensifolium**
subsp. **haematodes** Du Puy & Cribb
37. **C. munronianum** King & Pantling
38. **C. sinense** (Jackson in Andr.)Willd.
39. **C. kanran** Makino
section **Maxillarianthe** Schltr.
40. **C. cyperifolium** Wall. ex Lindley
subsp. **cyperifolium**
subsp. **indo chinense** Du Puy & Cribb

41. C. faberi Rolfe
var. **faberi**
var. **szechuanicum** (Y.S. Wu & S.C. Chen) Y.S. Wu & S.C. Chen
42. C. goeringii (Reichb.f.)Reichb.f.
var. **goeringii**
var. **serratum** (Schltr.)Y.S. Wu & S.C. Chen
var. **tortisepalum** (Fukuyama)Y.S. Wu & S.C. Chen

section **Geocymbidium** Schltr.
43. C. lancifolium Hook.

section **Pachyrhizanthe** Schltr.
44. C. macrorhizon Lindley

Key to the subgenera of *Cymbidium*

1. Lip fused at the base to the base of the column for about 3–6 mm .... **B.** *subgenus* **Cyperorchis**
   Lip free, attached to the base of the column, or rarely to a short column-foot **2**
2. Pollinia 4, in two unequal pairs; lip with callus ridges converging at the apex and forming a short tube at the base of the mid-lobe .... **C** *subgenus* **Jensoa**
   Pollinia 2, each deeply cleft, or rarely 4; callus ridges not as above .... **A.** *subgenus* **Cymbidium**

Key to the sections in subgenus *Cymbidium*

1. Pollinia 4, in two pairs . *section* **Borneense** (p. 80)
   Pollinia 2, deeply cleft behind **2**
2. Leaves elliptic, petiolate; callus ridges reduced to two small swellings at the base of the mid-lobe of the lip .... *section* **Bigibbarium** (p. 110)
   Leaves ligulate to linear-elliptic, without a distinct petiole; callus not as above **3**
3. Scape erect; callus ridges crenulate .... *section* **Austrocymbidium** (in part) (p. 88)
   Scape suberect to pendulous; callus ridges not crenulate **4**

4.* Leaves thick, coriaceous, somewhat rigid, narrowly ligulate with an obtuse to emarginate and unequally bilobed apex .... *section* **Cymbidium** (p. 62)
   Leaves usually thinner, linear-elliptic, acute to obtuse, usually oblique at the apex **5**
5. Pseudobulbs replaced by an elongated stem growing indeterminately; scape lateral, from within the axils of the leaves .... *section* **Austrocymbidium** (in part) (p. 88)
   Pseudobulbs present, produced annually; scape from the base of the pseudobulb **6**
6. Leaves long, slender, linear-elliptic, acute to acuminate; callus ridges pubescent, with glandular hairs .... *section* **Himantophyllum** (p. 83)
   Leaves not as above; callus ridges not usually pubescent, and without glandular hairs **7**
7. Flowers usually yellow-green, occasionally heavily marked with red-brown on the sepals and petals; sepals broad, rounded, concave; lip with narrow, reduced side-lobes; callus a glistening depression, or in two ridges which are either hairy or crenulate; pollinia ellipsoidal .... *section* **Austrocymbidium** (in part) (p. 88)
   Flowers usually red-brown; sepals oblong-elliptic, margins recurved; lip with broad side-lobes which weakly clasp the column; callus in two smooth, minutely papillose ridges; pollinia triangular .... *section* **Floribundum** (p. 105)

Key to the sections in subgenus *Cyperorchis*

1. Pseudobulb not produced annually, each shoot growing indeterminately for two or more years; scape from within the axils of the leaves; pollinia quadrangular .... *section* **Eburnea** (p. 132)
   Pseudobulbs produced annually; scape from near the base of the pseudobulb; pollinia triangular or clavate, rarely quadrangular **2**

* *C. chloranthum* (section *Austrocymbidium*) has somewhat similar leaves, but they are thinner and less rigid. It may be further distinguished by its erect scape with yellow flowers and crenulate callus ridges.

2. Lateral sepals curved downwards, giving the flower a narrow, triangular appearance; callus ridges united apically into a single, 3-lobed, cuneate (wedge-shaped) structure; column strongly hairy beneath, bright deep pink . . . . *section* **Annamaea** (p. 142)

   Lateral sepals porrect or spreading; callus ridges 2, not fused at the apex; column not strongly hairy beneath, not as above **3**

3. Leaves 2–4, short, to 17(22) cm, elliptic, at the apex of an exposed, lens-shaped pseudobulb; side-lobes of the lip almost uniformly purple-brown; mid-lobe with transverse dashes of purple-brown . . . . *section* **Parishiella** (p. 152)

   Leaves 5 or more, longer than above, slender, linear-elliptic, their bases covering an ovoid, usually lightly compressed pseudobulb; lip markings not as above **4**

4. Flowers large, usually 8–15 cm across; lateral sepals and petals spreading; anther-cap without an obvious backwards-pointing beak; pollinia triangular . . . . *section* **Iridorchis** (p. 114)

   Flowers smaller, less than 4 cm across, usually pendulous; sepals and petals porrect, giving the flower a bell-shaped appearance (but see *C. sigmoideum*); column and anther-cap with a pronounced backwards-pointing beak (rostellum); pollinia clavate to almost quadrangular . . . . *section* **Cyperorchis** (p. 145)

Key to the sections in subgenus *Jensoa*

1. Plant saprophytic, subterranean, without any true leaves . . . . *section* **Pachryhizanthe** (p. 190)

   Plant autotrophic, with two to several green leaves **2**

2. Leaves lanceolate, petiolate; pseudobulbs narrowly fusiform, with a lateral scape . . . . *section* **Geocymbidium** (p. 186)

   Leaves linear-elliptic, without a distinct petiole; pseudobulbs often inconspicuous, ovoid, with a basal scape **3**

3. Leaves 2–4 on each shoot . . *section* **Jensoa** (p. 156)

   Leaves 5–13* on each shoot . . . . *section* **Maxillarianthe** (p. 173)

Artificial key to the species of *Cymbidium*

1. Margin of the lip fused at the base to the base of the column for c. 3–6 mm, forming a short sac . . . **2**

   Lip free, attached to the base of the column or occasionally a short column-foot **20**

2. Callus composed of a single, wedge-shaped ridge **3**

   Callus in two parallel ridges **4**

3. Flowers usually 4–8; lip strongly veined with deep red; side-lobes of the lip with long hairs in lines over the veins; callus tapering into the mid-lobe of the lip, cream with pink mottling; column bright pink, with a dense indumentum of long hairs beneath; pollinia triangular . . . . **30. C. erythrostylum**

   Flowers usually solitary; lip white, sometimes with a few pink spots; side-lobes of the lip papillose or minutely hairy; callus ending abruptly behind the mid-lobe base, bright yellow; column white, almost glabrous beneath; pollinia quadrangular . . . . **26. C. eburneum**

4. Leaves 2–4 per pseudobulb, short, to 17(22) cm long, elliptic, at the apex of an exposed, lens-shaped pseudobulb; side-lobes of the lip almost entirely purple-brown . . . . . . **35. C. tigrinum**

   Leaves 5 or more, usually much longer than above, slender, linear-elliptic, their broad, sheathing bases covering an ovoid, somewhat compressed pseudobulb; side-lobes not coloured as above **5**

5. Sepals, petals and lip green, spotted with dark or purple-brown, glossy; flower small, the dorsal sepal 2.7–2.9 cm long; mid-lobe of the lip 2–2.5 mm broad, strap-shaped, glabrous, waxy . . . . **34. C. sigmoideum**

**C. goeringii* occasionally has 4 leaves, but can be distinguished by its usually single-flowered inflorescence.

Sepals, petals and lip not as above; flower often much larger, the dorsal sepal at least 3.2 cm long, and frequently much longer; mid-lobe of the lip 4–30 mm broad, not as above **6**

6. Lip white, with numerous fine maroon or red-brown spots; callus a broad, glabrous, slightly raised ridge behind, slightly swollen at the margins and forming two raised ridges which become strongly inflated and confluent at the apices, white . . . . **33. C. whiteae**
Lip and callus not as above **7**

7. Flowers pendulous, bell-shaped, the sepals and petals not or slightly spreading; anther-cap with a distinct backwards-pointing beak; pollinia clavate **8**
Flowers held erect, the lip and column almost horizontal, not bell-shaped; anther-cap without a distinct beak; pollinia triangular or quadrangular **9**

8. Leaves up to 1.4(2) cm broad; flower cream or yellowish with a cream to pale green lip, not or occasionally sparsely spotted with pale pink; mid-lobe strap-shaped at the base, expanded into two lobes apically and with an emarginate apex; callus ridges to the base of the lip, usually with a pair of auricles at the base which form a trough-shaped depression . . . . **31. C. elegans**
Leaves narrower, up to 0.8 cm broad; flower greenish-brown with a yellowish lip covered with numerous, fine, red spots; mid-lobe cordate to elliptic, mucronate; callus ridges short, quickly tapering from an inflated apex and terminating well behind the base of the lip, and lacking auricles at the base . . **32. C. cochleare**

9. Scape produced from the axils of the leaves; leaf apex usually finely forked, often with a short mucro in the sinus; sepals and petals white or pink; pollinia quadrangular **10**
Scape produced from near the base of the pseudobulb; leaf apex acute; sepals and petals white, green or brown; pollinia triangular (section *Iridorchis*) **12**

10. Stem not pseudobulbous, growing indeterminately for many years, eventually forming an elongated, strap-shaped base covered by scarious, persistent leaf bases with fresh leaves present only towards the apex; scape often with 5–10 flowers; petals narrow, 0.5–0.7 cm broad; mid-lobe of the lip up to 1.3 cm broad . . . . **29. C. mastersii**
Stem weakly pseudobulbous, growing for 2–4 years before a new growth is produced, not as above; scape usually with 2–5 flowers; petals wider, 1.1–1.3 cm broad; mid-lobe larger, 1.3–1.7 cm broad **11**

11. Flower smaller, dorsal sepal 4.4–4.8 cm long; side-lobes of the lip with rounded to subacute apices, shortly pubescent . . . . **28. C. roseum**
Flower larger, dorsal sepal about 5.9 cm long; side-lobes of the lip with acute, porrect apices, with minute, papillose hairs . . . . **27. C. parishii**

12. Sepals and petals white or pinkish; side-lobes of the lip with broadly rounded apices; pseudobulbs weakly compressed; scape erect to suberect; bracts up to 15 mm long; capsule almost spherical **13**
Sepals and petals yellowish to green, sometimes heavily lined and stained red-brown; side-lobes of the lip with triangular apices; pseudobulbs strongly bilaterally compressed; scape suberect to subpendulous, arching; bracts less than 5 mm long; capsule cylindrical **14**

13. Scape 100–150 cm long, with a very long, erect peduncle up to 125 cm long, and the flowers closely spaced in the apical portion; mid-lobe of the lip acute with maroon veining and spotting; terrestrial . . . . **24. C. insigne**
Scape 30–50 cm long, with a shorter, suberect peduncle about 20 cm long; mid-lobe of the lip rounded with heavy maroon blotches; epiphytic . . . . **25. C. sanderae**

14. Flowers small, less than 8 cm across; dorsal sepal less than 11 mm broad; petals very narrow, up to 7 mm broad, ligulate and strongly curved; leaves long, narrow, usually about 1 cm, but up to 1.6 cm broad; scape very slender; mid-lobe of the lip white or cream with sparse, irregular, red-brown spots . . . . **19. C. erythraeum**

Flowers lacking the above combination of characters, more than 8 cm across; dorsal sepal more than 11 mm broad; petals more than 7 mm broad, usually slightly dilated towards the apex and less strongly curved; leaf wider, up to 4 cm broad; scape more robust; mid-lobe of the lip strongly marked with red or red-brown spots or a single V-shaped blotch **15**

15. Mid-lobe of the lip with a large, V-shaped, red to pale chestnut, submarginal blotch, porrect, almost flat, margin slightly undulating, indumentum in two zones with short, silky hairs at the base and in the centre, with the V-shaped mark densely covered with minute hairs; side-lobes with right-angled apices; flower lacking an obvious scent **16**

Mid-lobe of the lip spotted or blotched with red or red-brown, recurved, margin strongly undulating; indumentum not in two distinct zones as above, sometimes with 2–3 lines of long hairs in the centre; side-lobes with acute apices; flowers sweetly scented **17**

16. Side-lobes of the lip strongly veined red-brown; sepals and petals yellow-green, striped red-brown on the veins; flowers about 8 cm across; dorsal sepal 4.5–4.9 cm long, 1.4–1.6 cm broad; callus long, two-thirds of the length of the disc .. .. **23. C. schroederi**

Side-lobes of the lip unmarked; sepals and petals usually clear green, lightly veined or shaded red-brown; flowers 8–10 cm across; dorsal sepal 4.8–5.7 cm long, 1.6–1.8 cm broad; callus short, about half of the length of the disc .... **22. C. lowianum**

17. Callus on the lip with long cilia which continue in 2 or 3 lines well into the mid-lobe; sepals and petals heavily marked red-brown; leaves up to 4 cm broad, plants very large; flowering August to December **18**

Callus on the lip papillose to pubescent, but the hairs not continuing beyond the apices of the callus ridges; sepals and petals clear green, or very lightly flushed with red-brown; leaves up to 2.5 cm broad, plants less robust; flowering January to April **19**

18. Petals sickle-shaped, often reflexed; lip large, prominent, cream, irregularly spotted and dashed red-brown; side-lobes fringed with cilia more than 1 mm long, and with hairs confined to the veins; callus of two ciliate ridges with a third line of cilia between; flower very large, 12–15 cm across; dorsal sepal 5.7–7.8 cm long, 1.4–2.9 cm broad, column 3.4–4.4 cm long, evenly winged to the base .... **17. C. tracyanum**

Petals lightly curved, spreading; lip not prominent, yellow, marked with a submarginal ring of red-brown spots; side-lobes fringed with short hairs less than 1 mm long, and with an even indumentum of short hairs; callus of two hairy ridges only; flower smaller, about 9–10 cm across; dorsal sepal 4.4–4.7 cm long, 1.2–1.8 cm broad; column 2.6–2.9 cm long with wings which taper to the base .... **18. C. iridioides**

19. Side-lobes of the lip spotted dark red-brown along the veins, fringed on the margins with cilia more than 1 mm long; mid-lobe large, striking; flower up to 15 cm across; dorsal sepal 5.6–6 cm long, 1.7–1.9 cm broad .... **20. C. hookerianum**

Side-lobes of the lip veined red-brown, with some spots towards the margins, fringed with short hairs; mid-lobe smaller, not prominent; flowers smaller, 9–10 cm across; dorsal sepal 4.4–5.7 cm long, 1.2–1.9 cm broad ...... **21. C. wilsonii**

20. Callus ridges strongly convergent in the apical half forming a short tube at the base of the mid-lobe of the lip **21**

Callus ridges not as above **30**

21. Plant saprophytic, entirely lacking leaves .... **44. C. macrorhizon**

Plant autotrophic, with 2 to several green leaves **22**

22. Leaves elliptic, petiolate; pseudobulb cigar-shaped; scape from near the middle of the pseudobulb .. .. **43. C. lancifolium**

Leaves linear-elliptic, without a distinct petiole; pseudobulbs ovoid, often inconspicuous; scape from near the base of the pseudobulb **23**

23. Flowers solitary, rarely paired .. **42. C. goeringii**
Flowers usually 3–26 .... **24**

24. Leaves 2–4 on each pseudobulb **25**
Leaves 5–10 or more on each pseudobulb **28**

25. Leaves dark green, glossy above; flowers large, dorsal sepal usually longer than 30 mm; petals porrect and covering the column; mid-lobe large, (10)12–16 mm long **26**
Leaves mid-green, not glossy, flower smaller, dorsal sepal usually less than 26(30) mm long; petals somewhat spreading, not covering the column; mid-lobe shorter, 6–10(12) mm long **27**

26. Leaves narrow, usually less than 1.5 cm broad; bracts long, equalling the pedicel and ovary in the lower flowers; sepals very slender, about 7 times as long as broad, finely acuminate .... **39. C. kanran**
Leaves broader, usually greater than 2 cm broad; bracts shorter, usually shorter than the pedicel and ovary in the lower flowers; sepals broader, about 4–5 times as long as broad, acute .... **38. C. sinense**

27. Scape usually with 3–8 flowers; sheaths overlapping, cymbiform, somewhat spreading from the peduncle; mid-lobe of the lip almost as broad as the side-lobes when the lip is flattened, usually 6–10 mm broad; column 10–15(18) mm long ... . **36. C. ensifolium**
Scape usually with 8–13 flowers; sheaths distant, amplexicaul, closely clasping the peduncle; mid-lobe of the lip small, much narrower than the side-lobes when the lip is flattened, 4.2–7.1 mm broad; column 7–11 mm long .... **37. C. munronianum**

28. Mid-lobe of the lip covered with glistening, inflated papillae; mid-lobe margin minutely fimbriate and strongly undulating .... **41. C. faberi**
Mid-lobe of the lip minutely papillose or with some inflated papillae; mid-lobe margin entire, lightly kinked or weakly undulating **29**

29. Lamina of the leaf 0.9–1.5 cm broad; sepals and petals green or yellowish, often stained with red-brown on the veins, especially towards the base of the mid-vein .... **40. C. cyperifolium**
Lamina usually less than 0.9 cm broad; sepals and petals usually cream, with pale greenish veins; side-lobes of the lip with broadly rounded apices .... **42. C. goeringii**

30. Callus ridges replaced by a viscid stripe **31**
Callus ridges not as above **32**

31. Plant with large, ovoid pseudobulbs; flowers distantly spaced on a 30–80 cm long, pendulous scape, produced from the pseudobulb base; mid-lobe of the lip narrowly obovate, with a truncated apex .... **11. C. madidum**
Plant with an elongated stem; flowers densely spaced on a c. 15–24 cm long, arching scape, produced towards the stem apex, usually in the axils of fallen leaves; mid-lobe of the lip ovate, with an obtuse to subacute apex .. **12. C. suave**

32. Callus ridges reduced to 2 small swellings at the base of the mid-lobe of the lip **33**
Callus ridges not as above **34**

33. Leaves elliptic, narrowing to a distinct petiole, 3.5–6.2 cm broad; scape sharply pendulous, with c. 15–35 flowers; lip rhomboid when flattened, the side-lobes not sharply demarked from the mid-lobe; mid-lobe of the lip with 2 large, rich purple blotches near the base; pollinia 2, cleft behind .... **16. C. devonianum**
Leaves linear-ligulate 0.5–2.1 cm broad; scape suberect, with 3–5 flowers; lip distinctly 3-lobed, the side-lobes angled at their apices; mid-lobe of the lip lightly blotched with maroon; pollinia 4, in two pairs .... **6. C. borneense**

34. Callus ridges strongly S-shaped and often broken in the middle, swollen towards the base and at the apex* **35**
Callus ridges entire, straight or slightly curved .. **36**

*See also *C. rectum* which differs from *C. bicolor* and *C. aloifolium* in its erect or suberect rather than pendulous or downward arching scape.

11. *Cymbidium suavissimum.* Burma, *Seth* 95, cult. Kew

12. *Cymbidium devonianum.* N India, *Seth* 56, cult. Kew

13. *Cymbidium tracyanum.* Burma, *Holford* s.n., cult. Kew

14. *Cymbidium iridioides.* N India, *Sainsbury* s.n., cult. Kew

35. Side-lobes of the lip not exceeding the column and anther-cap, finely mottled with maroon or red-brown . . . . **2. C. bicolor**
Side-lobes of the lip longer than the column and anther-cap, veined with maroon . . . . **1. C. aloifolium**

36. Callus ridges and mid-lobe of the lip pubescent **37**
Callus ridges and mid-lobe of the lip glabrous or minutely papillose **38**

37. Leaves 3–4 on each pseudobulb, broad, rigid and very coriaceous; scape usually with 20–60 flowers; dorsal sepal 11–25 mm long, obtuse to subacute; mid-lobe of the lip white, pale pink or pale green, lightly spotted with red or purple . . . . **8. C. canaliculatum**
Leaves usually 5–8 on each pseudobulb, slender, flexible and only slightly coriaceous; scape usually with 5–15 flowers; dorsal sepal usually 25–34 mm long, acute to shortly acuminate, often mucronate; mid-lobe of the lip deep maroon, with a basal, pale yellow, triangular patch . . . . **7. C. dayanum**

38. Dorsal sepal 25–36 mm long; mid-lobe of the lip 10–14 mm broad; column 15–18 mm long . . **39**
Dorsal sepal 11–25 mm long; mid-lobe of the lip 5–10 mm broad; column 6–15 mm long **41**

39. Leaves well-spaced along an elongated stem, with slightly oblique apices; scape with c. 1–5 flowers, produced from the axils of the upper leaves; side-lobes of the lip not angled at the apex, confluent with the base of the mid-lobe . . . . **13. C. elongatum**
Leaves covering a short, swollen or pseudobulbous stem, usually with strongly oblique to bilobed apices; scape with (7)10–33 flowers, produced from within the cataphylls at the base of the stem; side-lobes of the lip angled at the apex, making the lip strongly 3-lobed **40**

40. Side-lobes of the lip with porrect, triangular, acute apices, which are longer than the column and anther-cap; callus ridges strongly raised and terminating abruptly at the base of the mid-lobe of the lip . . . . **4. C. finlaysonianum**
Side-lobes of the lip with truncated, obtuse apices which are shorter than the column; callus ridges weakly raised, tapering gradually into the base of the mid-lobe of the lip . . . . **5. C. atropurpureum**

41. Scape pendulous or arching downwards; callus ridges often with a short, oblique groove near the apex . . . . **2. C. bicolor**
Scape erect, suberect, or arching upwards; callus ridges not as above **42**

42. Callus ridges crenulate; sepals and petals olive-green to bright yellow **43**
Callus ridges smooth; sepals and petals usually red-brown, with a narrow to broad cream margin **44**

43. Plant epiphytic; leaves 23–38 mm broad, oblique to unequally bilobed at the apex; peduncle 15–22 mm long, covered by overlapping sheaths; petals narrower than the sepals; side-lobes of the lip with subacute, porrect apices . . . . **10. C. chloranthum**
Plant terrestrial; leaves 9–15 mm broad, acute, slightly hooded at the apex; peduncle 35–60 cm long, incompletely covered by distant sheaths; petals almost as broad as the sepals; side-lobes of the lip rounded at the apex . . . . **9. C. hartinahianum**

44. Mid-lobe of the lip 5–5.5 mm broad, ligulate, pale yellow with a single maroon spot near the apex; column 8 mm long; pollinia connate at the top, forming a single pollinium . . . . . **3. C. rectum**
Mid-lobe of the lip 7–8 mm broad, ovate, white or cream with several maroon or pink blotches; becoming deep red on pollination or with age; column 12–15 mm long; pollinia 2, not connate **45**

45. Leaves narrow, up to 1.5(2) cm broad; cataphylls green; dorsal sepal 1.8–2.1 cm long; column lacking auricles at the base . . . . **14. C. floribundum**
Leaves broader, 3–3.8 cm broad; cataphylls purple; flower slightly larger, the dorsal sepal 2–2.5 cm long; column with small auricles at the base . . . . **15. C. suavissimum**

Subgenus Cymbidium

Cymbidium *subgenus* **Cymbidium**. Type: *C. aloifolium* (L.)Sw.

The type subgenus contains six sections which are characterised by having rather small (usually less than 4 cm diameter), relatively simple flowers. The sepals, petals, lip and column are all free, although there is occasionally a short column-foot. The two pollinia are deeply cleft behind, triangular to ellipsoid, and lightly compressed, attached by short, elastic caudicles to the rostellum (figure 14). In section *Borneense* there are two pairs of unequal pollinia.

Great variation exists within this subgenus, especially in the leaf morphology and anatomy, and in the structure of the callus.

Section **Cymbidium** *P. Hunt* in Kew Bull. 24: 94 (1970); Seth in Kew Bull. 37: 397–402 (1982); Seth & Cribb in Arditti (ed.), Orchid Biol., Rev. Persp. 3: 295 (1984). Type: *C. aloifolium* (L.)Sw.

section *Eucymbidium* Lindley, Gen. Sp. Orch. P1.: 162 (1833); Schltr. in Fedde, Repert. 20: 103 (1924).

The species in this section are distinguished by their thick, often rigid, ligulate leaves with obtuse to emarginate, bilobed apices, and pendulous to arching or rarely suberect scapes with well-spaced flowers. The flowers are cream or greenish, with red or brownish markings. The leaf anatomy and leaf surface morphology are very distinct and highly characteristic for this section. The mesophyll contains elongated, palisade-like cells in its upper layers, and the subepidermal fibre bundles are linked together by a complete subepidermal layer of lignified sclerenchyma (photograph 7; figure 5). The characters of the epidermis are shared by section *Borneense*, both sections having elliptical stomatal coverings, which have narrow, slit-shaped apertures (figures 7&8).

1. C. aloifolium *(L.)Sw.* in Nov. Act. Soc. Sci. Upsal. 6: 73 (1799); Lindley, Gen. Sp. Orchid. Pl.: 165 (1833) & in J. Linn. Soc. Bot. 3: 27 (1858); Hook.f., Fl. Brit. India 6: 10 (1891), *pro parte*; King & Pantling in Ann. Roy. Bot. Gard. Calcutta 8: 189–90, t.252 (1898); J.J. Smith, Orchid. Java: 482 (1905) & Figuren-Atlas: t.367 (1911); Hayata, Ic. Pl. Formos. 4: 74–6, f.37 (1914); Hara, Stearn & Williams, Enum. Flow. Pl. Nepal 1: 37 (1978); Pradhan, Indian Orchids 2: 475 (1979); Comber in Orchid Dig. 44: 164 (1980); Seth in Kew Bull. 37: 399, t.2a (1982); Seidenfaden in Opera Bot. 72: 77, f.41, t.5a (1983). Lectotype: Illustration in *Rheede*, Hortus Indicus Malabaricus 12(8): t.8 (1703) (chosen by Seth, 1982).

Epidendrum aloifolium L., Sp. Pl.: 953 (1753).
Epidendrum pendulum Roxb., P1. Coast Coromandel 1: 35, t.44 (1795). Type: Illustration in *Roxburgh, loc. cit.* (lectotype chosen by Seth, 1982).
Epidendrum aloides Curtis, Bot. Mag. 11: t.387 (1797), *sphalm.* for *E. aloifolium*.
C. pendulum (Roxb.)Sw., in Nov. Act. Soc. Sci. Upsal. 6:73 (1799); Cogniaux & Goossens, Dict. Ic. Orchid. 9: *Cymbidium*, t.6 (1899); Hu, Gen. Orchid. Hong Kong: 96 (1977).
Aerides borassii Buch.-Ham. ex J.E. Smith in Rees, Cyclop. 39: Addend. Aerides 8 (1819). Type: India, Mysore, *Buchanan-Hamilton* (holotype BM).
C. erectum Wight, Icon. P1. Ind. Or. 5: 21, t.1753 (1851); Reichb.f. in Walpers, Ann. Bot. 6: 623 (1864). Type: India, Iyamally Hills, Coimbatore, *Wight* (holotype BM?).
C. simulans Rolfe in Orchid Rev. 25: 175 (1917); Holttum, F1. Malaya 1: 519 (1957); Seidenfaden & Smitinand, Orchids of Thailand 1: 508 (1961); Wu & Chen in Acta Phytotax. Sin. 18: 300 (1980). Type: India, Sikkim, *Pantling* 268 (lectotype K!, chosen by Seth, 1982).
C. atropurpureum sensu Yen, Icon. Cymbid. Amoyens, A4 (1964) *non* (Lindley)Rolfe.
C. intermedium H.G. Jones in Reinwardtia 9: 71 (1974). Type: India, Bombay, cult. *Jones* C/85 (holotype herb. Jones).

Map 3: The distribution of the species in section *Cymbidium*:

A *C. aloifolium*
B *C. bicolor*
C *C. rectum*
D *C. finlaysonianum*
E *C. atropurpureum*
and in section *Borneense*: F *C. borneense*

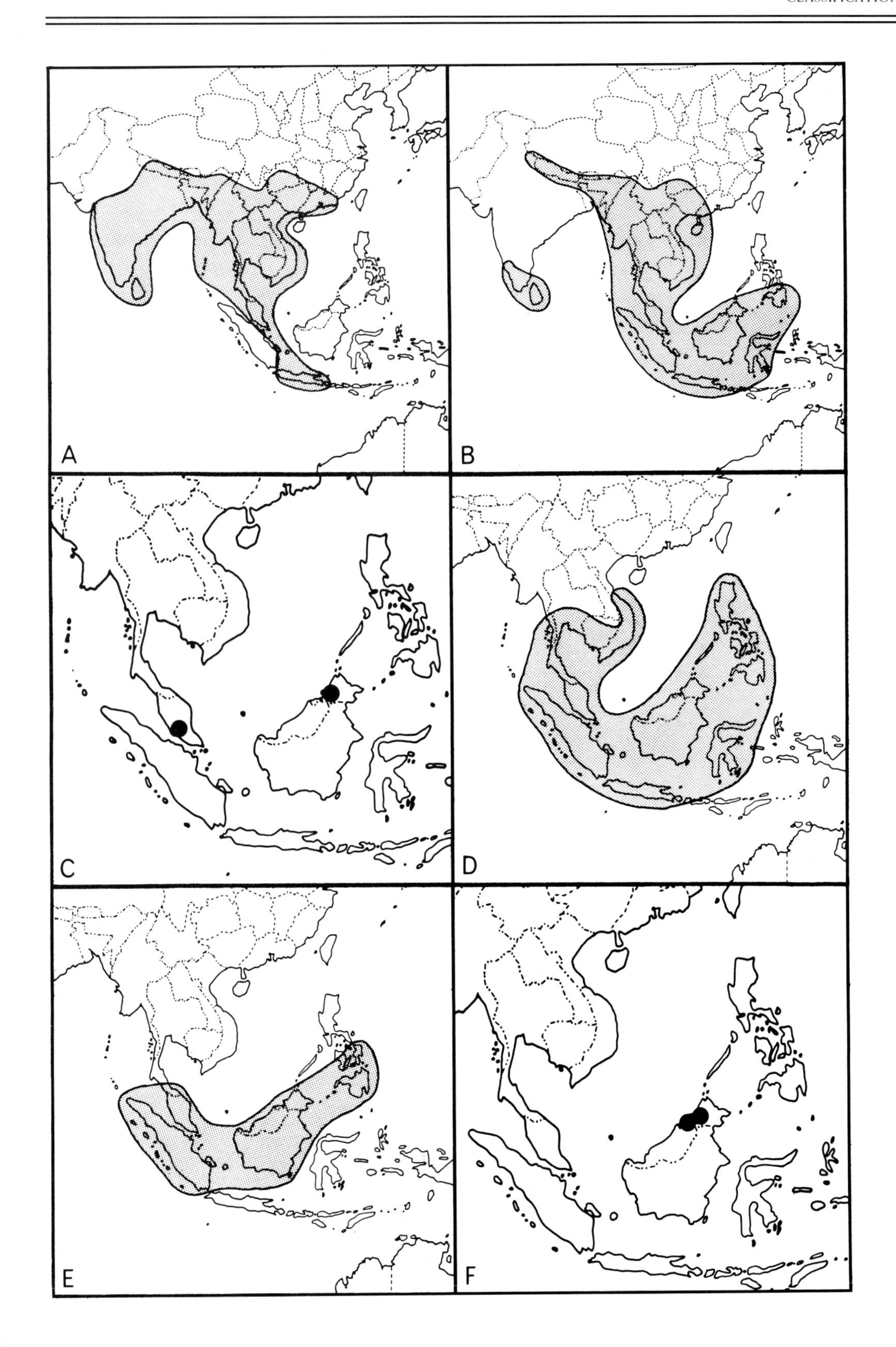
A
B
C
D
E
F

A medium-sized, perennial, epiphytic *herb*. *Pseudobulbs* usually strongly inflated, 6–9 × 3–4 cm, ovoid, bilaterally flattened, enclosed in the persistent leaf bases and 6–7 cataphylls. *Leaves* 4–5(6) per pseudobulb, (27)40–100 × 1.5–4.5(6.3) cm, ligulate, obtuse to emarginate and strongly unequally bilobed at the apex, strongly coriaceous, rigid, arching, articulated to a 4–18 cm long, broadly sheathing base; cataphylls up to 14 cm long, becoming scarious and eventually fibrous with age. *Scape* 30–70(90) cm long, from within the cataphylls, strongly pendulous except at the base, with (14)20–45 flowers; peduncle short, 4–14 cm, covered basally by 5–7 overlapping, cymbose, acute, spreading sheaths up to 3.5–7 cm long; rhachis 24–76 cm long; bracts 2–5 mm long, triangular. *Flowers* (3)3.5–4.2(5.5) cm across; lightly scented; rhachis, pedicel and ovary pale green, sometimes stained red-brown; sepals and petals pale yellow to cream with a broad, central, maroon-brown stripe, often with darker streaks; lip white or cream, side-lobes and mid-lobe veined maroon, mid-lobe yellow at the base; callus ridges and disc yellow; column dark maroon; anther-cap yellow. *Pedicel and ovary* 11–25 mm long. *Dorsal sepal* (17)19–24(28) × 5–8.5 mm, narrowly oblong to narrowly ligulate-elliptic, obtuse, somewhat mucronate, erect; lateral sepals similar, slightly oblique, spreading. *Petals* (15.5)18–23(26) × 5–8.5 mm, narrowly elliptic, obtuse to acute, sometimes mucronate, porrect and almost parallel above the column. *Lip* 15–23 × 10–14 mm when flattened, 3-lobed, saccate at the base, minutely papillose to minutely pubescent; side-lobes erect, acute to acuminate, porrect, weakly clasping the column, exceeding the column and the anther-cap; mid-lobe (7)8.4–12 × 6.5–9 mm, ovate, acute to obtuse, often mucronate, usually strongly recurved, margin entire; callus usually in two strongly sigmoid ridges which are often broken in the middle and inflated only towards the base and the apex. *Column* 10–12 mm long, arching, weakly winged towards the apex, with a very short column-foot; pollinia 2, triangular, deeply cleft, on a broadly triangular viscidium drawn into two short thread-like appendages at the tips. *Capsule* 4–7 × 2–3 cm, oblong-ellipsoidal with a short stalk and an apical beak about 1 cm long.

ILLUSTRATIONS. Plate 1; photographs 48, 49; figure 15.1.

DISTRIBUTION. Sri Lanka, Andaman Islands, India, Sikkim, Nepal, Bangladesh, S China, Hong Kong, Burma, Thailand, Cambodia, Laos, Vietnam, W Malaysia and Java (map 3A); 0–1500 m (4920 ft).

HABITAT. In the forks and hollows of large branches and tree trunks, usually in open forest in partial shade provided by the leaf canopy; flowering (March)April–June(August).

Although this species has a wide distribution, it is much less variable than the closely related *C. bicolor*, and the variation is more or less continuous. Leaf breadth is variable, but except in the Malay Islands and in Indo-China where the leaves are somewhat narrower, the broadest leaves on the plant exceed

Figure 15

1. ***C. aloifolium*** (Kew spirit no. 45061/46664)
   a Perianth, × 1
   b Lip and column, × 1
   c Pollinarium, × 3
   d Pollinium (reverse), × 3
2. ***C. bicolor*** subsp. ***bicolor*** (Kew spirit no. 45048)
   a Perianth, × 1
   b Lip and column, × 1
   c Pollinarium, × 3
   d Pollinium (reverse), × 3
3. ***C. bicolor*** *subsp.* ***obtusum*** (Kew spirit no. 48296)
   a Perianth, × 1
   b Lip and column, × 1
   c Pollinarium, × 3
   d Pollinium (reverse), × 3
4. ***C. rectum*** (Kew spirit no. 48674)
   a Perianth, × 1
   b Lip and column, × 1
   c Pollinarium, × 3
   d Pollinium (reverse), × 3
5. ***C. bicolor*** subsp. ***pubescens*** (Kew spirit no. 47389)
   a Perianth, × 1
   b Lip and column, × 1
   c Pollinarium, × 3
   d Pollinium (reverse), × 3
6. ***C. atropurpureum*** (Kew spirit no. 42895)
   a Perianth, × 1
   b Lip and column, × 1
   c Pollinarium, × 3
   d Pollinium (reverse), × 3
7. ***C. finlaysonianum*** (Kew spirit no. 48379)
   a Perianth, × 1
   b Lip and column, × 1
   c Pollinarium, × 3
   d Pollinium (reverse), × 3

1a
1b
1c
1d
2a
2b
2c
2d
3a
3b
3c
3d
4a
4b
4c
4d
5a
5b
5c
5d
6a
6b
6c
6d
7a
7b
7c
7d

2.5 cm (1 in) and, in the extreme case of the specimens from Yunnan in S W China, may reach 6.3 cm (2.5 in). Flower size is also variable, usually being 3–4.2 cm (1.2–1.6 in) in diameter, but some specimens may have flowers up to 5.5 cm (2.2 in) in diameter.

C. aloifolium can be most easily distinguished from *C. bicolor* by the markings on the lip, and by the length and shape of the side-lobes of the lip. The lip of *C. aloifolium* is always strongly veined with maroon, both on the mid- and side-lobes of the lip. Its side-lobes have long, acute or acuminate apices which extend beyond the anther-cap, while in *C. bicolor* they never exceed the anther-cap (figure 16).

Several other characters can usually be used to distinguish these species (table 10.2), but the ranges of variation overlap to some extent, and they cannot be relied upon individually. *C. aloifolium* is usually a more robust plant, producing larger pseudobulbs. Seth (1982) states that the leaf of *C. aloifolium* (more than 2.5 cm (1 in) broad) is broader than that of *C. bicolor* (less than 2.5 cm, 1 in). This is often the case, but in Thailand they have leaves of a similar width, about 2.3–2.4 cm (0.9 in) (Seidenfaden, 1983). In Java, the breadth of the leaf of both species can be about 1.5 cm (0.6 in). Other useful vegetative characters include the leaf number, which is usually 4–5 per growth in *C. aloifolium* and 5–7 in *C. bicolor*, and in the shape of the leaf apex which is unequally bilobed in both species, but obtuse in *C. bicolor*, and almost truncate or emarginate in *C. aloifolium*. The scape of the latter is always pendulous, whereas that of *C. bicolor* may be arching, although it can also be pendulous but is then always much shorter than that of *C. aloifolium*. Seth (1982) uses the shape of the side-lobes as a distinguishing character, stating that while those of *C. aloifolium* are acute, those of *C. bicolor* are obtuse. However, both *C. bicolor* subsp. *bicolor* and subsp. *pubescens* have acute side-lobe apices. The callus ridges are mostly very strongly sigmoid and often broken in *C. aloifolium*, while in *C. bicolor* they are often only weakly sigmoid or almost straight.

The difficulty of distinguishing these species from herbarium specimens has led to great confusion between them. Seth (1982) and Seth & Cribb (1984) separated these two species, and their treatment is followed here. However, the literature cited must be used with caution as the species have been frequently confused. The synonymy given by Seth is followed here, with the exception of *C. wallichii* Lindley, which has been transferred to the synonymy of *C. finlaysonianum*.

C. aloifolium was the first species of *Cymbidium* known in Europe, and is the type of the genus. Linnaeus described it as *Epidendrum aloifolium* in his *Species Plantarum* in 1753, based on an illustration made by Rheede at the end of the seventeenth century, and published in 1703 in his *Hortus Malabariensis*. This plant came from south-western peninsular India, and was described as growing on trees of *Strychnos nux-vomica*. Swartz transferred it to the genus *Cymbidium* in 1799.

Roxburgh described *Epidendrum pendulum* in 1795, from a plant collected in S E India, and the illustration he published in his *Plants of the Coast of Coromandel* now serves as the type. This was also transferred to the genus *Cymbidium* by Swartz in 1799. Lindley (1833) maintained it as distinct from *C. aloifolium*, but J.D. Hooker (1891) considered them conspecific. King & Pantling (1898) upheld them as separate species, and so did Duthie (1906) and Rolfe (1917), all giving different distinguishing characters and including different synonyms. From an examination of the type material and the original descriptions, it must be concluded that *C. pendulum* is conspecific with *C. aloifolium*, although the former name has often been wrongly used for specimens of *C. bicolor* from northern India.

Another specimen from peninsular India was named *C. erectum* by Wight in 1851. The description indicates that the leaves were ligulate, and deeply and obliquely emarginate at the apex. The side-lobes of the lip were acute, and although the description states that the callus ridges were straight, the illustration shows them as interrupted and similar to those of *C. aloifolium*. The main distinguishing feature was its erect scape. The description and drawing were probably prepared from a preserved specimen because all of the flowers are inverted, strongly suggesting that the scape was pendulous in life, as in *C. aloifolium*.

In 1917, Rolfe made a study of the *C. aloifolium*/*C. bicolor* complex, and recognised five separate species. One of those he described as a new species, *C. simulans*, distinguished by its broader leaves, its striped lip, and its interrupted, curved callus ridges. These characteristics are typical of *C. aloifolium*, with which it is now considered to be conspecific. No type specimen was cited, but Seth (1982) selected a specimen collected by Pantling as the lectotype, since Rolfe included a reference to King & Pantling's work in his description. Recently, this name has been applied to rather pale-coloured variants of *C. aloifolium* (Menninger, 1961), but these do not warrant specific status.

Recently, H.G. Jones (1974) differentiated as *C. intermedium* a plant with straight, entire callus ridges, reminiscent of those of *C. finlaysonianum*. Otherwise the plant does not differ from *C. aloifolium*. Although this is an extreme variant, and is unusual for *C. aloifolium*, the shape of the callus ridges is variable and such variants are occasionally found. This variation is mirrored in the closely related *C. bicolor*, where almost straight callus ridges are commonly found in subsp. *obtusum*.

C. aloifolium extends to W Malaysia and Java (Holttum, 1957; Backer & Bakhuizen, 1968; Comber, 1980), but it has not been recorded from Sumatra and Borneo. This anomalous distribution suggests that further collections will extend the known range of *C. aloifolium* in Malesia.

C. aloifolium and *C. bicolor* grow in similar habitats, and colonise similar niches. Typically, they occur in forks or hollows in the trunks or larger branches of forest trees, usually preferring to grow close to the bole of the tree. This position utilises the partial shade which the foliage of the tree provides, although both species can tolerate full exposure to the sun as often occurs when the tree is deciduous. They often grow in rotting wood, and will form large clumps on dead trees where their extensive root system grows into the rotting wood. The roots form a dense, spongy mass, and *C. aloifolium* may produce short, slender, erect, aerial roots which trap leaves and other plant debris. *C. aloifolium* also occurs on the trunks of palm trees (photograph 48), on rocks, and even on the walls of brick houses (Barretto, 1980).

These orchids are commonly lowland plants, and they are amongst the few epiphytic species which are able to survive on the isolated trees which remain when land is cleared for agricultural use. *C. aloifolium* occurs in this type of habitat over much of peninsular India. It also appears to tolerate trampling when the trees are climbed to remove branches for firewood. Both species are able to withstand dry, exposed conditions, and their leaves incorporate several xerophytic adaptations, including their thick, rather succulent but highly coriaceous texture, the presence of a complete subepidermal layer of lignified cells, and the stomata which are less densely spaced than in other species and have a characteristic slit-shaped, rather than rounded, aperture in the stomatal cover which can close completely, sealing the entire leaf surface in a continuous cuticular layer.

Although *C. aloifolium* and *C. bicolor* occur sympatrically over much of their range, they do not appear to hybridise. This is probably because they have slightly different flowering seasons. This is particularly pronounced in northern Thailand where *C. bicolor* flowers during the hot, dry season, from December to March and *C. aloifolium* during the rainy season, from May to August, although specimens flowering during March are not uncommon, especially in peninsular Thailand (Seidenfaden, 1983). However, in northern India, the flowering times are reversed, *C. bicolor* flowering in May and June and *C. aloifolium* in April and May (King & Pantling, 1898; Pradhan, 1979).

Altitudinal separation may also contribute to maintaining genetic isolation. In Thailand, *C. aloifolium* often occurs at slightly higher altitudes in the deciduous hill forests rather than in evergreen lowland forest. In northern India the reverse is again encountered, with *C. bicolor* growing along the Himalaya at slightly higher altitudes than *C. aloifolium*.

2. C. bicolor *Lindley*, Gen. Sp. Orchid. P1.: 164 (1833) & in Bot. Reg. 25: misc. 47 (1839); Thwaites, Enum. Ceylon P1.: 308 (1861); Reichb.f. in Walpers, Ann. Bot. 6: 625 (1864); Hook.f., F1. Brit. India 6: 11 (1891) *pro parte, syn. excl.*; Trimen, Handb. F1. Ceylon 4: 179 (1898) *pro parte*; Seth in Kew Bull. 37: 397–402 (1982). Type: Sri Lanka, *Macrae* 54 (holotype K!).

C. aloifolium sensu Jayaweera in Dassanayake (ed.), F1. Ceylon 2: 183–5, f.82 (1981) *pro parte, non* (L.)Sw.

A medium-sized, perennial, epiphytic *herb. Pseudobulbs* usually not strongly inflated, up to 5 × 2.5 cm, elongate-ovoid, slightly bilaterally flattened, enclosed in the persistent leaf bases and 4–5 cataphylls. *Leaves* (4)5–7 per pseudobulb, up to 30–68(90) × (0.8)1.2–2.9(3.1) cm, narrowly ligulate, obtuse, unequally bilobed to oblique at the apex, cartilaginous, stiff, arching, articulated to a 3–12 cm long, broadly sheathing base; cataphylls up to 11 cm long, becoming scarious and eventually fibrous with age, the longest with an abscission zone near the apex and a short lamina. *Scape* 10–50(72) cm long, from within the cataphylls, arching to pendulous, with 5–26 flowers; peduncle short, 2–12(18) cm, covered towards the base by about 5 overlapping, cymbose, spreading sheaths up to 3.5–5.5 cm long; rhachis 8–33 cm long; bracts 1.5–4.5 mm long, triangular. *Flowers* 2.5–4.5 cm across; lightly fruit-scented; rhachis green, pedicel and ovary pale yellow to red-brown; sepals and petals pale yellow to cream, with a broad, weakly defined central stripe of maroon-brown; lip white or cream with a pale yellow patch at the base, the side-lobes finely mottled with maroon or purple-brown, the mid-lobe white to yellow at the base, variously spotted or blotched with maroon or purple-brown, with a narrow cream margin; callus ridges cream to yellow, often finely mottled behind with red-brown; column cream, pale green or yellow, variously shaded with maroon; anther-cap cream to pale yellow. *Pedicel and ovary* 9–42 mm long. *Dorsal sepal* 16.5–28 × (3.2)4–6.5 mm, narrowly oblong to narrowly obovate-ligulate, obtuse to subacute, often weakly mucronate, erect; lateral sepals similar, subacute, often weakly mucronate, erect; lateral sepals similar, slightly oblique, spreading. *Petals* (14)15–21.5 × 4–6.2 mm, narrowly oblong to narrowly elliptic, obtuse to acute, somewhat porrect but not closely covering the column. *Lip* 12.5–18 × (8.5)9.5–15.5 mm when flattened, 3-lobed, saccate at the base, minutely papillose to shortly pubescent; side-lobes erect, weakly clasping the column, shorter than, or as long as the column, but not exceeding the anther-cap, acute to obtuse, porrect to recurved; mid-lobe 5.2–8.7 × 6–9.6 mm, elliptic to ovate, rounded to obtuse and often weakly mucronate at the apex, usually strongly recurved, minutely papillose to shortly glandular-pubescent, sometimes with a minutely undulating margin; callus of two ridges which may be entire and almost parallel to weakly or strongly sigmoid, or broken in the middle and inflated only towards the base and the apex, minutely papillose to minutely pubescent. *Column* (8)9–12 mm long, arching, winged towards the apex, with a very short column-foot; pollinia 2, triangular, deeply cleft, on a triangular viscidium, drawn into two, short, thread-like appendages at the tips. *Capsule* 4–6 × 2–3 cm, oblong-ellipsoidial, with a short stalk and an apical beak about 8 mm long.

DISTRIBUTION. Sri Lanka, India, S China, Indo-China, W Malaysia, Java, Sumatra, Borneo, Sabah, Celebes, Philippines (maps 3B, 4); 0–1500 m (4920 ft).

HABITAT. In the forks and hollows of large branches and tree trunks, usually in open forest in partial shade provided by the leaf canopy; flowering (December)March-June. See *C. aloifolium* for a more detailed discussion of the habitats and ecology, and the individual subspecies for their essential characters and distributions.

In his type description, Lindley (1833) stated that although *C. bicolor* was related to *C. aloifolium*, it could be distinguished by the colour of the lip, and by the saccate lip base. Although *C. aloifolium* also has a saccate lip base, further differences between these species have been documented, and they are maintained as distinct species in this study. The simplest method of distinguishing *C. bicolor* is by the colour of the lip, especially of the side-lobes, which are mottled with maroon or red-brown rather than veined and streaked with maroon. These characters are difficult to ascertain from herbarium material lacking flower colour notes, and the large variation in *C. bicolor* in particular has, in the past, caused great taxonomic confusion. The misapplication of the name *C. pendulum* to the Himalayan specimens of *C. bicolor* is one prominent example of this. The literature and synonymy cited here is based mainly on the revision of section *Cymbidium* by Seth (1982), and Seth & Cribb (1984), with the exceptions of *C. pulchellum* Schltr. (1910) which is now considered to be conspecific with *C. chloranthum* Lindley, and *C. rectum* Ridley, which is a distinct species distinguished by its narrow, canaliculate leaves, its suberect scape, its short (8 mm, 0.3 in) column, its connate pollinia, and the distinctive single maroon spot at the apex of the pale yellow, ligulate mid-lobe. The literature and specimens cited, and the distributions given by the various authors, are often confused and incorrect, and should be used with caution.

C. bicolor is a widespread and rather variable species, varying particularly in the breadth of its leaves, the angle at which its scape is held, the length of its scape, the length of its pedicel and ovary, the size of its flowers, the shape of its side-lobe apices, the shape of its callus ridges and the lip indumentum. Three more or less distinct, geographically separated taxa can be distinguished. Intermediates may occur where the distributions meet, and the differences are not sufficient to warrant the recognition of these taxa as separate species. However, as there is a geographically linked pattern in this variation, these taxa are recognised here at subspecific rank. The most useful distinguishing characters separating these three subspecies and *C. aloifolium* are summarised in table 10.2.

Key to the subspecies of *C. bicolor*

1. Sepals 4–6.5 mm longer than the petals; petals usually porrect, connivent above the column . . . . *subsp.* **bicolor**
   Sepals as long as or up to 2 mm longer than the petals; petals weakly porrect and somewhat spreading **2**
2. Lip shortly, but densely pubescent, usually with some glandular hairs especially near the base of the mid-lobe; side-lobes acute (figure 16); scape strongly pendulous . . . . *subsp.* **pubescens**
   Lip papillose or weakly pubescent on the side-lobe tips or the apices of the callus ridges; side-lobes obtuse to subacute (figure 16); scape arching to pendulous . . . . *subsp.* **obtusum**

Table 10.2: A comparison of *C. aloifolium* and the three subspecies of *C. bicolor*

Character	*C. bicolor*			*C. aloifolium*
	subsp. *bicolor*	**subsp. *pubescens***	**subsp. *obtusum***	
Leaf breadth–max (mm)	(8)13–27	15–19(22)	13–29(31)	15–45(63)
Scape length (mm)	28–31 (approx)	10–39	18–50(72)	30–70(90)
Flower number	9–18	5–21	10–26	(14)20–45
Ovary length (mm)	18–38	9–21	9–42	11–25
Angle of the scape	arching	pendulous	arching or pendulous	pendulous
Dorsal sepal length (mm)	(18)20.5–28	17.5–22	16.5–21.5	17–28
Petal length (mm)	(14)15–21.5	17–21	15.5–19.5	15.5–26
Sepal minus petal length (mm)	4–6.5	0.5–1	1–2	1.5–2
Angle of the petals	connivent above the column	almost spreading	almost spreading	connivent above the column
Lip indumentum	papillose	pubescent, often with some glandular hairs	papillose or very weakly pubescent	papillose to weakly pubescent
Side-lobe apex shape	acute, long	acute, long	obtuse to subacute, short	acute to acuminate
Callus shape	sigmoid or broken	broken	sigmoid to straight	usually strongly sigmoid or broken
Side-lobe colour	mottled purple	mottled maroon	mottled maroon	striped maroon
Mid-lobe colour	lemon yellow at the base, with a large, submarginal purple blotch	pale yellow spotted and blotched with maroon	pale yellow or cream mottled and blotched with maroon or brown	striped maroon
Distribution	Sri Lanka and southern peninsular India	Malesia	Indo-China, to northern India, Nepal and China (Hainan)	Sri Lanka to N India, China, Indo-China, W Malaysia, Java

subsp. bicolor

This subspecies is characterised by an arching scape, a rather long pedicel and ovary, a long, slender dorsal sepal which exceeds the petals by 4–6.5 mm giving the flower a spidery appearance, petals which run parallel to each other or even cross above the column, a papillose lip with rather long, acute side-lobes (which do not, however, exceed the column) and a strongly sigmoid or broken pair of callus ridges (figure 16D). The lip is cream, with purple-mottled side-lobes, and a large, submarginal purple blotch towards the apex of the mid-lobe, and pale lemon yellow callus ridges, the colour of which extends into the base of the mid-lobe.

ILLUSTRATIONS. Plate 2 (right); photographs 50, 51; figure 15.2.

DISTRIBUTION. Sri Lanka, S India (map 4).

C. bicolor was described by Lindley in 1833, from a specimen collected by Macrae in Ceylon (Sri Lanka). Subsp. *bicolor* is confined to Sri Lanka and southern and south-western peninsular India (map 4), and its distribution appears not to overlap with that of subsp. *obtusum*.

C. bicolor *subsp.* **obtusum** *DuPuy & Cribb* **subsp. nov.** affine subspecibus aliis sed inflorescentia pendula vel arcuata, petalis patentibus sepalis aequantibus, labello lobis lateralibus obtusis vel sub-acutis, labello ad apicem loborum lateralium et calli sparse papilloso differt. Typus: Thailand, Uttaradit, *Menzies & Du Puy* 120 (holotypus K!).

C. crassifolium Wall. Cat. 7357 *nom.nud.*

C. mannii Reichb.f. in Flora 55: 274 (1872). Type: India, Assam, *Mann* (holotype W).

Map 4: The distribution of the subspecies of *C. bicolor*

Subspecies *bicolor* is isolated from the other two subgenera. Subspecies *obtusum* and subspecies *pubescens* overlap in southern Indo-China, and somewhat intermediate specimens occur.

Figure 16: A comparison of the lip shapes of *C. aloifolium* and the subspecies of *C. bicolor*

A & B *C. bicolor* subsp. *obtusum* is characterised by its short, obtuse side-lobe apices, its short, rounded mid-lobe and its continuous callus ridges. Specimen A, with almost straight callus ridges, originates in N India, and has a short, pendulous scape. Specimen B, with sigmoid callus ridges, originates in Thailand, and has a longer, arching scape with well-spaced flowers.

C *C. bicolor* subsp. *pubescens* has acute side-lobe apices, broken callus ridges and a minute indumentum.

D *C. bicolor* subsp. *bicolor* has acute side-lobe apices, broken callus ridges and a longer mid-lobe. The mid-lobe is distinctively coloured with a yellow base, and a large U-shaped purple blotch, rather than the numerous spots found in the other species.

E *C. aloifolium* differs from *C. bicolor* in its long, acute side-lobe apices, which exceed the column and anther-cap in length. The side-lobes of the lip are strongly veined with maroon, rather than mottled as in *C. bicolor*.

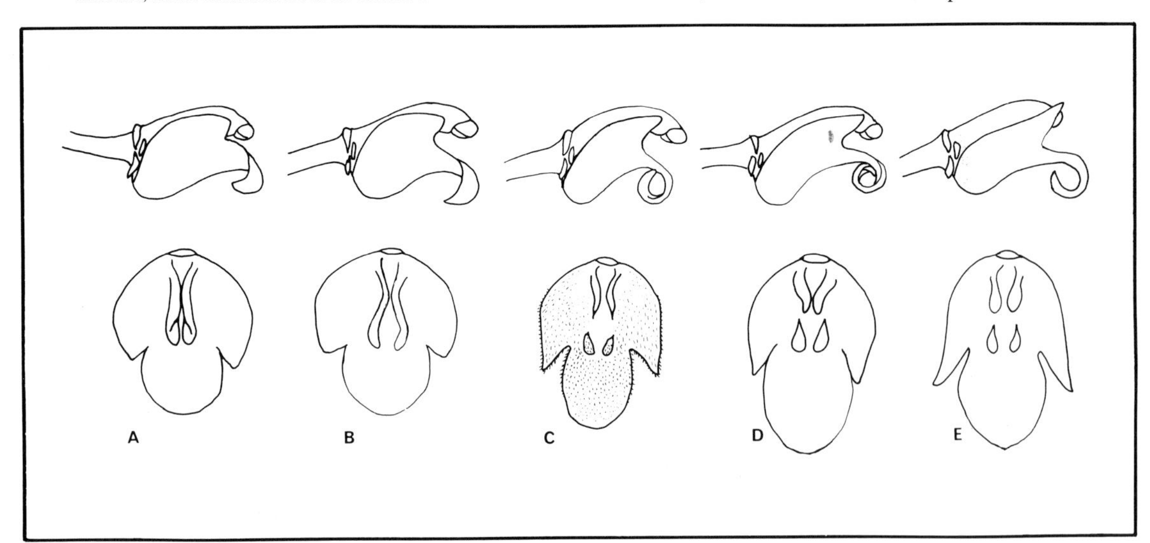

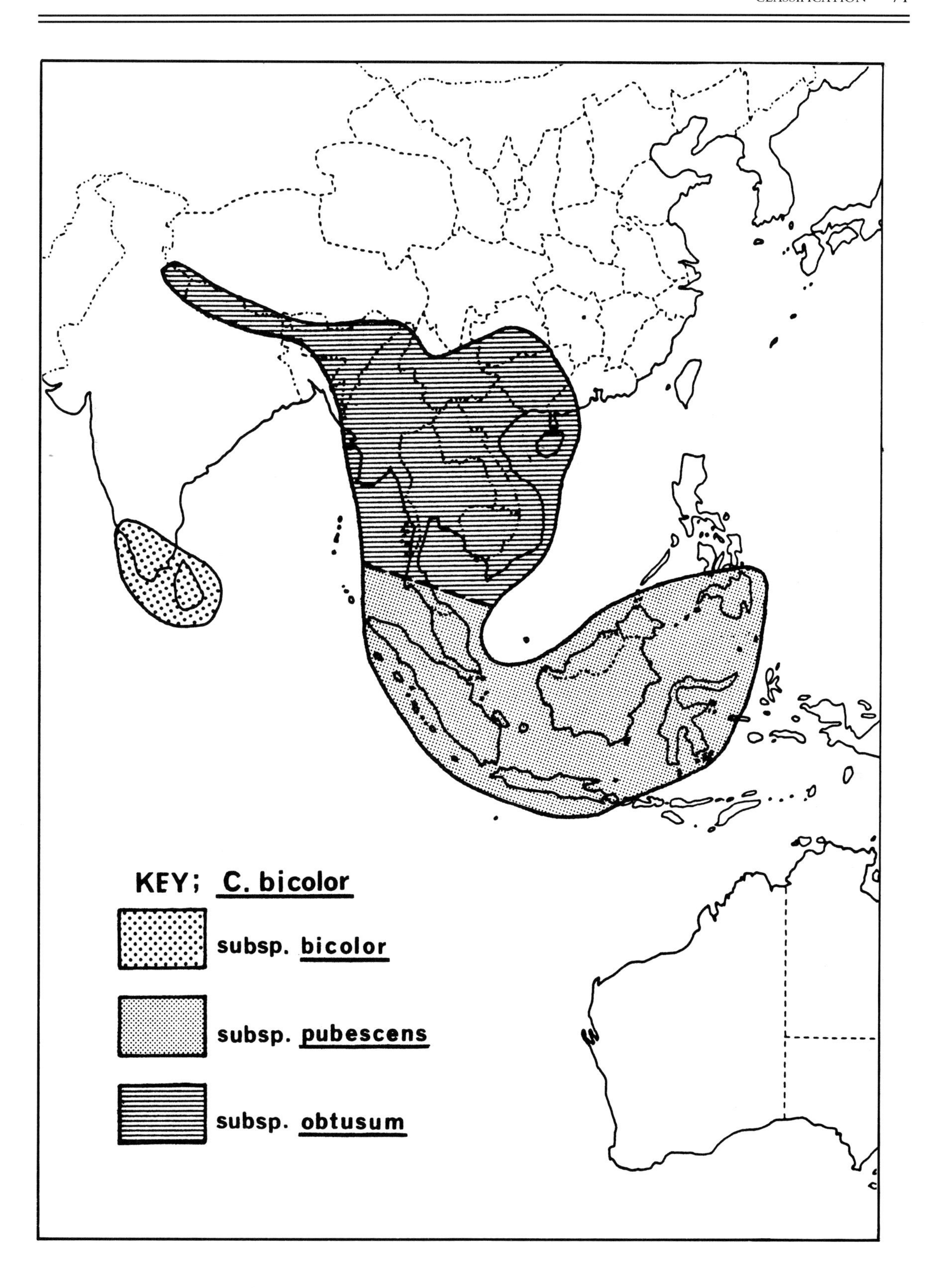
KEY; C. bicolor
subsp. bicolor
subsp. pubescens
subsp. obtusum

C. pendulum sensu King & Pantling in Ann. Roy. Bot. Gard. Calcutta 8: 188, t.251 (1898), *non* (Roxb.)Sw.
C. pendulum sensu Duthie in Ann. Roy. Bot. Gard. Calcutta 9: 136 (1906), *non* (Roxb.)Sw.
C. flaccidum Schltr. in Fedde, Repert. 12: 109 (1913) & in Fedde, Repert. Beih. 4: 267 (1919); Tso in Sunyatsenia 1: 154 (1933). Type: China, Kweichow, *Esquirol* 2728 (holotype B†)
C. pendulum sensu Schltr. in Fedde, Repert. Beih. 4: 271 (1919) *syn. excl.*, & in Fedde, Repert. 20: 104 (1924) *non* (Roxb.)Sw.
C. pendulum sensu Pradhan, Indian Orchids 2: 475 (1978) *non* (Roxb.)Sw.
C. pendulum sensu Hara, Stearn & Williams, Enum. Flow. P1. Nepal 1: 38 (1978) *non* (Roxb.)Sw. *syn. excl.*
C. pendulum sensu Wu & Chen in Acta Phytotax. Sin. 18: 300 (1980) *non* (Roxb.)Sw.
C. bicolor sensu Seidenfaden in Opera Bot. 72: 81, f.44, pl. 5b (1983), *syn. excl.*

This subspecies often has rather broad leaves, up to about 25–30 mm wide. The scape is often longer than in the other subspecies, and varies from arching to pendulous. Its petals are spreading, and its lip is papillose or weakly pubescent, especially at the tips of the side-lobes. The side-lobes of the lip are short, and usually obtuse at the apex (figure 16A, B). The mid-lobe of the lip is characteristically broadly elliptic and rounded at the apex, appearing circular in outline. Its callus ridges are entire, varying in shape from sigmoid to almost straight and parallel with a short furrow near the apex. The lip is cream, the side-lobes mottled with maroon, and the mid-lobe spotted and mottled with maroon, sometimes with a brighter yellow base.

ILLUSTRATIONS. Photographs 52, 53; figure 15.3.

DISTRIBUTION. Nepal, N India (Meghalaya, Sikkim, Assam, Arunachal Pradesh), Bhutan, S W China, Burma, Indo-China (map 4).

Some differences can be seen between the Indo-Chinese specimens and those from China and the Himalaya. The latter have rather narrower leaves, a shorter and pendulous rather than arching inflorescence, shorter pedicel and ovary lengths, and the callus ridges are usually straight not sigmoid. These are all rather variable characters with many intermediates, and the flowers appear to be almost identical in shape and colour (figure 16A, B). The decision to maintain all of this variation in a single subspecies is reinforced by the otherwise anomalous occurrence of plants resembling the N Indian variant in S China. Further investigation may indicate that these should be treated as distinct varieties, but lack of material makes this inadvisable at present.

In 1898, King & Pantling misapplied the name *C. pendulum* (Roxb.)Sw. to specimens of *C. bicolor* from Sikkim. Subsequent authors have followed this, including Duthie (1906), Schlechter (1919), Pradhan (1978) and Hara, Stearn & Williams (1978). *C. pendulum*, however, was described from a specimen collected in southern peninsular India, and is conspecific with *C. aloifolium.*

C. mannii Reichb.f. (1872) was described from northern India, and was considered by Seth (1982) to be a synonym of *C. bicolor.* However, Hooker f. (1891) and Seidenfaden (1983) included it as a synonym of *C. aloifolium.* The description is rather poor, but suggests an affinity with *C. bicolor* rather than *C. aloifolium*, especially in the shape of the callus ridges which are described as straight and parallel or sigmoid, sometimes with a sulcate apex.

C. flaccidum was collected by Esquirol in China, the locality being given originally as Sichuan, but is now thought to be Kweichow, although the type specimen has since been destroyed. It was described by Schlechter (1913, 1919), who indicated an affinity with *C. bicolor*, and described the plant as having leaves 2 cm (0.8 in) broad, a very short ovary, side-lobes of the lip with obtuse apices, the mid-lobe of the lip broadly ovate and obtuse, and the callus ridges only slightly curved. It was included by Seth (1982) as a synonym of *C. bicolor*, and this treatment is followed here.

Tso (1933) collected further specimens of this variant in Guangdong, his description closely matching the N Indian specimens of *C. bicolor.* Further specimens have been collected in Hainan. Wu & Chen (1980) give the distribution of *C. bicolor* in China as Guangdong, Guangxi, Guizhou, Hainan and Yunnan (map 4).

subsp. **pubescens** *(Lindley) Du Pay and Cribb* **comb. et stat. nov.** Type: Singapore, *Cuming*, cult. *Loddiges* (holotype K!).

C. aloifolium sensu B1., Bijdr. F1. Nederl. Indie: 378, t.19 (1825), *non* (L.)Sw., *sphalm* for *C. aloifolium*.
C. pubescens Lindley in Bot. Reg. 26: misc. 75 (1840) & in Bot. Reg. 27: t.38 (1841); Hook.f., F1. Brit. Ind. 6: 11 (1891); J.J. Smith, Orchid. Java: 483 (1905) & Figuren-Atlas: t.368 (1911); Ames & Quisumbing in Philippine J. Sci. 49: 491, t.2 (4 & 5), 8, 21, 22 (1932); Holttum, F1. Malaya, ed. 3, 1: 522 (1964); Comber in Orchid Dig. 44: 164 (1980).
C. aloifolium var. *pubescens* (Lindley)Ridley in J. Roy. As. Soc. Str. Br. 59: 196 (1911).
C. pubescens Lindley var. *celebicum* Schltr. in Fedde, Repert. 10: 190 (1911). Type: Sulawesi, Minahassa Peninsula, Lansot, *Schlechter* 20627 (holotype B†).
C. celebicum (Schltr.)Schltr. in Fedde, Repert. 21: 197 (1925).

This has narrow leaves, rarely more than 2 cm broad. It has a sharply pendulous, often rather short, few-flowered scape, and a short pedicel and ovary. The petals are usually somewhat spreading. The lip is shortly pubescent, and usually some of these hairs are glandular (swollen at the tips), especially towards the base of the mid-lobe. The side-lobe apices are usually acute, but shorter than the column, and the callus ridges are broken in the middle (figure 16C). The lip is cream, mottled with maroon on the side-lobes, and spotted and blotched with maroon or red-brown on the mid-lobe which is yellow towards the base. The callus ridges are also yellow.

ILLUSTRATIONS. Plate 2 (left); photographs 54, 55; figure 15.5.

DISTRIBUTION. W Malaysia, Java, Sumatra, Borneo, Celebes, Philippines (Mindanao) (map 4).

C. celebicum, described by Schlechter (1911, 1925) from Celebes, and tentatively included here as a synonym of subsp. *pubescens*, differs somewhat from the above description in that the inflorescence is arching, the side-lobes of the lip are rather blunt and the callus ridges are not broken. Specimens described by Ames & Quisumbing (1932) and Quisumbing (1940, 1954) from Mindanao have a large red-brown submarginal blotch on the mid-lobe. These are somewhat intermediate between subsp. *obtusum* and subsp. *bicolor* respectively. Specimens from peninsular Thailand have the lip shape of subsp. *obtusum*, but are more strongly pubescent than usual.

3. C. rectum *Ridley* in J. Roy. Asiat. Soc. Str. Br. 82: 198 (1920); Holttum, F1. Malaya 1: 516 (1953); Du Puy & Lamb in Orchid Rev. 92: 335, f.297 (1984). Type: Malaya, Negri Sembilan, *Genyns-Williams* s.n. (holotype SING.!).

A medium-sized, perennial, epiphytic *herb*. *Pseudobulbs* not strongly inflated, about 5 × 2 cm, elongate-ovoid, enclosed in the sheathing leaf bases and 2–3 cataphylls. *Leaves* 7–9 per pseudobulb, up to 60 × 0.8–1.4 cm, narrowly ligulate, unequally bilobed to oblique at the tip, strongly V-shaped in section, coriaceous, very stiff, arching, articulated 1–8 cm from the pseudobulb. *Scape* up to 40 cm long, suberect to horizontal, often pendulous in fruit, with up to 17 flowers; peduncle short, up to 9 cm, covered towards the base by about 5 sheaths; sheaths up to 4 cm long, cymbose, overlapping, spreading; bracts 2–3 mm long, triangular. *Flowers* 3–4 cm across; lightly fruit-scented; rhachis, pedicel and ovary pale green, stained red-brown; sepals and petals pale yellow or cream with a broad central stripe of maroon-brown extending to the tip; lip white with a pale yellow patch at the base, side-lobes lightly spotted maroon, mid-lobe lightly spotted maroon at the base, with a broad primrose yellow central band and a single maroon spot at the apex; callus ridges white with yellow apices; column cream, strongly stained maroon above, white and speckled maroon below; anther-cap pale yellow. *Pedicel and ovary* 1.5–3 cm long. *Dorsal sepal* 17–20 × 7–8 mm, narrowly oblong to elliptic, obtuse or weakly mucronate, erect, margins lightly revolute; lateral sepals similar, slightly oblique, spreading. *Petals* narrower, 17–18 × 6 mm, narrowly elliptic, acute, weakly porrect, but not covering the column. *Lip* 15 mm long, 3-lobed; side-lobes erect, weakly clasping the column, obtusely angled and appearing truncated at the apex, shorter than the column; mid-lobe 8–9 × 5–5.5 mm, ligulate, acute, weakly recurved, minutely papillose, margin undulate towards the base; callus of two entire sigmoid

FIGURE 17: Floral dissection of *C. rectum*

A Dorsal sepal (× 1.75)
B Lateral sepal (× 1.75)
C Petal (× 1.75)
D Lip, flattened (× 1.75)
E Lip and column, lateral view (× 1.75)
F Lip, ventral view (× 1.75)
G Pollinarium (× 5)

Specimen: Sabah, Tenom district, *Lamb* C.C.L. No. 1 (herbarium and spirit preserved material, K).

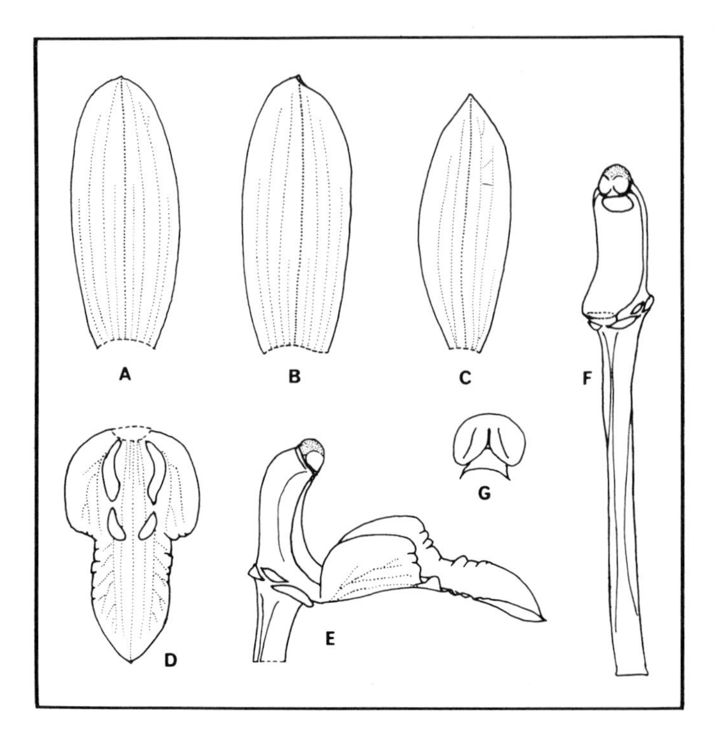

ridges, less inflated in the centre than at either the base or apex. *Column* short, 8 mm long, weakly winged at the apex, with a very short column-foot; pollinia 2, deeply cleft, connate at the top and forming one single pollinium, viscidium crescent-shaped, drawn into two, short, thread-like appendages at the tips.

ILLUSTRATIONS. Plate 3, photographs 56, 57; figures 15.4, 17.

DISTRIBUTION. Sabah, W Malaysia (map 3C); 450–500(800) m (1475–1640(2625) ft).

HABITAT. Epiphytic in full sun or light shade in open, damp forest; flowering sporadically throughout the year, but probably with a peak during September–December.

C. rectum was known as early as 1902, and had been in cultivation in Singapore Botanic Garden for many years before Ridley described it in 1920. It had been known in the gardens as *C. 'erectum'* (a name already used by Wight (1852) for another species), the specific epithet referring to its upright scape, which is an unusual character in section *Cymbidium*. A water-colour sketch of *C. rectum* by Ridley in the collection at Kew, taken from the specimen which flowered in 1902 (*Ridley* 11370), is mistakenly labelled as 'near *C. acutum*'. This specimen was thought to have been collected in Perak, from the hills behind Taiping. It was not until the type specimen was collected by Genyns-Williams in 1916 that its occurrence in Malaya could be confirmed as near Siliau, in the Negri Sembilan Hills. Holttum (1953) gives a second locality as Selangor, also in Malaya.

Subsequently, the species was lost to cultivation and the name fell out of usage. Seth (1982) placed it in his synonymy of *C. bicolor*. Recently, however, *C. rectum* was rediscovered in Sabah by Mr A. Lamb, who sent a living plant to Kew, where it flowered in September 1984. This plant corresponds closely with Ridley's sketch of 1902 and the description has been drawn up from the living material.

All of the specimens so far examined have an unusual pollinarium structure (figure 17). The pollinia are similar to other species of section *Cymbidium* except that they are connate at the top, the two cleft pollinia being fused into a single structure.

C. rectum may also be distinguished from other allied species by its narrow, strongly channelled leaves (up to 1.4 cm (0.6 in) broad), its suberect to horizontal scape, its mid-lobe of the lip which is ligulate, has an undulating margin and is characteristically coloured cream with a primrose-yellow central stripe, a single apical maroon spot and usually some sparse maroon spots at the base. The callus ridges are almost straight and are unbroken, the column is short (8 mm (0.3 in) long), and the anther cap is unusually small, about 1.5 mm (.06 in) across (figure 17).

The general habit of the plant and the scape are very similar to other species in sect. *Cymbidium*. The thick, fleshy, coriaceous leaf, with an unequally bilobed apex and a continuous subepidermal layer of sclerenchymatous cells linking the subepidermal fibre strands, is characteristic of this section. The leaf may be confused with narrow-leaved specimens of *C. bicolor* or *C. atropurpureum*, but it is normally narrower than in those species, and more strongly V-shaped in section. The flower shape and colour are also typical of sect. *Cymbidium*, closely resembling *C. aloifolium* and *C. bicolor*.

C. rectum is also, in some respects, close to *C. chloranthum* in sect. *Austrocymbidium*. They both have erect inflorescences, broad, elliptic floral segments, an undulating mid-lobe margin and a short column with a small anther.

C. rectum has not been rediscovered in Malaya, and is currently only known from the Sook Plain in Sabah, at about 450–500 m (1475–1640 ft) altitude. This area is being cleared for the development of cattle farms and much of its habitat has already been lost, making *C. rectum* a species in imminent danger of extinction.

It grows epiphytically in *Baeckia frutescens* forest, on podsolic soils, and is usually seen on small, stunted trees less than 6 m (20 ft) from the ground, and often lower. It receives only light shade, and the root system is full of biting ants which feed on the nectar exuded behind the sepals and at the base of the pedicel. In return, the ants protect the flowers from herbivorous insects. The rather beautiful forest where it grows also contains the lovely 'sealing wax' palm *Cyrtostachys renda* (syn. *C. lakka*). *C. rectum* also grows at slightly higher altitudes in adjacent valleys which contain more swampy forest, where it appears to grow more luxuriantly.

4. C. finlaysonianum *Lindley*, Gen. Sp. Orchid. P1.: 164 (1833); Hook.f., F1. Brit. India 6: 11 (1891); J.J. Smith, Orchid. Java 6: 481 (1905) & Fig. Atlas: t.366 (1911); Quisumbing in Philippine J. Sci. 72: 483–5, t.1 (5 & 6), t.3 (1940); Holttum, Orchid. Malaya, ed. 3,: 520, t.150 (1964); Backer & Bakhuizen, F1. Java 3: 396 (1968); Comber in Orchid Dig. 44: 165, + fig. (1980); Seth in Kew Bull. 37: 401, pl.27b (1982); Seidenfaden in Opera Bot. 72: 76–7, f.41, pl.4c (1983); Du Puy & Lamb in Orchid Rev. 92: 355, f.301 (1984). Type: Cochinchina, Turon Bay (Tourane?), *Finlayson* in *Wallich* 7358 (holotype K!).

C. pendulum sensu Blume, Bijdr. 379 (1825), *non* (Roxb.)Sw.

C. wallichii Lindley, Gen. Sp. Orchid. P1.: 165 (1833). Type: Malaya, Penang, *Porter* (lectotype K!, lectotype chosen here), **syn. nov**.

C. pendulum sensu Lindley, Gen. Sp. Orchid. P1.: 165 (1833) *syn.excl.*, & in Bot. Reg. 26: t.25 (1840), *non* (Roxb.)Sw.

C. pendulum (Roxb.)Sw. var. *brevilabre* Lindley in Bot. Reg. 30: t.24 (1842); Reichb.f. in Walpers, Ann. Bot. 6: 624 (1864). Type: Philippines, cult. *Loddiges*, *Cuming* (holotype K!).

C. tricolor Miq., Choix P1. Buitenz. t.19 (1863). Type: Java, cult. Buitenzorg (Bogor) Bot. Gard. (holotype U).

C. pendulum sensu Vidal, Phan. Cuming Philippin.: 150 (1885), *non* (Roxb.)Sw.

C. aloifolium sensu Guillaumin in Lecompte, F1. Gen. Indo-Chine. 6: 415 (1932) *syn. excl.*, *non* (L.)Sw.

A very large, perennial, epiphytic or lithophytic *herb*. *Pseudobulbs* up to 8 × 5 cm, ovoid, bilaterally flattened, usually obscure and weakly inflated, enclosed in the persistent leaf bases and 5–8 cataphylls which become scarious and eventually fibrous with age. *Leaves* distichous, 4–7 per pseudobulb, up to (36)50–85(100) × (2.7)3.2–6 cm, ligulate, obtuse to emarginate and strongly unequally bilobed at apex, very coriaceous and rigid, almost erect, articulated to a broadly sheathing base up to 8–16 cm long, the shortest sometimes reduced to cataphylls with an abscission zone near the apex and a very short lamina. *Scape* (20)30–115(140) cm long, from within the cataphylls, sharply pendulous, with (7)12–26 well-spaced flowers; peduncle short, 5–12 cm, covered by 6–8 overlapping, cymbose, acute, spreading sheaths up to 3.7–8 cm long; rhachis about 25–110 cm long, usually becoming slender and fractiflex towards the apex: bracts small, 1–3 mm long, triangular. *Flowers* 4–5.7 cm across; usually weakly fruit-scented; rhachis, pedicel and ovary green, stained red-brown; sepals and petals dull green to straw-yellow, usually suffused with red-brown, especially towards the tips of the sepals and along the centre of the petals; lip white, side-lobes suffused and strongly veined with purple-red, mid-lobe yellow in front of the callus ridges and with a large, submarginal, U-shaped, purple-red blotch towards the apex and often some other reddish spotting; callus ridges bright yellow in front, stained purple-red behind; column deep purple at the tip, becoming yellowish towards the base; anther-cap cream. *Pedicel and ovary* 14–45 mm long. *Dorsal sepal* 25–33 × 7–11 mm, narrowly ligulate-elliptic, obtuse, erect, margins revolute; lateral sepals similar, slightly oblique, spreading. *Petals* 24–30 × 7–11 mm; narrowly elliptic to ovate, obtuse to subacute, weakly porrect, margins weakly revolute, 7–9 veined. *Lip* 24–28 × 14–18 mm when flattened, 3-lobed, usually broadest across the side-lobes, papillose or with some minute hairs; side-lobes erect, longer than the column and usually also exceeding the anther-cap, upper margins involute, apices triangular, acute to acuminate, porrect; mid-lobe large, 11–14 × (9)11–14 mm, broadly elliptic, obtuse to emarginate, mucronate, recurved, margin undulate; callus of 2 parallel, strongly raised and well-defined ridges which terminate abruptly at the base of the mid-lobe. *Column* 15–18 mm long, arching, lightly winged; pollinia 2, about 2 mm long, triangular, deeply cleft, on a broadly triangular viscidium drawn into two short appendages at the tips. *Capsule* 5–10 × 3–4 cm, oblong-ellipsoidal, narrowing to a short pedicel and a 15–18 mm long apical beak formed by the persistent column.

ILLUSTRATIONS. Plate 4; photograph 58, figure 15.7.

DISTRIBUTION. S Vietnam, Cambodia, S Thailand, W Malaysia, Sumatra, Java, Borneo, Sabah, the Philippines and Sulawesi (map 3D); 0–300 m (0–985 ft).

HABITAT. On trees in open lowland forest or secondary forest, usually near the coast, or on exposed coastal rocks, sometimes colonising rubber, palm and other lowland tree crops; flowering all year, with seasonality in some regions (June-September(October) in cultivation in Europe and N America.)

The type specimen of *C. finlaysonianum* was collected by Finlayson in Vietnam (probably at Tourane Bay, now Da Nang), and an isotype was included in the Wallich herbarium (no. 7358). However, the species was described by Lindley (1833), and the holotype is in his herbarium. This species had previously been described by Blume (1825) as *C. pendulum*, but this name is invalidated by its previous application by Swartz (1799) to a different taxon, now included in *C. aloifolium.* The inclusion by Seth (1982) of *C. finlaysonianum* as a distinct species in section *Cymbidium* is followed here, and his synonymy is accepted with the addition of *C. wallichii* Lindley, which he included as a synonym of *C. aloifolium*, and of *C. pendulum sensu* Vidal, which he included as a synonym of *C. atropurpureum.*

Lindley (1833) cited three specimens when he described *C. wallichii.* They are all in the Wallich herbarium, numbered 7352a-coll. Finlayson; 7352b-Penang, Porter; 7352c-Attran R., Wallich. There is a fourth specimen in the Lindley herbarium which appears to be the same collection as 7352b. Of these, 7352c is a specimen of *C. aloifolium* (L.)Sw. Unfortunately, Seth (1982) selected 7352c as the lectotype of *C. wallichii*, and accordingly placed *C. wallichii* in the synonymy of *C. aloifolium.* The description by Lindley includes the phrase 'Lamellis continuis parallelis', and a drawing on his herbarium sheet of 7352b shows these two straight callus ridges. The specimen 7352c has the strongly sigmoid, interrupted keels characteristic of *C. aloifolium*, and therefore does not agree with the type description. Seth's lectotypification is therefore rejected in accordance with article 8 of the I.C.B.N., and the specimen in the Lindley herbarium (coll. Penang, Porter, Wallich 7352b) is here selected instead as the lectotype of *C. wallichii.* This specimen is identical with *C. finlaysonianum.*

C. pendulum var. *brevilabre* was also described by Lindley, in the *Botanical Register* (1842), where it was

Cymbidium hookerianum. Young s.n., cult. Kew

16. *Cymbidium lowianum.* R.B.G. Kew no. 000-73-13582, cult. Kew

17. *Cymbidium insigne.* Thailand, *Menzies & Du Puy* 500 (right) and 501 (left)

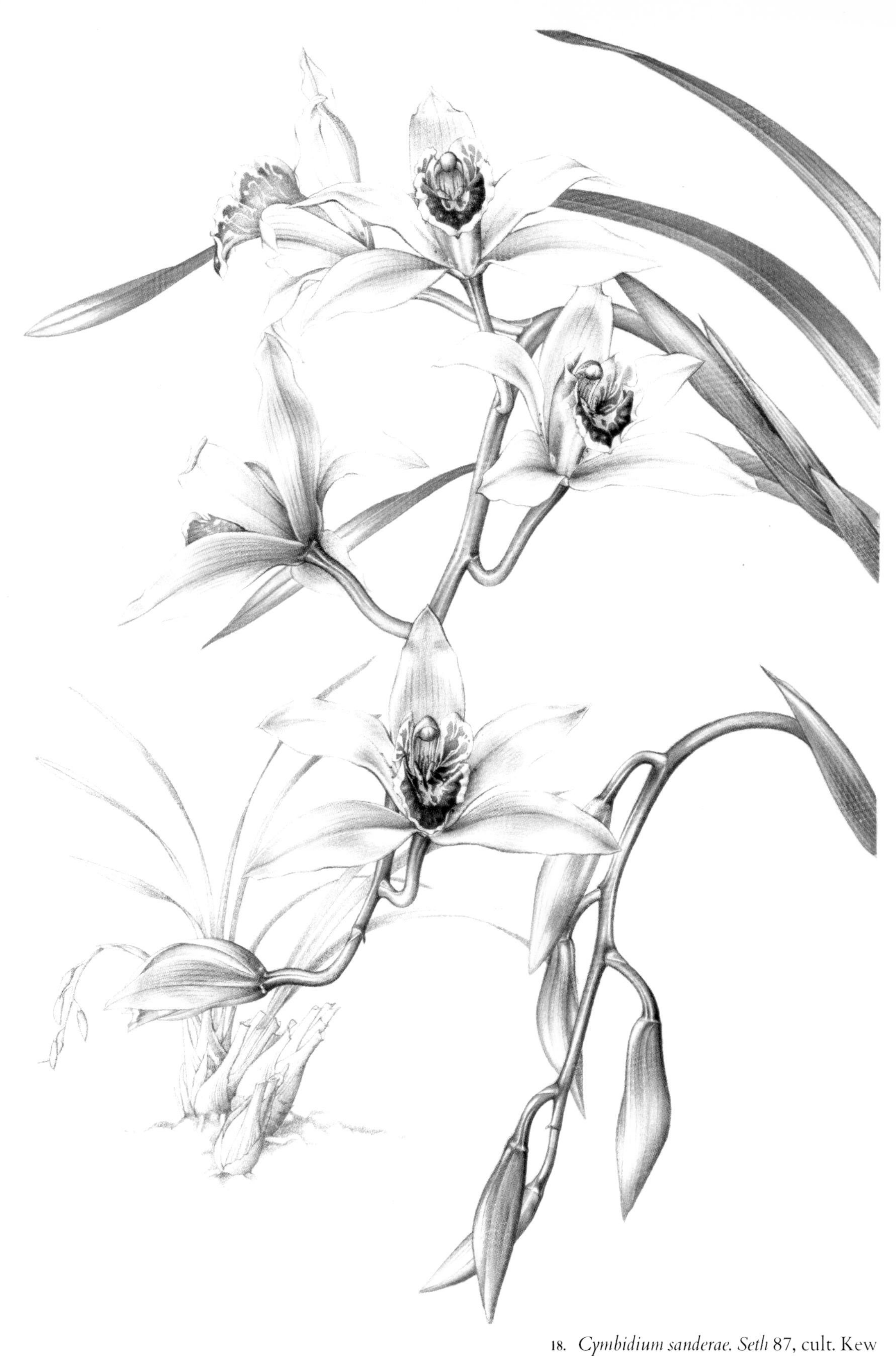

18. *Cymbidium sanderae. Seth* 87, cult. Kew

19. *Cymbidium eburneum.* N India, *Andrew* s.n., cult. Kew

20. *Cymbidium mastersii.* Bhutan, *Grierson & Long* s.n., cult. Edinburgh

simultaneously figured. He distinguished it from the related species by its broader, nearly round mid-lobe of the lip. The plate, type description and type specimen are all indistinguishable from *C. finlaysonianum*, under which species this taxon is now included as a synonym.

The leaf breadth and scape length vary somewhat in *C. finlaysonianum*, but this is often dependent on the age and general size of the specimen. It is a relatively uniform species, although some variation in flower colour and in flower size has been described. Du Puy & Lamb (1984) note that two colour variants occur in different habitats in Sabah, and that the differences are maintained when individuals are transplanted to the same environment. The variant which occurs as an epiphyte or lithophyte in lowland areas has greenish sepals and petals stained with red-brown, while the variant which colonises rocks on the coast has golden to straw-coloured sepals and petals, with little red staining. Seidenfaden (1983) noted that in Thailand the specimens which occur towards the north of the distribution of this species have slightly smaller flowers (dorsal sepal about 25 mm (1 in) long) than those which grow in the more southerly, peninsular area.

C. finlaysonianum is closely related to *C. atropurpureum* (see figure 18, and the discussion of the latter for the distinguishing characters) but it is more common, and has a wider distribution. It occurs in southern Indo-China. Thailand, peninsular Malaysia, Sumatra, Borneo, widely in the Philippines (Quisumbing, 1940, 1954), Sulawesi and W Java (where the climate is less seasonal, Backer & Bakhuizen, 1968; Comber, 1980).

It is a lowland species, often growing near the sea and rarely above 300 m (985 ft), and is therefore more commonly encountered than the other tropical *Cymbidium* species. The very robust habit of this species, and its ability to form enormous clumps, make it a conspicuous element in the lowland flora. In common with *C. aloifolium* and *C. tracyanum* it has the ability to form slender, erect roots which trap dead leaves and other plant debris. It often harbours biting ants, bees and even snakes. Its large size makes it suited to forks in the trunks and sturdy branches of trees in open forest, and its tolerance of exposed sites has made it an efficient coloniser of secondary forest. This latter attribute has also allowed it to colonise economic tree crops such as rubber in Sabah (Du Puy & Lamb, 1984) and Java (Comber, 1980), and *Borassus* (Toddy) palms and other fruit trees in Thailand, where it can become a pest. It interferes with the morning tapping of rubber trees by retaining and dripping overnight rain. Its tolerance of exposed habitats is demonstrated by its ability to form large colonies on rocks on the coast in Sabah and W Malaysia, where it is exposed to full sun and salt-bearing winds. In Sarawak and Java, it has also been reported growing on trees in humid mangrove swamps (Backer & Bakhuizen, 1968). In Sabah, it has been seen to be pollinated by large bees (A. Lamb, pers. comm.).

The flowering season varies throughout the range of this species, and seems to be dependent on climate fluctuations. In more northerly and seasonal regions of the Philippines it usually flowers between February and May, while in Indo-China it flowers during April to September. In the more equatorial regions flowering occurs all year round. In cultivation in Europe and N America it is a summer- and autumn-flowering species, usually between June and September.

5. C. atropurpureum *(Lindley)Rolfe* in Orchid Rev. 11: 190–1 (1903); Ames, Orchid., 2: 218 (1908), & 5: 199 (1915); J.J. Smith in Fedde, Repert. 32: 336 (1933); Quisumbing in Philippine J. Sci. 72: 481–3, t.1 (1 & 2), t.4 (1940); Holttum, Fl. Malaya 1: 516, t.151 (1953); Seth in Kew Bull. 37: 400, pl.26B (1982); Seidenfaden in Opera Bot. 72: 77, pl.4D (1983); Du Puy & Lamb in Orchid Rev. 92: 358, f.302 (1984). Type: ?Java, cult. *Rollissons* (neotype K!).

C. pendulum (Roxb.)Sw. var. *atropurpureum* Lindley in Gard. Chron.: 287 (1854); Hook.f. in Bot. Mag. 94:t.5710 (1865). Types: Philippines, *Cuming* & cult. *Knowles* (syntypes not located).

C. pendulum (Roxb.)Sw. var. *purpureum* W. Watson, Orchids: 151 (1890), *sphalm.* for var. *atropurpureum.*

C. finlaysonianum Wallich ex Lindley var. *atropurpureum* (Lindley) Veitch, Man. Orchid. Pl. 2: 16 (1894).

C. atropurpureum (Lindley)Rolfe var. *olivaceum* J.J.Smith, Orchid. Java in Bull. Dep. Agric. Ind. Neerland. 43: 60–2 (1910); Backer & Bakhuizen, Fl. Java 3: 396 (1968) Type: Java, Leomadjang, *Connell* (holotype L!) **syn. nov.**

A large, perennial, epiphytic or rarely lithophytic *herb*. *Pseudobulbs* up to 10 × 6 cm, ovoid, often obscurely and weakly inflated, bilaterally flattened, enclosed in the persistent leaf bases and about 4 scarious cataphylls. *Leaves* usually 7–9 per pseudobulb, up to 50–90(125) × 1.5–4 cm, ligulate, obtuse and strongly unequally bilobed at the apex, coriaceous, rather rigid, arching, articulated to a broadly sheathing base up to 15–20 cm long, the shortest reduced to cataphylls with an abscission zone near the apex and a short lamina. *Scape* 28–75 cm long, from within the cataphylls, arching to strongly pendulous, with (7)10–33 flowers; peduncle short, 5–16 cm, covered basally by 6–8 overlapping, cymbose, acute, spreading sheaths up to 7 cm long; rhachis 20–55 cm long, pendulous; bracts 1–4 mm long, triangular. *Flowers* 3.5–4.5 cm across; usually strongly coconut-scented; rhachis, pedicel and ovary pale green, often flushed with purple; sepals deep maroon to dull yellow-green with strong maroon staining; lip white, becoming yellow with age, side-lobes stained maroon-purple, mid-lobe yellow in front of the callus ridges and blotched with maroon; callus ridges bright yellow in front, stained maroon behind; column deep maroon, sometimes paler in front, anther-cap white to pale yellow. *Pedicel and ovary* 15–26 mm long. *Dorsal sepal* 28–33 × 7–10 mm, narrowly ligulate-elliptic, obtuse, suberect, margins strongly revolute; lateral sepals similar, slightly falcate and oblique, somewhat pendulous and porrect. *Petals* 25–30 × 7.5–11 mm, narrowly elliptic, subacute, weakly porrect, margins sometimes revolute. *Lip* 21–25 × 13–15 mm when flattened, 3-lobed, usually broadest across the mid-lobe, minutely papillose to minutely pubescent, the longest hairs on the side-lobe tips; side-lobes erect, much shorter than the column, apices obtuse and appearing truncated; mid-lobe large, 11–13 × 13–14 mm, broadly ovate to rhomboid, obtuse to emarginate, weakly recurved, margin entire; callus of two slightly sigmoid, weakly raised ridges which are rounded and somewhat confluent at the apices and merge gradually with the base of the mid-lobe. *Column* 16–18 mm long, arching, narrowly winged; about 3.5 mm across, pollinia 2–2.5 mm long, triangular, deeply cleft, on a broadly triangular viscidium drawn into two acuminate tips.

ILLUSTRATIONS. Plate 5; photograph 59; figures 15.6, 18.

DISTRIBUTION. S Thailand, W Malaysia, Sumatra, Java, Borneo and the Philippines (map 3E); 0–2200 m (0–7220 ft).

HABITAT. In the forks of forest trees, and occasionally on rocks, usually in lowland forests and often near the sea; flowering March–May (and not uncommonly at other times).

Lindley (1854) first described this species as a variety of *C. pendulum*, a treatment which J.D. Hooker followed in Curtis's *Botanical Magazine* of 1865. Veitch (1894) included it as a variety of *C. finlaysonianum*. It was first treated as a distinct species by Rolfe in 1903, and subsequent authors have accepted this status (Ames, 1908; Merrill, 1925; J.J. Smith, 1933; Holttum, 1953; Seidenfaden, 1983). Seth (1982) included this species in his revision of section *Cymbidium*, and his treatment is followed here with the exception of *C. pendulum sensu* Vidal, *non* (Roxb.)Sw., which is a synonym of *C. finlaysonianum*. In the type description, Lindley cited two specimens, one cultivated by Knowles, the other collected by Cuming in the Philippines. The latter specimen seems to have been the one he used to prepare the description. However, this specimen is not present in the Lindley herbarium, and has not been located elsewhere. Seth gives the type as cult. Rollissons, a specimen which is said to have come from Java, although this locality is doubtful. This specimen is the plant figured in the *Botanical Magazine* (Hooker f., 1865) and also mentioned by Rolfe (1903).

C. atropurpureum is a variable species distributed from the Philippines to the Malay Peninsula, through Borneo and Sumatra. It varies principally in its leaf breadth, its scape length and its flower colour. The scape length varies in individual specimens, although apparently at random within populations. The most distinctively coloured variant of *C. atropurpureum* is found in the Philippines and in lowland Sabah. It has deep wine-red sepals and petals, and narrow leaves up to 2.5 cm (1 in) broad. This was the variant described by Lindley (1854), J.D. Hooker (1865), Rolfe (1903) and Quisumbing (1940). However, plants from the more western part of its distribution

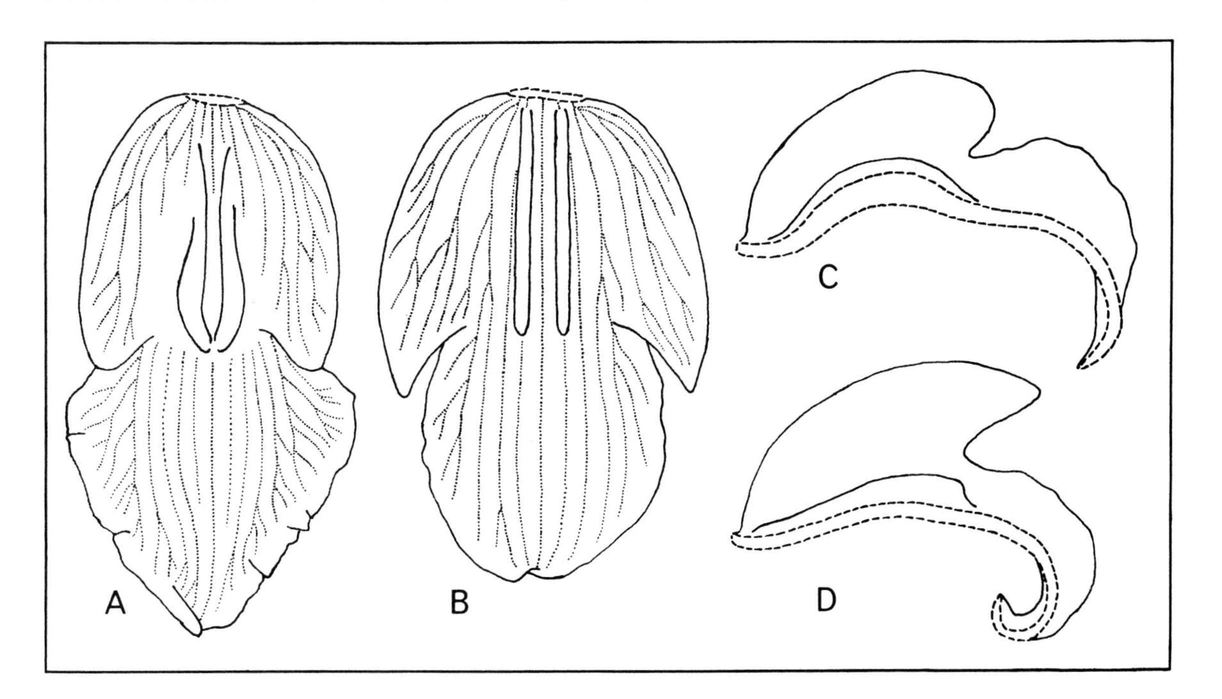

Figure 18: A comparison of the lips of *C. atropurpureum* and *C. finlaysonianum*

A & C *C. atropurpureum*, illustrating its short, truncated, rather narrow side-lobes, and obscurely defined, rounded callus ridges

B & D *C. finlaysonianum*, illustrating its long, acute, broad side-lobes, and strongly raised, parallel callus ridges which are angled at the apices

[A & B are flattened lips, while C & D are longitudinal sections through the centre of the lip.]

(W Malaysia, S Thailand, Sumatra) have greenish sepals, although the petals and the column remain deep maroon and may be assignable to var. *olivaceum* described by J.J. Smith (1910) from Java. Du Puy & Lamb (1984) reported that a greenish-flowered specimen with purple petals was collected in Sabah at higher altitudes than the more common deep purple-flowered variant. This specimen also had broader leaves than the lowland variant (up to 4 cm (1.6 in) broad), suggesting that further investigations might show that a variant with greenish sepal colour and broader leaves might be distinguished.

Holttum (1953) and Seidenfaden (1983) both state that this species can be distinguished from *C. finlaysonianum* by its narrower leaves. This is often the case, but some specimens have broad leaves, up to 4 cm (1.6 in), leading to confusion between the two taxa. The most reliable characters which distinguish them are the length and shape of the side-lobes of the lip, and the shape of the callus ridges (figure 18). In *C. atropurpureum* the side-lobes are much shorter than the column, and their apices are obtuse to rounded, appearing to be somewhat truncated, while in *C. finlaysonianum* their apices are triangular, acute and strongly porrect, exceeding both the column and the anther-cap. The callus ridges in *C. atropurpureum*, usually weakly defined, are rounded and somewhat confluent at their apices, merge into the base of the mid-lobe of the lip and are distinct from those of *C. finlaysonianum* which are strongly raised and terminate abruptly at the base of the mid-lobe.

J.J. Smith (1910) reported a single collection of *C. atropurpureum* on Java, but Backer & Bakhuizen (1968) suggest that this may have been a cultivated plant. J. Comber (pers. comm.) has recently recorded this species from western Java. In Thailand, it is confined to the peninsula, near the W Malaysian border (Seidenfaden, 1983). Quisumbing (1940, 1954) gives its distribution in the Philippines as Luzon, Leyte and Mindanao, although he suggests that it may have been introduced into Luzon and is not native there. It is a lowland forest species in the northern parts of its distribution, but it has been collected at 1200 m (3940 ft) in Sabah, and at 2200 m (7220 ft) in Sumatra. It usually flowers between March and May, although flowering at other times appears to be relatively common.

Section **Borneense** *Du Puy & Cribb*, **sect. nov.** (subgenus *Cymbidium*) affine sectio typico sed callo redacto et polliniis quatuor differt; e subgenero *Jensoa*, similtudine polliniis quatuor, sed stomatibus ellipticis similibus rimis, cellulis laevigatis epidermidis, et anatomia foliorum distinguendo. Type: *C. borneense* J.J. Wood.

The recent discovery of *C. borneense*, a highly unusual terrestrial species, has necessitated the establishment of a separate section. It is characterised by its callus ridges which are reduced to two small swellings at the base of the mid-lobe of the lip, and the presence of four pollinia. This latter character would normally indicate an affinity with the species in subgenus *Jensoa*. However, it lacks the other characteristics of this subgenus, and the leaf surface morphology indicates a close relationship with section *Cymbidium*. The elliptical stomatal coverings with long, narrow, slit-shaped apertures are characteristic of sections *Cymbidium* and *Borneense*, and the papillose epidermal cells characteristic of subgenus *Jensoa* are absent. The leaf anatomy differs from section *Cymbidium* in that there are no palisade-like mesophyll cells, and the subepidermal fibre bundles are not linked by a continuous subepidermal layer of sclerenchyma, preventing its inclusion in that section. The affinities of section *Borneense* are more fully discussed in the following species account.

6. C. borneense *J.J. Wood* in Kew Bull. 38: 69–70, t.1 (1983); J.J. Wood & Du Puy in Orchid Dig. 48: 115–16, + figs. (1984); Du Puy & Lamb in Orchid Rev. 92: 352, f.392–3 (1984). Type: Borneo, Sarawak, *Lewis* 314 (holotype K!, isotype SAR).

A medium-sized, perennial, terrestial *herb*. *Pseudobulbs* 8 × 1.5 cm, fusiform, with 6–13 distichous leaves and covered by the sheathing leaf bases with a 2 mm broad membranous margin, and occasionally 2–3 scarious or fibrous cataphylls on young specimens. *Leaves* (12)40–79 × 0.5–2.1 cm, linear-ligulate, acute, arching, somewhat coriaceous, articulated 8–12 cm from the base, not constricted into a petiole. *Scape* 16–18 cm long, suberect, arising from within the lower leaf bases, with 3–5 flowers produced in the apical third of the scape and held below the leaves; peduncle arching upwards, with 3–5 distant, amplexicaul sheaths up to 1.5 cm long; bracts 0.5–1.5 cm long, narrowly ovate, acute. *Flowers* small, c. 4 cm across; coconut-scented; rhachis green, pedicel and ovary pale olive-green, stained with red-brown; sepals and petals cream with a narrow white margin, strongly stained and blotched with maroon-purple, especially in the centre; lip white, side-lobes speckled with maroon, mid-lobe with some maroon blotches and a pale yellow patch at the base; callus ridges pale yellow; column cream above, stained maroon-purple below; anther-cap pale yellow. *Pedicel and ovary* 2–3.5 cm long. *Dorsal sepal* 2.2–2.8 × 0.5–0.6 cm, narrowly oblong-elliptic, apiculate, erect, margins somewhat revolute; lateral sepals similar, subacute, slightly curved, spreading. *Petals* 2–2.4 × 0.6–0.8 cm, narrowly ovate, oblique, slightly broader than the sepals, porrect, but not forming a hood over the column. *Lip* about 1.5 cm long when flattened; side-lobes erect, rounded, subacute at the apex, minutely papillose; mid-lobe 7–8 × 7 mm, ovate to oblong, obtuse to subacute, weakly recurved, minutely papillose, margin entire; callus very reduced, composed of two small swellings at the base of the mid-lobe. *Column* 1.2–1.3 cm long, arching, narrowly winged, minutely papillose; pollinia 4, ovate-elliptic, in two unequal pairs; viscidium triangular, with two short processes at the lower corners.

ILLUSTRATIONS. Plate 6; photographs 60–62; figures 19, 22.2.

DISTRIBUTION. Borneo (Sarawak, Sabah) (map 3F); 150–1300 m (490–4265 ft).

HABITAT. Rainforest, in deep shade, in humus-rich soils over limestone or ultra-basic rocks, often near streams; flowering March-April in Sabah, October in Sarawak.

C. borneense was described by J.J. Wood (1983) from material collected in northern Sarawak. It has since been found in central Sabah (Du Puy & Lamb, 1984). It appears to be endemic to northern Borneo. Specimens from Sabah flowered at Kew in March 1984.

FIGURE 19: Floral dissection of *C. borneense*

A Dorsal sepal (× 1.75)
B Lateral sepal (× 1.75)
C Petal (× 1.75)
D Column and ovary (× 1.75)
E Lip, flattened (× 1.75)
F Lip, lateral view (× 1.75)
G Lip, front view (× 1.75)
H Pollinia, a single pair (× 5)
I Viscidium (× 5)
J Floral bract (× 1.75)

Specimen: Borneo, Sarawak, *Lewis* 314 (K) – holotype.

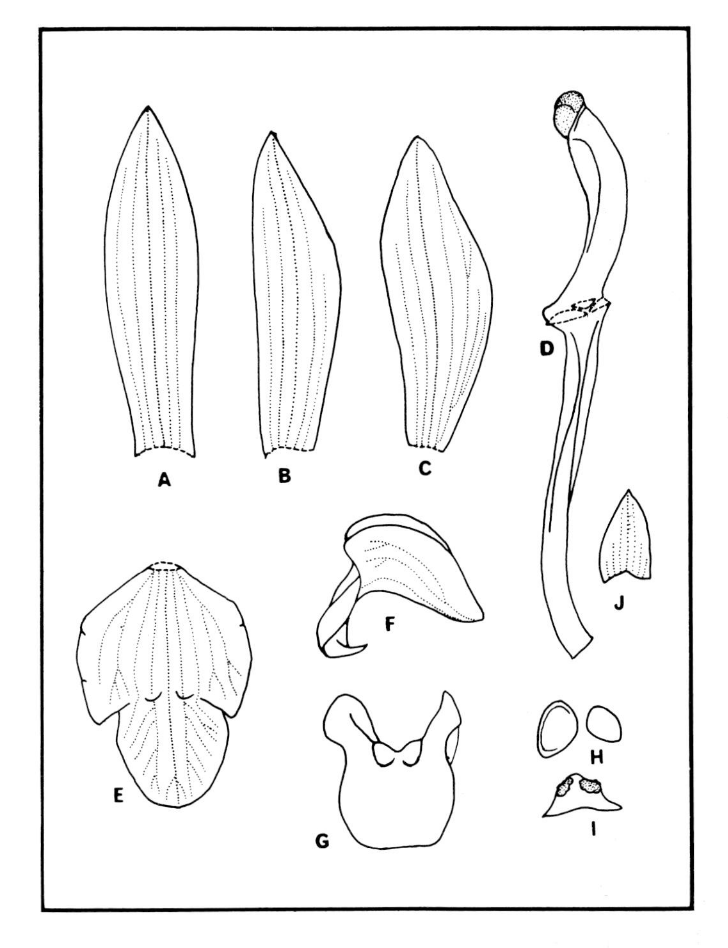

In Sarawak, *C. borneense* is found in lowland rain forest (150–250 m, 490–820 ft), but in Sabah it grows at higher altitudes in lower montane ridge-top forest (about 1200 m, 3940 ft) with *Casuarina sumatrana*. It prefers humus-rich soils over limestone or ultra-basic rocks, and has been found growing amongst serpentine boulders in Sabah. It is a forest-dwelling species, growing in damp, shady conditions, often near streams.

The taxonomic position of *C. borneense* is somewhat complex. It has some characters which are normally diagnostic of subgenus *Jensoa*, and some of subgenus *Cymbidium*. Subgenus *Jensoa* is usually distinguished by the presence of four pollinia in two pairs, while the other two subgenera have two cleft pollinia. This would therefore place *C. borneense* in subgenus *Jensoa*, close to *C. cyperifolium* because of the numerous crowded, distichous leaves (Wood & Du Puy, 1984). However, the characteristically complex callus structure of that subgenus is absent in *C. borneense*, its callus being reduced to two small swellings (figure 19).

Studies of the leaf surface show that species in subgenus *Jensoa* have strongly raised, hemispherical stomatal covers with a circular aperture and papillose epidermal cells. *C. borneense* lacks this, having smooth epidermal cells and ellipsoidal stomatal covers with slit-shaped apertures, a pattern characteristic of species in section *Cymbidium*. The texture of the leaf is more coriaceous than the species in subgenus *Jensoa*, more like that of section *Cymbidium*. However, *C. borneense* leaves lack the unequally bilobed apex of section *Cymbidium*. Transverse sections of the leaf show that the leaves do not have a subepidermal layer of lignified cells, and that the leaves are thinner and lack the layer of palisade-like mesophyll cells of section *Cymbidium*. These comparisons are summarised in table 10.3.

Table 10.3: A comparison of *C. borneense* with subgenus *Jensoa* and section *Cymbidium*

Character	subgenus *Jensoa*	*C. borneense*	subgenus *Cymbidium*, section *Cymbidium*
Leaf texture	leaf not coriaceous	leaf coriaceous	leaf very coriaceous
Stomatal cover shape	stomatal covers hemispherical	stomatal covers ellipsoid	stomatal covers ellipsoid
Stomatal aperture shape	stomatal aperture circular	stomatal aperture narrow, slit-shaped	stomatal aperture narrow, slit-shaped
Epidermal cell surface	epidermal cells papillose	epidermal cells smooth	epidermal cells smooth
Subepidermal strands	subepidermal sclerenchyma strands separate	subepidermal sclerenchyma strands separate	subepidermal sclerenchyma strands linked by a continuous subepidermal layer of sclerenchyma
Mesophyll	mesophyll cells all spherical	mesophyll cells all spherical	mesophyll cells in the upper layers elongated, palisade-like
Callus shape	callus in two ridges, convergent towards the tip, forming a short tube at the base of the mid-lobe	callus reduced to two small swellings	callus in two ridges, parallel, S-shaped or interrupted
Pollinia	four pollinia in two unequal pairs	four pollinia in two unequal pairs	two deeply cleft pollinia
Fragrance	often perfumed	coconut-scented	weakly fruit-scented, except *C. atropurpureum* which is strongly coconut-scented

The number of pollinia is usually considered to be a conservative character, and has been used by Dressler (1981) to differentiate the tribe Cymbidieae from the other Vandoid orchids. All of the genera related to *Cymbidium* have two, usually cleft, pollinia (figure 14). This suggests that the four pollinia of species in subgenus *Jensoa* are an advanced character in the tribe Cymbidieae, probably evolving from the two deeply cleft pollinia by the loss of the fusion along the inner margin. It seems possible that this step could occur twice in *Cymbidium*, giving rise to four pollinia in subgenus *Jensoa*, and again separately in *C. borneense*. The leaf epidermis and stomatal characters strongly suggests a link with section *Cymbidium*, and the conclusion drawn is that the ancestor of *C. borneense* was a species with two deeply cleft pollinia, and leaves similar in epidermal and stomatal morphology to those found in section *Cymbidium*. If it were proposed that *C. borneense* should be placed in subgenus *Jensoa*, on the basis of pollinium number, the dissimilarities in stomatal cover shape, stomatal aperture shape and epidermal cell surface type would have to be accounted for (see table 10.3, figure 20). It would be necessary to postulate the separate evolution of this complex set of leaf characters in both section *Cymbidium* and in *C. borneense*, and moreover to suggest the loss of the characteristic leaf surface features found throughout subgenus *Jensoa*. This alternative seems to be so unlikely that *C. borneense* is included here in subgenus *Cymbidium*, close to section *Cymbidium* (see figure 20 and Du Puy, 1986 for a cladistic analysis of this group of taxa).

Although *C. borneense* is related to section

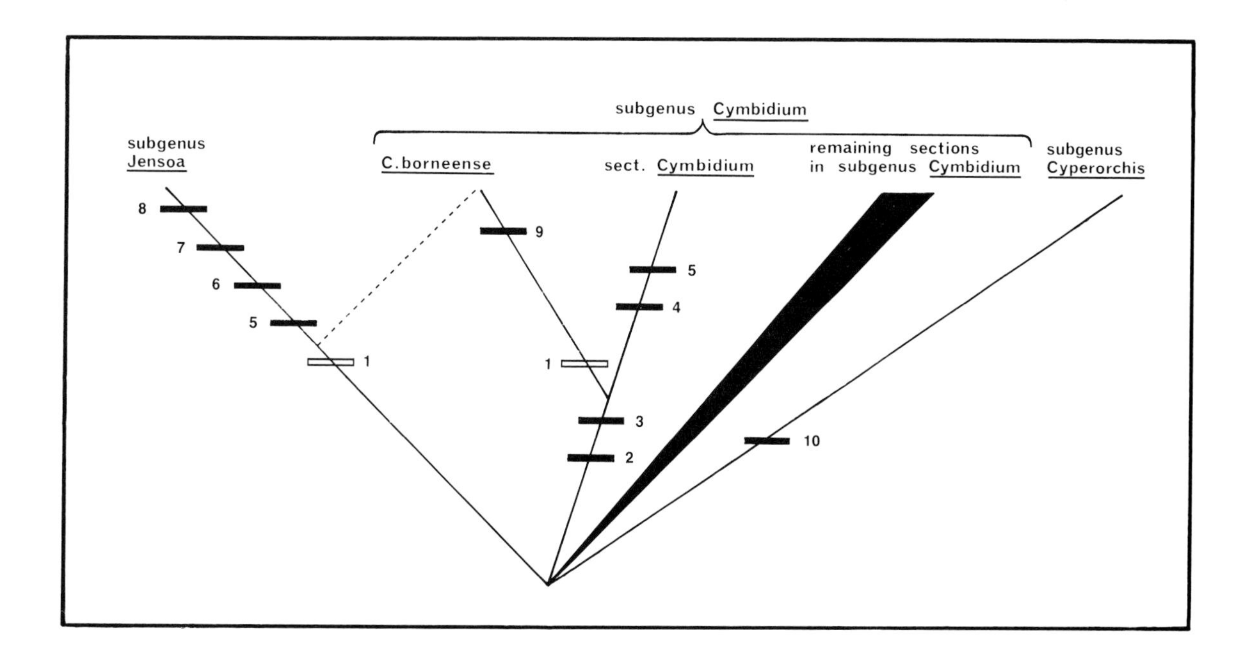

FIGURE 20: Cladogram showing the probable relationship between *C. borneense* and subgenus *Cymbidium.* The dotted line indicates where *C. borneense* would be placed if it were included in subgenus *Jensoa.* Hybridisation between species in these two subgenera could also be proposed as the origin of this species.

Synapomorphies indicated:

1 Four pollinia.
2 Oblong-ellipsoidal covering over the stomata.
3 Slit-shaped stomatal aperture.
3a Circular stomatal aperture.
4 Mesophyll cells elongated, forming a palisade-like layer.
5 Continuous subepidermal layer of sclerenchymatous cells.
6 Callus ridges convergent towards their apices, forming a short tube at the base of the mid-lobe of the lip.
7 Epidermal cells papillose.
8 Stomatal cover raised well above the rest of the epidermis.
9 Callus reduced to two small swellings at the base of the mid-lobe of the lip.
10 Column and lip fused at the base.

Cymbidium, it has four pollinia and its callus is very reduced. It has thinner leaves and the apex is not strongly unequal or bilobed. It lacks the palisade-like mesophyll cells and the subepidermal sclerenchymatous layer in the leaves. It also has longer floral bracts. These differences are enough to justify a new section, section *Borneense*, to accommodate this one species.

The possibility that *C. borneense* may be a hybrid between species in subgenus *Jensoa* and subgenus *Cymbidium* must also be considered. The most likely parents in Borneo would be *C. ensifolium* subsp. *haematodes* and *C. atropurpureum*. This might explain the unusual mixture of characters found in *C. borneense*. However,there is currently little evidence to support this hypothesis, and the following reasons argue strongly against this possibility:

1. sterility barriers exist between these two subgenera and primary hybrids are sterile;
2. the specimens of *C. borneense* do not display a range of variation intermediate between the proposed parental species;
3. the distribution of the species is too great to be the result of a single hybridisation event.

Section **Himantophyllum** *Schltr.* in Fedde, Repert, 20: 103 (1924); P. Hunt in Kew Bull. 24: 94 (1970); Seth & Cribb in Arditti (ed.), Orchid Biol., Rev. Persp. 3: 297 (1984). Type: *C. dayanum* Reichb.f. (lectotype chosen by P. Hunt, 1970).

This section is related to section *Cymbidium* and contains a single distinctive species. This affinity is indicated by its pendulous scape, its two, triangular, cleft pollinia, and the anatomy of its leaves which have many small subepidermal fibre bundles directly subjacent to the epidermis and broadly rounded margins.

This section is characterised by its slender, acuminate leaves which resemble those of *C. ensifolium* (section *Jensoa*) but are rather thicker and do not have protruding vascular bundles. The maroon and white flowers are also distinctive, with sharply acute sepals and petals, the porrect petals tending to form a hood over the column and two strongly pubescent callus ridges. The hairs are glandular, with swollen tips, a character which is highly unusual in the genus.

7. C. dayanum *Reichb.f.* in Gard. Chron.: 710 (1869); Ridley, F1. Malay Peninsula 4: 146 (1924); Quisumbing in Philippine J. Sci. 72: 488–9, pl. 2,8 (1940); Holttum, F1. Malaya: 515 (1953); Liu & Su, F1. Taiwan 5: 940 (1978); Wu & Chen, in Acta Phytotax. Sin. 18: 301, f.3 (1980); Seidenfaden in Opera Bot. 73: 82–3, t.45 (1983); Du Puy in Orchid Rev. 91: 366 (1983); Du Puy & Lamb in Orchid Rev. 92: 353, f.296 (1984). Type: Assam, cult. *Day* (holotype W., ?isotype K!).

C. leachianum Reichb.f. in Gard. Chron. n.s. 10: 106 (1878); Rolfe in J. Proc. Linn. Soc. 36: 30 (1903); Schltr. in Fedde, Repert, Beih. 4: 265 (1919); Hu in Quart. J. Taiwan Mus. 26: 138 (1973). Type: Taiwan, *Corner*, cult. *Leach* (holotype W).

C. pulcherrimum Sander in Gard. Chron. ser. 3, 10: 712 (1891). Type: cult. *Sander* (holotype not preserved).

C. eburneum var. *dayana* Hook.f., F1. Brit. India 6: 12 (1891), *non* var. *dayi* Jennings, Orchids: t.16 (1875) (= *C. eburneum*), **syn. nov.**

C. simonsianum King & Pantling in J. Asiatic Soc. Bengal 64(2): 338-9 (1895) & in Ann. Roy. Bot. Gard. Calcutta 8: 188, t.250 (1898); Hook.f. in Bot. Mag. 128: t.7863 (1902); Hayata, Icon. P1. Formosana 4: 82–3, f.39 (1914); Hu in Quart. J. Taiwan Mus. 26: 138 (1973); Pradhan, Indian Orchids 2: 474 (1979). Type: Sikkim, Teseta Valley, *Pantling* 51 (holotype K!, isotype P).

C. acutum Ridley in J. Linn. Soc. 32: 334 (1896) & Mat. Fl. Malay Penins. 1: 140 (1907). Type: Malaya, Perak, Waterloo Estate, *Elphinstone* (holotype SING!, isotype K!).

C. alborubens Makino in Bot. Mag. Tokyo 16: 11 (1902); Schltr. in Fedde, Repert, Beih. 4: 265 (1919); Hu in Quart. J. Taiwan Mus. 26: 134 (1973); Mark, Ho & Fowlie in Orchid Dig. 50: 15 (1986). Type: Japan, Musashi, *Makino*, cult. Tokyo Bot. Gard. (holotype MAK).

C. simonsianum f. *vernale* Makino in Iinuma, Somoku-Dzusetsu ed. 3, 4(18): 1186 (1912), **syn. nov.**

C. angustifolium Ames & Schweinfurth in Ames, Orchid. 6: 212–14 (1920); Masamune, Enum. Phaner. Bornearum: 150 (1942). Types: Sabah, Kiau, *Clemens* 74, 39, 82 (syntypes AMES).

C. sutepense Rolfe ex Downie in Kew Bull.: 382–3 (1925); Seidenfaden & Smitinand, Orchids of Thailand: 512 (1961); Seidenfaden in Opera Bot. 72: 83 (1983). Type: Thailand, Doi Suthep, *Kerr* 113 (holotype K!), **syn. nov.**

C. poilanei Gagnepain in Bull. Mus. Nat. Hist. Paris, ser. 2, 3: 681 (1931); Guillaumin in Lecompte, Fl. Gen. Indo-Chine: 418–19, f.38 (1932); Seidenfaden in Opera Bot. 72: 73 (1983). Type: Cambodia, Montagnes de l'Elephant, *Poilane* 316 (holotype P), **syn nov.**

C. dayanum Reichb.f. var. *austro-japonicum* Tuyama in Nakai, Icon. P1. As. Orient. 4: 363–5, t.118 (1941); Garay & Sweet, Orch. S Ryukyu Islands: 144 (1974); Lin, Native Orchids Taiwan: 103, + figs (1977); Seidenfaden, Opera Bot. 72: 83 (1983). Type: Japan, Kyushu, *Hurusawa*, cult. Koisikawa Bot. Gard. (holotype not located), **syn. nov.**

C. eburneum var. *austro-japonicum* (Tuyama) Hiroe, Orchid Fl. 2: 96 (1971). Type: as above, **syn nov.**

A medium-sized, perennial, epiphytic *herb*. *Pseudobulbs* small, about 4 × 2.5 cm, fusiform, slightly bilaterally compressed, usually covered by scarious remains of the persistent leaf bases and the cataphylls; cataphylls to 18 cm long, the longest becoming leaf-like with an abscission zone near the apex; becoming scarious and eventually fibrous with age. *Leaves* distichous, (4)5–8 per pseudobulb, up to (30)40–95(115) × 0.7–1.6(2.4) cm, linear-elliptic, acute to acuminate at the tip, with a slightly oblique apex, erect or arching towards the tip, somewhat thickened and coriaceous, dark green, the mid-vein prominent below, articulated 3–8 cm from the pseudobulb. *Scape* 18–30(35) cm long, peduncle short, suberect to horizontal, covered in pink-veined sheaths up to

8 cm long, inflated and slightly spreading in the apical half, cylindrical at the base, with 5–15(20) flowers; bracts 2–10 mm long, triangular, acute, purplish at the base. *Flowers* 4–5 cm across; not usually scented; rhachis, pedicel and ovary pale green; sepals and petals white or cream with a central maroon stripe which does not reach the apex, or occasionally suffused wine-red with a deeper central stripe, and a narrow whitish margin which may be absent except towards the base; lip white, strongly marked with maroon, with an orange or yellow spot at the base; side-lobes white, veined maroon, with a maroon margin; mid-lobe deep maroon with a basal, triangular, pale yellow stripe which does not reach the apex; callus ridges white or cream; column very dark maroon above and below; anther-cap pale yellow. *Pedicel and ovary* 2–3.5(4) cm long. *Dorsal sepal* (21)25–34 × (5)6–8 mm, narrowly elliptic to oblong-lanceolate, tapering to an acute or shortly acuminate apex, margin slightly erose towards the base, erect; lateral sepals similar, weakly oblique and slightly pendulous. *Petals* 18–28 × 5–7.5 mm, narrowly oblong to elliptic, obtuse to acute, apiculate, considerably shorter and narrower than the sepals, porrect and tending to cover the column. *Lip* 3-lobed, 15–22 × 10–15 mm when flattened; side-lobes well defined, erect and weakly clasping the column, as long as the column, with strongly porrect, triangular tips which are subacute at the apex and minutely papillose; mid-lobe 8–12 × (6)7–9 mm, ovate, entire, strongly recurved, apex subacute, mucronate to shortly acuminate, minutely papillose, often with minute hairs towards the base and with two lines of longer, often glandular, hairs extending from the callus tips to the centre of the mid-lobe; callus ridges 2, well defined, parallel, from the base of the lip to the base of the mid-lobe, covered in glandular hairs about 0.2 mm long. *Column* 11–14 mm long, arching, very weakly winged; pollinia 2, triangular, about 1.2–1.5 mm long, deeply cleft, on an oblong viscidium usually with minute extensions at the lower corners. *Capsule* 4–6 × 1.5–2 cm, ellipsoidal, tapering at the base to a stalk and at the apex to a short (1 cm) beak, formed by the persistent column.

ILLUSTRATIONS. Plate 7; photographs 63, 64, 65; figure 22.3.

DISTRIBUTION. N India (Sikkim, Assam), China, Taiwan, Ryukyus, Japan, Philippines (Luzon), Thailand, Cambodia, W Malaysia, Sumatra, Sabah (map 6C); (0) 300–1800 m ((0)985–5900 ft).

HABITAT. A light position in evergreen forest, in hollows in trees and on fallen, rotting logs, often in damp and rotting wood; flowering August-November(December), and sporadically throughout the year in tropical latitudes.

C. dayanum is a beautiful species with elegant, white and wine-red flowers on a pendulous inflorescence, and with graceful foliage. It is placed in subgenus *Cymbidium* as it has two cleft pollinia and lacks any fusion between the margins of the base of the lip and the base of the column. Superficially, however, it resembles some of the species in subgenus *Jensoa*, especially in the vegetative habit and the slender, acute, arching leaves. It can be recognised from those species in its vegetative state by its rather thicker leaves with a somewhat rounded margin and a slightly unequal tip, and its epiphytic habit. It can be easily distinguished from the other species in subgenus *Cymbidium* by its narrow, acute leaves, its sharply pointed sepals, petals and lip, and its well defined, parallel, white callus ridges which are covered in short, white glandular hairs which extend in two lines beyond the callus tips into the mid-lobe of the lip. These differences are sufficient to place it in a section of its own, a classification further justified by its unusual leaf anatomy.

John Day of Tottenham, London, was the first person to flower this species in Europe, in 1869, having imported it from Assam in 1865. Reichenbach described it in 1869, and John Day's drawing of the type specimen is preserved in his scrapbooks at Kew.

King & Pantling (1895, 1898) described and illustrated *C. simonsianum* from material collected by Pantling in the Teesta Valley, Sikkim (although Pradhan, 1979, states that the locality is actually Darjeeling). Neither the type specimen nor the description differs from *C. dayanum*, and following Ridley (1924), Holttum (1953), Ohwi (1965), Liu & Su (1978), Wu & Chen (1980), Seidenfaden (1983) and Seth & Cribb (1984) it is included here as a synonym of *C. dayanum*.

Sander (1891) named and described a cultivated specimen from Assam as *C. pulcherrimum.* Although a type specimen does not appear to have been preserved, the description is unquestionably very close to *C. dayanum,* and it is therefore treated as a synonym of this species.

J.D. Hooker (1891) included *C. dayanum* in his *Flora of British India* as *C. eburneum* var. *dayana.* These two species are very distinct and are classified in separate subgenera. The confusion arose through the current use at that time of the name *C. eburneum* var. *dayi* Jennings (1875) for a colour variant of *C. eburneum* which had a red-spotted lip. Hooker also noted that, at the time of writing, he had not seen a specimen of *C. dayanum.*

Although the first collections of *C. dayanum* were from northern India, it has a wide distribution, and it was not long before collections were made from other localities. The first collection from Taiwan, made by A. Corner and cultivated by Leach, was named by H.G. Reichenbach (1878) as *C. leachianum.* He distinguished this variant from *C. dayanum* by its broader leaf, its less acute sepals, its shorter mid-lobe of the lip and its interrupted callus ridges. This description does appear to differ somewhat from *C. dayanum,* but it is part of the population in Taiwan which has been studied by several later authors as *C. dayanum* var. *austro-japonicum* (Tuyama, 1941; Ohwi, 1965; Garay & Sweet, 1974; Lin, 1977), *C. alborubens* (Makino, 1902; Schlechter, 1919; Hu, 1973; Mark *et al.,* 1986) or *C. simonsianum* (Makino, 1912; Hayata, 1914). Wu & Chen (1980), Seidenfaden (1983) and Seth & Cribb (1984) have all included *C. leachianum* as a synonym of *C. dayanum.*

Makino described *C. alborubens* in 1902 from a specimen collected in southern Japan. He did not compare it with *C. dayanum* or *C. simonsianum.* Although Schlechter (1919) recognised this name, Makino (1914) himself had previously included it in *C. simonsianum* instead. Its description does not differ from *C. dayanum,* and following the work of many authors on the floras of Japan and Taiwan, it is placed in the synonymy of *C. dayanum* (Tuyama, 1941; Ohwi, 1965; Garay & Sweet, 1974; Lin, 1977; Liu & Su, 1978; Wu & Chen, 1980). In 1914, Makino also published the name *C. simonsianum* f. *vernale.* This was a spring- rather than autumn-flowering variant. Despite the observation by Nagano (1955) of these two variants in cultivation, it has proved impossible to distinguish them morphologically, and they are not given any formal taxonomic recognition here.

As noted previously, several authors accept that there are differences between *C. dayanum* from northern India and the variant found in Japan and Taiwan. The most common classification for the latter is as *C. dayanum* var. *austro-japonicum,* named by Tuyama (1941) from material collected in the south of the island of Kyushu, Japan. He noted that it was morphologically indistinguishable from the specimens from Taiwan. He included *C. alborubens* Makino and *C. simonsianum sensu* Hayata, *non* King & Pantling as synonyms of his new variety. He differentiated var. *austro-japonicum* from the type variety by its narrower leaves (1.1–1.3 cm (0.4–0.5 in) broad), and its narrower corolla lobes (sepal 7.5 mm (0.29 in) broad, petal 6 mm (0.24 in) broad, lip 21 × 15 mm, 0.8 × 0.6 in). These sizes are, in fact, typical for *C. dayanum,* and cannot be used to differentiate the taxa. The rest of the description is also in accord with that of *C. dayanum.* Of the other authors who have used this varietal name, only Lin (1977) gives any differences between it and the type variety. He states that var. *austro-japonicum* differs in the apex of the mid-lobe of the lip, which is rounded rather than acute, and the inflorescence, which is pendulous rather than suberect. However, the inflorescence of *C. dayanum* is usually pendulous except where the crowded pseudobulbs and leaves artificially direct the scape upwards. Descriptions of *C. dayanum* from Taiwan and Japan vary somewhat. Garay & Sweet (1974) and Ohwi (1965) describe the apex of the mid-lobe of the lip as cuspidate; Liu & Su (1978) as acute; Hayata (1914) as cuspidate-acuminate; Tuyama (1941) as acuminate to apiculate; and Seidenfaden (1983) describes a specimen from Taiwan as having an acute mid-lobe. This strongly contradicts the statement by Lin concerning the lip shape, and there is no evidence that the mid-lobe is any different from the specimens collected from other parts of the distribution. Specimens from China also have an abruptly apiculate apex. There does not seem to be any consistent difference between the populations, and var. *austro-japonicum* is not therefore recognised here as distinct.

Shortly before his death, Rolfe examined Kerr's collections from Thailand, and had named and described, but not yet published, several new species.

Downie (1925) subsequently published these descriptions. Rolfe distinguished his new species, *C. sutepense*, from *C. dayanum* by its larger flowers. However, the dimensions given fall within the range of *C. dayanum*. Seidenfaden (1983) concluded that *C. sutepense* cannot be maintained as a distinct taxon, a view which is followed here.

C. poilanei, from Cambodia, was described by Gagnepain (1931) as a relative of *C. ensifolium*, differing in having white flowers and prominent side-lobes to the lip. Seidenfaden (1983) recognised it as a separate species, and added that it could be further distinguished, from the other species in section *Jensoa*, by a row of hairs along the crest of the callus ridges. These characters are all strongly characteristic of*C. dayanum*, which also closely resembles *C. ensifolium* in its vegetative habit. Furthermore, Gagnepain noted that it has the same flower colour as *C. dayanum*. He distinguished *C. poilanei* from *C. dayanum* by its bilobed leaves, its pendulous scapes, its larger flowers and its acuminate lip apex. Further differences evident in the description are the absence of callus ridges and the four pollinia of *C. poilanei*. However, Seidenfaden commented on the hairy callus of the type specimen, and the observed number of pollinia is uncertain – Gagnepain giving it tentatively as four in his description, but following it with a question mark. The leaves of *C. dayanum* are not bilobed, and appear shortly acuminate. Closer examination shows that they are in fact slightly oblique at the tip. The scape tends to be pendulous when possible, but may not be able to assume this habit because of the closely packed and congested pseudobulbs and leaves. The flowers are slightly smaller than would be expected in *C. dayanum*, but the difference is marginal. The description of shape of the lip of *C. poilanei* is incongruous, but further examination may show it to be apiculate. Furthermore, *C. poilanei*, like *C. dayanum*, is an epiphyte.

Guillaumin (1932) worked with Gagnepain and both authors collaborated to produce the chapter on the Orchidaceae for Lecompte's Indo-Chinese Flora. The illustration given there of *C. poilanei* closely resembles *C. dayanum*. *C. poilanei* is therefore treated here as a synonym of *C. dayanum*. Seidenfaden noted that later specimens identified by Guillaumin as *C. poilanei* should be placed in*C. ensifolium*.

Holttum (1953) described *C. dayanum* from Malaya as having either white or mauve petals and sepals with a median purple band. In Sabah, the extreme of this variation is found, the sepals and petals being deep wine-red in the centre and flushed with maroon to the margin. Although the leaves are narrow on this variant, they are not excluded from the range of expected variation in *C. dayanum*. It is otherwise identical in habit and floral morphology (Du Puy & Lamb, 1984). There appears to be a gradual darkening of the flower colour in a cline towards the southern and south-eastern extremes of the distribution of *C. dayanum*, with darkest being found on Sabah. Specimens from Luzon (Philippines) have the white sepals and petals with a central stripe typical of *C. dayanum* from the more northerly parts of its distribution.

C. acutum Ridley, collected in Perak, Malaya, was described by Ridley as having the petals and sepals as 'nearly white ... with a medium band of purple'. In 1924 he included it in the synonymy of *C. dayanum*. This name has been used for the slightly darker variants found in Malaya and Sumatra.

Ames & Schweinfurth (1920) named dark red-flowered specimens, collected by Clemens on Mt Kinabalu, Sabah, as *C. angustifolium*. Although they did not compare them with *C. dayanum*, they differentiated them from *C. acutum* by the shorter leaves, the short, suberect scape and differently marked lip of the latter. The leaf length in *C. dayanum* is very variable, and the variation includes the lengths of both. A suberect scape is not unusual for *C. dayanum*, as the closely set pseudobulbs with their dense covering of leaf bases and cataphylls often direct the scape upwards before it can adopt its usual pendulous habit. A short scape may become suberect under these circumstances. The lip of *C. acutum* is described as dark carmine with golden spots. The lip of specimens from all over the range of *C. dayanum* are distinctly coloured deep wine-red with whitish streaks and red veins on the side-lobes, a central yellow stripe on the mid-lobe and white callus ridges. The lip of the Sabah variant is almost identical in colour with that of specimens from Sumatra, Malaya and further north.

C. dayanum occurs in northern India (Sikkim, Assam), although it does not appear to occur in Meghalaya. Its distribution extends east through southern China to Taiwan and Japan (southern

Kyushu). Wu & Chen (1980) give its distribution in China as the southernmost provinces; Fujian, Guangdong, Guangxi, Hainan, Taiwan and Yunnan. It has also been collected in the mountains of northern Luzon, in the Philippines. It occurs in Thailand and Cambodia, and probably also in Laos and Vietnam. Its most southerly localities are in W Malaysia, Sumatra and Sabah, but it has not been recorded from Java.

Holttum (1957) and Du Puy & Lamb (1984) note that in the southern portions of the distribution of *C. dayanum*, it prefers the cooler, higher altitudes of about 1000–1500 m (3280–4920 ft). Similarly, in Thailand and northern parts of W Malaysia it is often found at about 600–1200(1600) m (1970–3940(5250) ft) altitude, in China (Hainan) at 1700 m (5580 ft) and in Sikkim between 300–1300 m (985–4265 ft). In Taiwan it occurs at lower altitudes below 1000 m (3280 ft) (Lin, 1977).

This species usually flowers between August and November(December), although out-of-season flowering is not uncommon in cultivation. In the more equatorial regions of its distribution (Sumatra, Sabah) it tends to flower during August and September, but due to the lack of strong seasonality in the climate, it is not unusual for it to flower at other times of the year. There is a variant in cultivation in Japan which is reputed to flower in the early spring.

The habitat of *C. dayanum* is almost exclusively in hollows on the limbs or trunks of tall trees or on rotten trees or logs in situations where the plant is in light shade, although it occasionally occurs on humus-covered rocks. In Thailand we have collected it from rotting, felled tree trunks in evergreen forest, where its roots formed a strong network between the bark and the damp, decaying core which some of the roots also penetrated (Du Puy, 1983). In Sabah, we have again seen it growing on rotten logs and dead trees, the roots penetrating into the rotting bark, in positions where it receives full sun for at least part of the day (Du Puy & Lamb, 1984).

Section **Austrocymbidium** *Schltr.* in Fedde, Repert. 20: 104 (1924); P. Hunt in Kew Bull. 24: 94 (1970); Seth & Cribb in Arditti (ed.), Orchid Biol., Rev. Persp. 3: 298 (1984). Type: *C. canaliculatum* R.Br. (lectotype chosen by P. Hunt, 1970).

When Schlechter (1924) established this section he only included the Australian species, distinguished by their green or yellowish-green flowers often marked with red. He noted that some species had strongly inflated pseudobulbs, while others had a shortly elongated stem. Seth & Cribb (1984) extended the limits of this section to include the two Malaysian species *C. chloranthum* and *C. hartinahianum*, and further characterised the section by its broad, often obovate sepals and petals. This section is further increased here by the inclusion of *C. elongatum*, a newly described species from Borneo.

This section is rather difficult to characterise. The species usually have leaves articulated close to the pseudobulbs, and many densely crowded flowers on the scape. The flowers colour in yellow or greenish, except for some variants of *C. canaliculatum* which are greenish but are heavily marked with red-brown. The sepals are characteristically broad, obovate to oblong-elliptic and rounded or obtuse at the apex, and are usually much broader than the petals. The lip has small, narrow side-lobes. The column is short, and the two pollinia are ellipsoidal rather than triangular, and are parallel to each other (figure 14B).

The species in this section can be divided into two groups on the basis of their vegetative habit. *C. canaliculatum*, *C. hartinahianum*, *C. chloranthum* and *C. madidum* have large, ovoid, bilaterally flattened pseudobulbs which are produced annually. Conversely, *C. suave* and *C. elongatum* lack pseudobulbs and have instead a slender stem which grows indeterminately, giving these species an almost monopodial habit. This division appears to be artificial, however, when the flowers are examined. The flowers of *C. suave* and *C. madidum* are similar in colour and shape, and they both have a highly unusual single, glistening, somewhat viscid, ligulate depression on the lip in place of the usual paired callus ridges. This latter character is so distinctive that it could also be used to

Map 5: The distribution of the species in section *Austrocymbidium*:

A *C. chloranthum*
B *C. hartinahianum*
C *C. madidum*
D *C. canaliculatum*
E *C. suave*
F *C. elongatum*

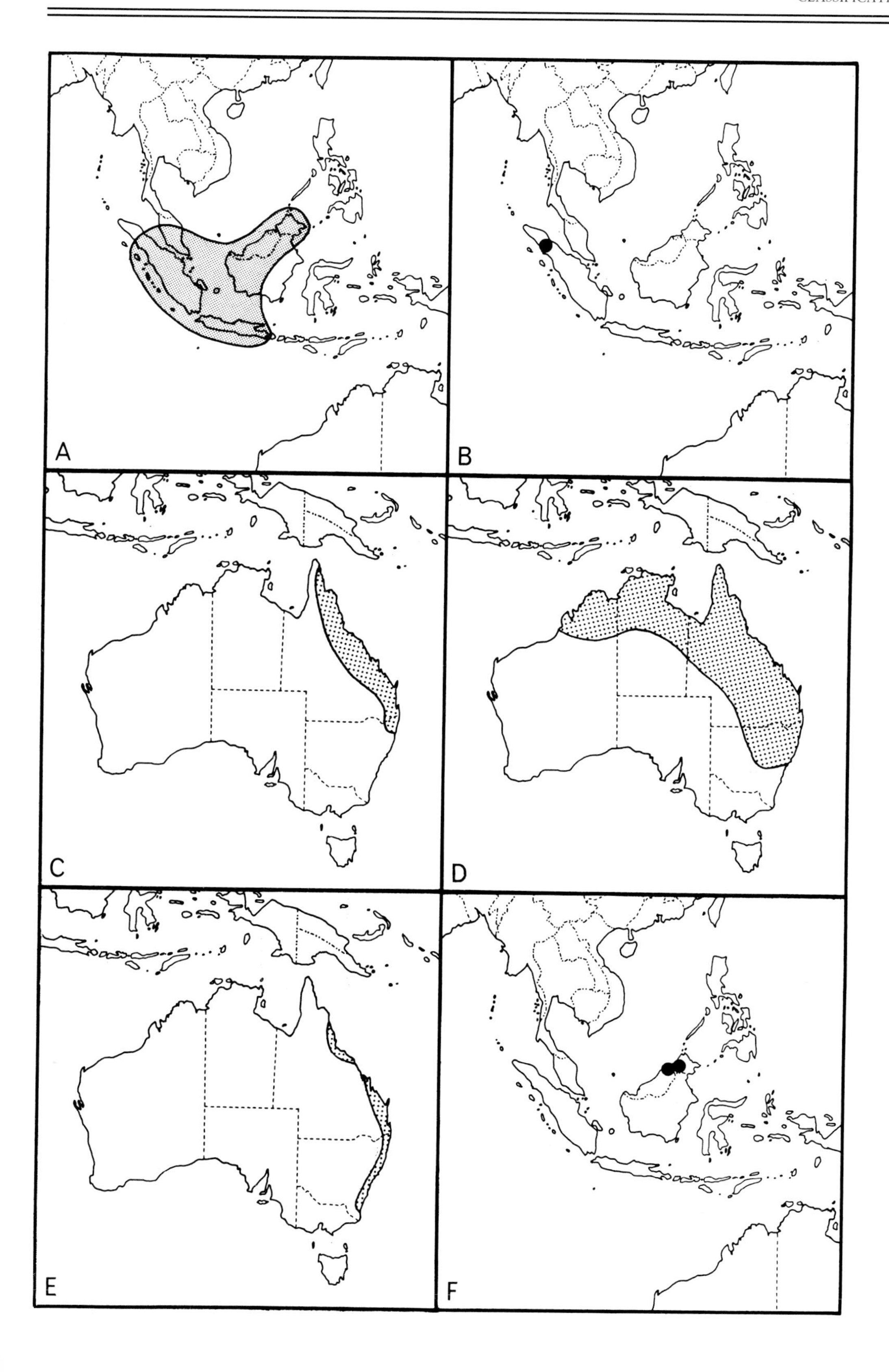
A
B
C
D
E
F

subdivide the section. The crenulate callus ridges of *C. chloranthum* and *C. hartinahianum* are also very distinctive. Furthermore, the unusual flower shape, the hairy callus ridges, the thick, rigid leaves and other extreme xerophytic adaptations of *C. canaliculatum* could also be used to separate this species in a section of its own. The most satisfactory treatment therefore seems to be to maintain this group as a single section.

8. C. canaliculatum *R.Br.*, Prodr. Fl. Nov. Holl.: 331 (1810); Hook.f. in Bot. Mag. 96: t.5851 (1870); Bailey, Fl. Queensland 5: 1547 (1902); Rupp in Proc. Linn. Soc. New South Wales 59: 94-100, f.1-3 (1934); Dockrill in Australian Pl. 3: 293-5, + fig. (1966) & Australian Indig. Orchids 1: 630-2, + fig. (1969). Type: Australia, *R. Brown 5303* (holotype BM!, isotype K!).

C. hillii F. Muell. in Regel, Gartenflora 28: 138-9 (1879); Bailey, Queensland Fl. 5: 1547 (1902); Rupp in Proc. Linn. Soc. New South Wales 62: 299 (1937). Type: Australia, N Queensland, Mulgrave Mts., *Hill*, cult. Brisbane Botanic Gardens (holotype MEL).

C. sparkesii Rendle in J. Bot. 36: 221 (1898) & in l.c. 39: 197 (1901). Type: Australia, N E Queensland, *Jones*, cult. *Sparkes* (holotype BM!, isotype K!).

C. canaliculatum R. Br. var. *sparkesii* (Rendle) Bailey in Compr. Cat. Queensland Pl.: 845 (1931); Rupp in J. Linn. Soc. New South Wales 59: 98-9, f.3 (1934).

C. canaliculatum R. Br. var. *canaliculatum* f. *inconstans* Rupp in J. Linn. Soc. New South Wales 59: 98-9 (1934).

C. canaliculatum R.Br. var. *canaliculatum* f. *aureolum* Rupp in J. Linn. Soc. New South Wales 59: 99 (1934).

C. canaliculatum R.Br. var. *marginatum* Rupp in J. Linn. Soc. New South Wales 59: 99 (1934).

C. canaliculatum R.Br. var. *marginatum* Rupp f. *fuscum* Rupp in J. Linn. Soc. New South Wales 59: 99 (1934).

C. canaliculatum R.Br. var. *marginatum* Rupp f. *purpurascens* Rupp in J. Linn. Soc. New South Wales 59: 99 (1934).

C. canaliculatum R.Br. var. *barrettii* W.H. Nicholls in Australian Orchid Rev. 7: 40 (1942) Type: Australia, Northern Territory, Groote Eylandt, *Barrett* (holotype not located).

A medium-sized, perennial, epiphytic *herb. Pseudobulbs* large, up to 12 × 3.5 cm, elongate-ellipsoidal, bilaterally compressed, covered by persistent, sheathing leaf bases and 3–4 cataphylls up to about 8 cm long, the longest of which have an abscission zone near the apex and a very short lamina, soon becoming scarious and eventually fibrous. *Leaves* 3–4 per pseudobulb, distichous, 15–46(65) × 1.5–3(4) cm, narrowly elliptic, tapering to an acute, slightly cucullate apex, coriaceous and stiff, canaliculate, glaucous, narrowing and becoming conduplicate towards the base, articulated close to the pseudobulb to a broadly sheathing base 2–10 cm long, with a 2–4 mm broad, membranous margin. *Scape* (15)25–55 cm long, suberect to horizontal, arching, usually from the axils of the upper cataphylls, often more than one per pseudobulb; peduncle 12–24 cm long, with 5–8 distant, closely sheathing, scarious, often purple-tinted sheaths up to 3.5 cm long; rhachis (3)13–35 cm long, with (13)20–60 closely spaced flowers; bracts 2–4(8) mm long, triangular, acute to acuminate. *Flowers* 1.8–4 cm across, very variable in colour; not scented; rhachis, pedicel, ovary and bracts usually stained purple-brown; sepals and petals greenish or brown to almost black on the outer surface, dull to golden-yellow or green within, usually spotted and blotched or uniformly coloured with red-brown to deep magenta, usually with a narrow greenish margin, or occasionally uniformly deep reddish-black; lip white or cream, sometimes tinged with green or pink, lightly spotted with red or purple; callus ridges cream to pale green; column yellowish-green or green, often blotched with red-brown, to uniformly dark maroon, paler at the base, white streaked maroon below, anther-cap cream or green to dark red-brown or maroon. *Pedicel and ovary* very slender, 2.3–3.5 cm long. *Dorsal sepal* 11–25 × 4.5–6(10) mm, narrowly oblong-elliptic, obtuse to subacute, often mucronate, erect; lateral sepals similar, slightly broader, somewhat oblique, spreading. *Petals* slightly shorter and narrower than the sepals, 10–20 × 4.5–5.5(8) mm, narrowly elliptic, obtuse to acute, often mucronate, slightly oblique, weakly porrect or spreading. *Lip* 9–15 × 4–10 mm when flattened, 3-lobed, minutely pubescent; side-lobes well-defined, with acute to obtuse, porrect apices; mid-lobe 5–7(9) × 5–7.5(10) mm, broadly elliptic to ovate, with a rounded, obtuse to shortly acuminate apex, decurved, margin entire or weakly undulating; callus of 2, parallel, well-defined, slightly pubescent ridges, from the base of the lip to the base of the mid-lobe. *Column* 6–13 mm long, winged towards the apex; pollinia 2, about 1 mm long, ellipsoidal, deeply cleft; viscidium quadrangular, about 0.7 mm across, extending into two, short, thread-like appendages at the lower corners. *Capsule* 4–6 × 1.5–2.5 cm, fusiform-ellipsoidal, tapering at the base to a short pedicel, and at the apex into a short beak formed by the persistent column.

ILLUSTRATIONS. Plate 8; photographs 66, 67; figure 21.1.

DISTRIBUTION. Australia, from northern Western Australia to Cape York, Queensland, and south to central New South Wales (map 5D); sea level to about 1000 m (3280 ft).

HABITAT. Epiphyte on *Eucalyptus* and *Melaleuca* trees, growing in rotting wood in hollow trees and in hollows formed by fallen branches, often in very dry areas, usually in partial shade; flowering September-December.

C. canaliculatum is vegetatively a very distinctive species, described by R. Brown in 1810, from a specimen probably collected at Broad Sound, Queensland. It can form large clumps and, as each new pseudobulb often produces more than one scape with 20–50 or more densely crowded flowers each, a single large plant may carry several hundred flowers. The size of the flowers and the length of the bracts is variable. The sepal, petal and lip apices vary from obtuse or subacute to acute, and the lip apex may be shortly acuminate. The breadth of the side-lobes of the lip and the length of the column are also extremely variable. This variation appears to be continuous, and is significant even between different flowers on the same scape. Flower colour is also extremely variable from greenish, variously marked with red-brown, to almost black.

C. canaliculatum can be distinguished from the other Australian species by its highly characteristic leaves, its densely crowded scapes, and by the presence of two, distinct, parallel callus ridges on the lip. It can be further distinguished from *C. suave* by its inflated pseudobulbs.

The variation in flower colour is so great that it has led some authors to separate the more distinctive extremes at specific or varietal level. A discussion of Rupp's (1934) treatment serves to illustrate the variation involved. Rupp divided this species into three varieties, based mainly on the colour of the sepals and petals: the typical variety (var. *canaliculatum*), var. *marginatum* and var. *sparkesii*. He also recognised two forms of each of the first two of these. His var. *canaliculatum* has flowers with red-brown spots and blotches on the sepals and petals, over a dull green to bright yellow-green background colour. He recognised f. *inconstans*, which included most of the different shades of colour within this variety, and f. *aureolum* which included the brightest yellow-green variants. Var. *canaliculatum* is widespread in New South Wales and southern Queensland, from the coast well inland over the Great Dividing Range, and perhaps into the Northern Territory and the north of Western Australia, with f. *aureolum* more or less confined to the western plains of New South Wales.

Var. *marginatum* included all specimens with more-or-less uniformly brown or reddish flowers, usually with a pale green margin. The two forms of this variety which Rupp recognised were f. *fuscum*, brown in colour, and f. *purpurascens* which was described by Rupp as 'bright or deep magenta'. He gave the distribution of this variety as northern Queensland and Cape York, and included the variant illustrated by J.D. Hooker in Curtis's *Botanical Magazine* (1870). Rupp depicted these varieties and forms, and also included an intermediate with blotched sepals and uniformly coloured sepals. We have seen a specimen with the opposite colour combination of blotched petals and uniformly coloured sepals. In reality, there seems to be a complete intergradation between these taxa, and Rupp himself states that the taxa are 'more or less connected by intermediates'.

A very striking extreme of the colour variation described in var. *marginatum* was recognised by Rupp as var. *sparkesii*, which had previously been described as *C. sparkesii* by Rendle (1898, 1901), and later reduced to varietal status within *C. canaliculatum* by Bailey (1913). Rendle distinguished this variant by 'the deep, dark crimson colour of the flower, which in reflected light appears almost black', and by its slightly longer perianth segments (sepals 20 × 5 mm

(0.8 × 0.2 in), petals 16 × 4.5 mm (0.6 × 1.8 in)). Bailey noted that he could not distinguish it from *C. canaliculatum*, except by the dark red flower colour, and certainly the dimensions given by Rendle are no different from those commonly found in *C. canaliculatum.* The lip of var. *sparkesii* is usually suffused with pink on the margins and green on the disc. and Rupp suggested that the lip is usually smaller than normal, although this variation is, none the less, continuous. Var. *sparkesii* occurs in tropical Queensland, north of Townsville. The colour remains its only distinguishing character, but the darker specimens included by Rupp in his var. *marginatum* f. *purpurascens* approach var. *sparkesii*, and are also found in northern Queensland. Dockrill (1966, 1969), recognising the continuous nature of this variation, sank all of these taxa into *C. canaliculatum*, and this treatment is followed here.

Rupp (1934) also included a note concerning several reports of an albino variant of *C. canaliculatum*, and indicated that this species also has a tendency to produce abnormal floral variants such as peloric and compound flowers where, for example, the backs of two lateral sepals from otherwise separate flowers are fused together.

In 1942, Nicholls described var. *barrettii* from the Northern Territory, distinguished by its pale greenish-yellow colour but lacking any brown markings on the sepals and petals, and occasionally with some basal red spotting at the back of the sepals. This, then, represents the opposite extreme of colour variation to the dark red variants recognised by Rupp as var. *sparkesii*.

M. Clements (pers. comm.) has reported specimens with exceptionally large flowers from the Northern Territory (Arnhem Land) and also perhaps from the Atherton Tableland. These, and other large-flowered variants, should be investigated cytologically, and might prove to be natural polyploids.

F. Mueller described *C. hillii* in 1879 from material cultivated in Brisbane, but originally collected by Hill from *Eucalyptus* trees in the coastal forests of the Mulgrave Range in northern Queensland, probably near Cairns. Mueller differentiated this from *C. canaliculatum* in several characters, including its thinner, less rigid, flat (not canaliculate) leaves with 3 prominent nerves below, its fewer-flowered inflorescences, its longer bracts, its shorter pedicel and ovary, its longer and more acute, oblong-lanceolate sepals and petals, its glabrous lip with weakly defined side-lobes, and its long, semi-lanceolate, acuminate mid-lobe which is nearly three times as long as broad and is much larger than the rest of the lip. Bailey (1902) repeated some of these characters, and maintained it as a separate species. Rupp (1937) claimed that it had been re-collected by Barrett on the Daly River in the Northern Territory. He described this new collection as being similar to *C. hillii* in its leaf with three prominent nerves, its longer perianth segments (sepals 21 × 4.5 mm (0.8 × 1.8 in), petals 19 × 4 mm (0.7 × 1.6 in)) and its very long lip which agreed well with the type description (lip 16.5 × 5 mm (0.7 × 0.2 in), mid-lobe 10 × 4 mm (0.4 × 0.16 in), acuminate). He added a further character, that of the capsule, which he described as 5.5 cm (2.2 in) long and 1.5 cm (0.6 in) broad at its broadest point, 2 cm (0.8 in) from the apex. Although rather narrow, this does not seem to differ greatly from typical *C. canaliculatum.* However, the leaf characters certainly appear to be distinct, and some of the lip characters such as the length/breadth ratio of the mid-lobe of the lip appear to be outside of the very wide range of variation known in *C. canaliculatum.* However, in most of its characters it agrees well with *C. canaliculatum*, and Dockrill (1969) considered it to be a synonym of that species. Further research may show this to be a separate taxon, or the result of hybridisation with *C. madidum* or *C. suave.*

The most southerly localities in the distribution of *C. canaliculatum* are the Hunter River near the coast, and Forbes on the western slopes of the Great Dividing Range, both in New South Wales. It is most commonly found on the drier western slopes of the Great Dividing Range, and the western plains, and is rather uncommon on the coast, which has a higher rainfall. Its distribution extends northwards through Queensland to the tip of Cape York, and although it is found in the coastal forest it does not occur in the rainforest. The capacity of this species to withstand drought is reflected in its wide range over the north of Australia into Arnhem Land and the Kimberleys, to Roebuck Bay in the north of Western Australia, where it occurs in tropical savanna and scrubland, but again is uncommon on the coast. Indeed Mueller (1879) noted that this is often the only orchid found over much of this arid area, such as near Sturt's Creek

21. *Cymbidium erythrostylum.* Vietnam, *Easton* s.n., cult. Kew

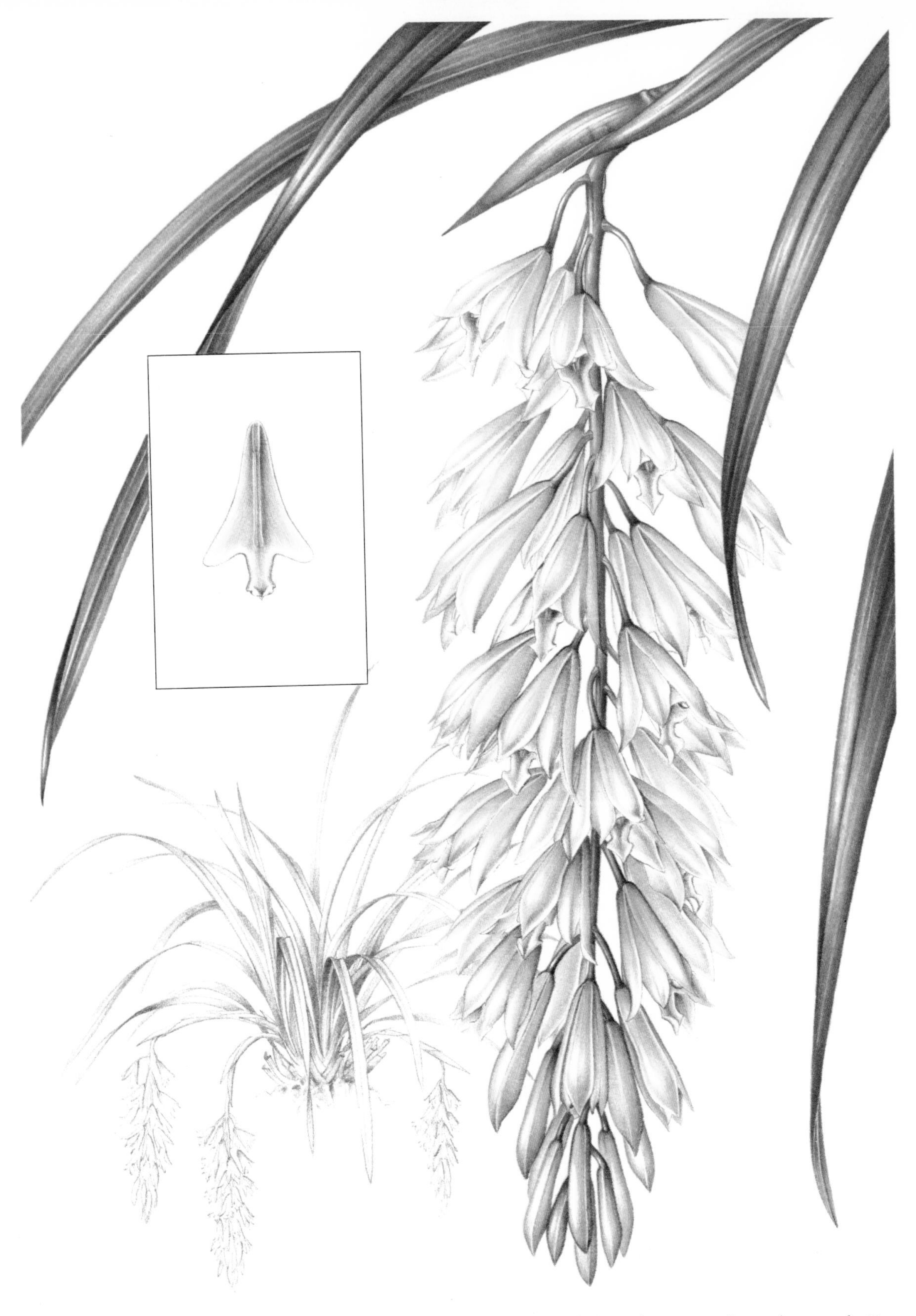

22. *Cymbidium elegans.* N India, *Seth* 109, cult. Kew

23. *Cymbidium tigrinum.* N E India, *Rittershausen* s.n., cult. Kew

24. *Cymbidium ensifolium* subsp. *haematodes*. Sabah, *Bailes & Cribb* 746, cult. Kew

and the Victoria River, where the annual rainfall is less than 50 cm (20 in).

Its drought tolerance is partly due to its thick, leathery leaves which are resistant to desiccation, and the shape of these leaves, which are erect, hooded at the apex and strongly V-shaped in section, and channel any available water directly to the base of the plant. Furthermore, Winter *et al.* (1983) have shown that the leaves are also xerophytically adapted in their physiology, using Crassulacean Acid Metabolism (CAM) to allow photosynthesis to continue without the usual debilitating loss of water.

C. canaliculatum is almost always found growing in rotting wood in hollows in tree trunks and branches of *Eucalyptus* and *Melaleuca* species which often have a decaying central core in the trunk (photograph 66). In this situation, the surrounding wood of the tree helps to maintain the moisture in the substrate into which the orchid forms an extensive root system, up to 12 m (39 ft) or more in length. Clements (pers. comm.) has successfully used a wood-rotting fungus in the symbiotic germination of seeds of this orchid. It is often found around creek beds, but this is probably due to the environmental preferences of the trees on which it grows, rather than as a consequence of any direct benefit to the orchid. It prefers some shade, but usually occurs in open woodland and will withstand exposure to strong, direct sunlight. When conditions are dry, such as in the plains of New South Wales where the annual rainfall is less than 50 cm (20 in), it can tolerate summer temperatures in excess of 35°C (95°F) and freezing winter night temperatures.

9. C. hartinahianum *Comber & Nasution* in Bull. Kebun Raya 3: 1–3, + fig. (1977) & in Orchid Dig. 42: 55–7, + fig. (1978). Type: Sumatra, Sidikalang, *Nasution & Bukit 6* (holotype BO!).

A medium-sized, perennial, terrestrial *herb. Pseudobulbs* conspicuous, up to 7 × 3.5 cm, ovoid, bilaterally compressed, covered by persistent, sheathing leaf bases and 3–4 cataphylls up to 9 cm long which became scarious and fibrous with age. *Leaves* 7–10 per pseudobulb, distichous, 13–30(60) × 0.9–1.5 cm, ligulate, acute, with a slightly hooded, oblique apex, somewhat coriaceous, V-shaped in section, articulated 2–6 cm from the pseudobulb to a broadly sheathing base. *Scape* 50–80(100) cm long, erect, produced in the axil of the cataphyll which is ruptured by the developing scape; peduncle 35–60 cm long, with 8–9 keeled, distant, cymbiform, loose sheaths up to 6 cm long, which are somewhat inflated and lack a cylindrical basal region; rhachis about 15–30 cm long, with (9)14–21 flowers; bracts 1–8 mm long, triangular, acute, scarious. *Flowers* about 3.5 cm in diameter; not scented; rhachis, pedicel and ovary green; petals and sepals olive-green to purple-brown, with some brownish staining towards the base; lip white, the side-lobes barred with red, and the mid-lobe sparsely blotched with red; callus ridges pale yellow; column dark purple-brown; anther-cap pale yellow. *Pedicel and ovary* slender, 2–3.5 cm long. *Dorsal sepal* 15–20 × (4)6–8 mm, oblong-elliptic to obovate, obtuse, erect; lateral sepals similar, spreading. *Petals* almost as broad as the sepals, 14–18 × 6–8 mm, oblong-elliptic, slightly oblique, spreading or weakly porrect. *Lip* about 21 mm long, 3-lobed, minutely pubescent; side-lobes erect, rounded, with rounded apices; mid-lobe short, 7 × 8–10 mm, oblong, with a broadly rounded or weakly mucronate apex, porrect or slightly decurved, margin undulating at the base; callus in 2 glabrous, somewhat crenulate ridges, extending from the base of the lip into the base of the mid-lobe, convergent towards their apices. *Column* 11–12 mm long, narrowly winged; pollinia 2, ellipsoidal, deeply cleft, on a crescent-shaped viscidium. *Capsule* about 3.2–4 × 1.8–2.5 cm, ellipsoidal, strongly winged and acute-angled in transverse section, stalked, pendulous, with an 11 mm long apical beak formed by the persistent column.

ILLUSTRATIONS. Photograph 75; figure 21.3.

DISTRIBUTION. Sumatra (map 5B); 1700–2700 m (5580–8860 ft).

HABITAT. In rough damp grassland with *Imperata cylindrica, Themeda villosa* and *Gleichenia sp.* ferns; flowering February-June.

C. hartinahianum is related to *C. chloranthum*, both having pseudobulbs of similar size and shape, erect scapes, and numerous small, slender-stalked, greenish flowers with similar broad, rounded floral segments

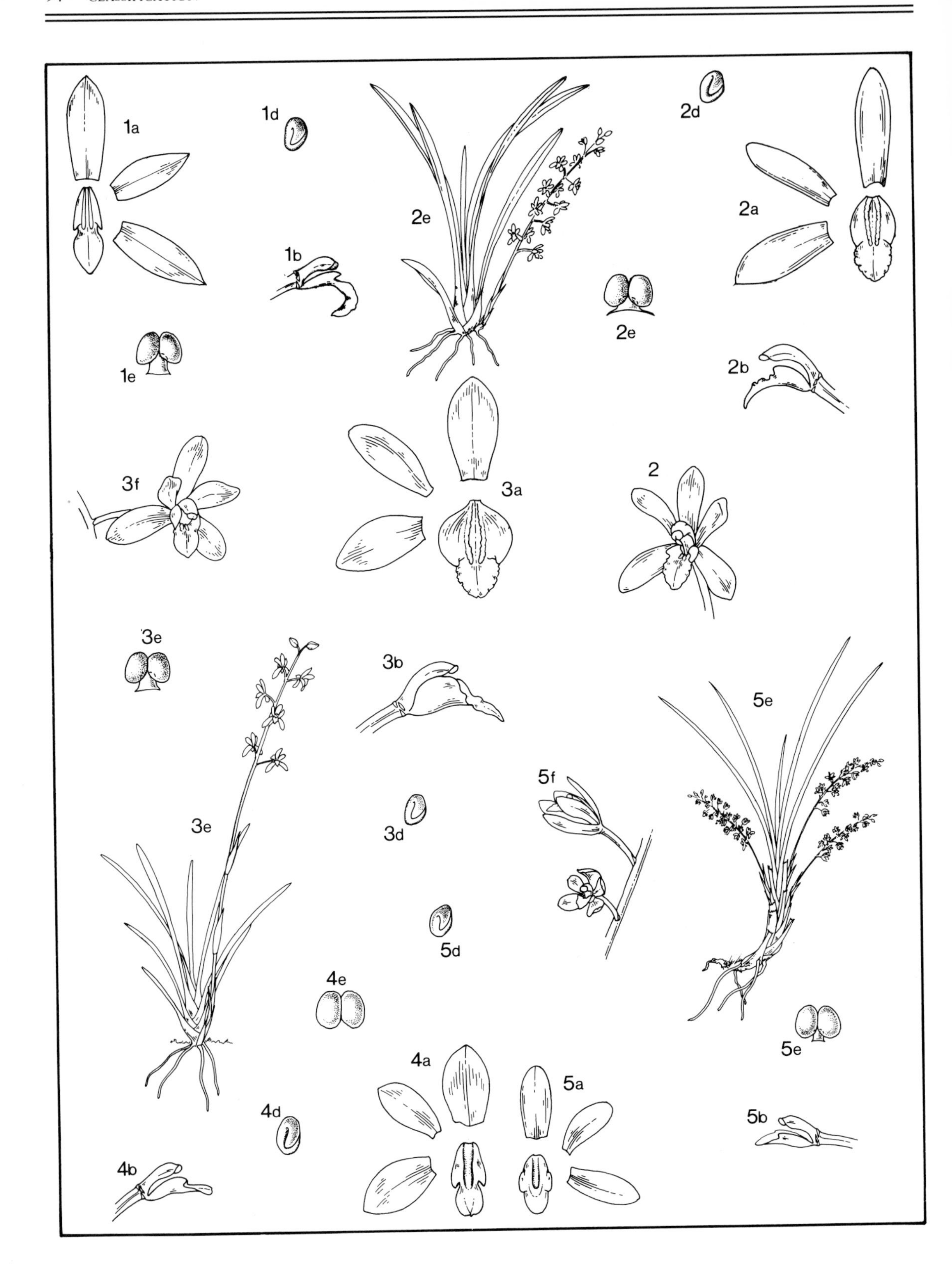
1a
1d
2e
2d
2a
1b
2e
1e
2b
3f
3a
2
3e
3b
5e
5f
3e
3d
5d
4e
5e
4a
5a
4d
5b
4b

Figure 21

1. ***C. canaliculatum*** (Kew spirit no. 47247)
   a Perianth, × 1
   b Lip and column, × 1
   c Pollinarium, × 3
   d Pollinium (reverse), × 3
2. ***C. chloranthum*** (Kew spirit no. 47268)
   a Perianth, × 1
   b Lip and column, × 1
   c Pollinarium, × 3
   d Pollinium (reverse), × 3
   e Flowering plant, × 0.1
   f Flower, × 0.8
3. ***C. hartinahianum*** (Kew spirit no. 39722)
   a Perianth, × 1
   b Lip and column, × 1
   c Pollinarium, × 3
   d Pollinium (reverse), × 3
   e Flowering plant, × 0.1
4. ***C. madidum*** (Kew spirit no. 45606)
   a Perianth, × 1
   b Lip and column, × 1
   c Pollinarium, × 3
   d Pollinium (reverse), × 3
5. ***C. suave*** (Kew spirit no. 8818)
   a Perianth, × 1
   b Lip and column, × 1
   c Pollinarium, × 3
   d Pollinium (reverse), × 3
   e Flowering plant, × 0.1
   f Flowers, × 1

and crenulate callus ridges. This latter, and their winged capsules, are characters found only in this species and in *C. chloranthum*. However, *C. hartinahianum* has narrower, usually shorter, more coriaceous leaves which are V-shaped in section and lack the bilobed apex usually found in *C. chloranthum*. *C. hartinahianum* also has a longer scape with a much longer, erect peduncle which is only partially covered by the distant sheaths. Its flowers are slightly larger than those of *C. chloranthum*, and are darker in colour, with a much darker column. The petals and sepals are almost equal in size, whereas in *C. chloranthum* the petals are narrower than the sepals. In *C. hartinahianum* the lip has broader, rounded side-lobes which are not angled at their apices, and a smaller, rather truncated mid-lobe, which has an undulating margin at the base only. Comber & Nasution (1978) illustrate flowers of both of these species.

The leaves of *C. hartinahianum* resemble those of *C. canaliculatum* but are much narrower. The similarity in vegetative morphology between these two species suggests that they are closely related to each other.

C. hartinahianum was described by Comber & Nasution in 1977. It appears to have a very restricted distribution, having been collected several times from a single locality on the western side of Lake Toba in northern Sumatra, on a plateau at about 2000 m (6560 ft), 1100 m (3610 ft) higher than the lake itself. The plateau has a vegetation of broken forest and rough, damp grassland, known locally as blang, which grows on poor soil, and is dominated by the ubiquitous tropical grass *Imperata cylindrica*, with tufts of *Themeda villosa* and patches of a coarse, bracken-like fern, *Gleichenia*. These are species which will tolerate very poor soils. *C. hartinahianum* grows in this grassland as a terrestrial. Although this habitat is unusual for a *Cymbidium* species, the environment resembles that in which many epiphytic species survive. The soil is so poor that it cannot support a strongly growing vegetation, nor a wide diversity of species, allowing the orchid to grow in damp conditions amongst small ferns and mosses, in good light, and without competition from other strongly growing plants.

10. C. chloranthum *Lindley* in Bot. Reg. 29: 68 (1843); Hook. in Bot. Mag. 82: t.4907 (1856); Rolfe in Orchid Rev. 27: 128–9 (1919); J.J. Smith, Enum. Orchids Sumatra: 336 (1933); Holttum, Fl. Malaya 1: 515 (1953); Backer & Bakhuizen, Fl. Java 3: 396 (1968); Comber in Orchid Dig. 44: 164, + fig. (1980); Du Puy & Lamb in Orchid Rev. 92: 353, f.295 (1984); Seth & Cribb in Arditti (ed.), Orchid Biol., Rev. Persp. 3: 298–9 (1984). Type: cult. *Loddiges* (holotype K!)

C. variciferum Reichb.f. in Bonplandia 4: 324 (1856); Hook.f., Fl. Brit. India 6: 14 (1891); Rolfe in Orchid Rev. 27: 129 (1919). Type: cult. *Booth* (holotype W).

C. sanguineolentum Teijsm. & Binnend. in Tijdschr. Neder. Ind. 24: 14–15 (1862); J.J. Smith, Orchid. Java: 479–81 (1905) & Figuren-Atlas: t.165 (1911); Backer & Bakhuizen, Fl. Java 3:396 (1968). Type: Java, Gunong Salak, *Teijsmann & Binnendijk* 902 (holotype BO!).

C. sanguineum Teijsm. & Binnend. in Cat. Hort. Bog.: 51 (1866) [sphalm. for *C. sanguineolentum* Teijsm. & Binnend.].

C. pulchellum Schltr. in Fedde, Repert. 8: 570–1 (1910) & in Fedde., Repert., Beih. 74: t.65, no. 260 (1934). Type: Borneo, Sarawak, Kuching, *Schlechter* 15846 (holotype B†), syn. nov.

A medium-sized, perennial, epiphytic *herb. Pseudobulbs* large, up to 11 × 4 cm; ovoid, strongly bilaterally flattened, covered by the persistent, sheathing leaf bases and surrounded by 3–5 cataphylls up to 10 cm long, the longest like the true leaves, with an abscission zone near the apex and a very short lamina, becoming scarious and eventually fibrous with age. *Leaves* 5–6(7) per pseudobulb, distichous, up to 40–60 × 2.3–3.8 cm, ligulate to ligulate-elliptic, obtuse, unequally bilobed at the apex, with a small mucro in the sinus, somewhat coriaceous, but thinner and more flexible than those of species in section *Cymbidium*, narrowing and becoming conduplicate towards the base of the lamina, articulated close to the pseudobulb to a broadly sheathing base 2–7 cm long, with a 2–4 mm broad, membranous margin. *Scape* (30)36–47 cm long, erect, usually produced in the axil of the lowest leaf at the base of the pseudobulb; peduncle 15–22 cm long, covered in the basal half by about 6 overlapping, keeled, cymbiform, acute sheaths, up to 5.5 cm long, spreading apically and with cylindrical bases; rhachis about 25–29 cm long, with (15)20–25 closely spaced flowers; bracts very short, about 1-2 mm long, triangular, acute. *Flowers* about 3 cm in diameter; not scented; rhachis, pedicel and ovary bright green; sepals and petals pale yellow to yellow-green, with some red speckling at the base of the petals; lip pale yellow-green, the side-lobes mottled and barred with red, especially towards the margin, the mid-lobe sparsely spotted with red and with a broad white margin, callus ridges yellow; column yellow-green, lightly speckled with red; anther-cap yellow. The whole flower, but especially the lip, becomes strongly suffused with crimson after pollination. *Pedicel and ovary* long, slender, (19)25–38 mm long. *Dorsal sepal* 17–20 × 5.5–7 mm, oblong-elliptic to obovate, obtuse, erect; lateral sepals similar, slightly broader, up to 8 mm broad, spreading. *Petals* narrower than the sepals, 17–19 × 4–5 mm, narrowly oblong-elliptic, subacute, slightly oblique, weakly porrect but not closely covering the column. *Lip* 12–14.5 mm long, 3-lobed, minutely pubescent; side-lobes small, erect, with porrect, subacute apices; mid-lobe (5.5)6.5–7.5 × (4)6–7.5 mm, broadly ovate with a broadly rounded apex, decurved, margin undulate; callus of 2 glabrous, crenulate ridges extending from the base of the lip into the base of the mid-lobe. *Column* short, 9–10 mm long, weakly winged towards the apex; pollinia 2, 1.2 mm across, ellipsoidal, deeply cleft, on a crescent-shaped viscidium extending into two, short, thread-like appendages at the tips. *Capsule* about 2.5 × 1.3 cm, ellipsoidal, strongly winged and acute-angled in transverse section.

ILLUSTRATIONS. Photographs 72–74; figure 21.2.

DISTRIBUTION. W Malaysia, Sumatra, Java and Borneo (map 5A); 250–1000 m (820–3280 ft).

HABITAT. On trees in evergreen tropical forest, in moist shade; flowering sporadically throughout the year.

This attractive epiphyte is unusual in that its many-flowered inflorescences are erect, while most of the other epiphytic species have arching or pendulous scapes. Another noticeable feature is the strong colour change in the flower induced by pollination, removal of the anther, or even by mechanical disturbance of the stigmatic surface, the flower becoming suffused with a strong carmine pink. This 'blushing' is well known in other species in this genus, such as those in section *Iridorchis* and in the modern, commercial hybrids, but it is particularly strong in *C. chloranthum*.

The pseudobulbs of *C. chloranthum* are large and slightly flattened, with the abscission zone of the leaf close to the pseudobulb, strongly resembling the pseudobulbs of the Australian *C. madidum* and *C. canaliculatum*. The strap-shaped leaves are unequally bilobed or strongly oblique at the apex, resembling those of the species in section *Cymbidium*, but are thinner, less rigid and lack a lignified layer of cells below the epidermis. In fact, *C. chloranthum* could be considered to be somewhat intermediate between section *Austrocymbidium* and section *Cymbidium*, but the flowers and scape are closer to the former. The flowers most strongly resemble those of *C. madidum* and *C. hartinahianum*. The former has a pendulous inflorescence, broader petals and a sticky depression instead of callus ridges on the lip. The latter, a closely related species which is rare and known only from a single locality in Sumatra, is a terrestrial, with narrower, acute leaves, petals and sepals of almost equal breadth, a narrower and shorter mid-lobe of the lip and a darker olive-green flower colour. *C. chloranthum* and *C. hartinahianum* both have distinctively gnarled and lumpy callus ridges, and winged capsules.

Historically, some confusion has surrounded the distribution of *C. chloranthum*. It was described by Lindley (1843) from a plant cultivated at Loddiges' Nursery which was said to have been originally collected in Nepal. This plant was later figured in the *Botanical Magazine* (W.J. Hooker, 1856). Since then, no specimens have been collected in or near the Himalaya of northern India, and this locality must be regarded as an error. Indeed the closest authentic locality appears to be in W Malaysia (Pahang, *Carr* 290).

C. variciferum Reichb.f. was also described from cultivated material, in 1856, but J.D. Hooker (1891) noted that it was conspecific with *C. chloranthum*, and suggested that it was Australian, a provenance which must also be regarded as erroneous. Rolfe (1919) also sank *C. variciferum* under *C. chloranthum* and gave the first confirmed locality for this species as Java, from a specimen imported and cultivated by Sander.

In the same publication, *C. sanguineolentum* Teijsmann & Binnendijk was also reduced to synonymy in *C. chloranthum*. This was originally described in 1862 from material collected on Gunong Salak, in Java. The habit of *C. sanguineolentum* was compared with that of *Iridorchis gigantea* Bl. (= *C. iridioides* D. Don), and the large, ovoid, bilaterally flattened pseudobulbs of *C. chloranthum* do indeed resemble those of *C. iridioides*, but the leaves, scape and flowers differ markedly. J.J. Smith (1905) accepted *C. sanguineolentum* in his *Orchid Flora of Java*, but did not discuss how it differed from *C. chloranthum*, which his detailed description and later figure (1911) closely match. Backer & Bakhuizen (1968) include *C. sanguineolentum* as a synonym of *C. chloranthum*, and this treatment is followed here. The name *C. sanguineum* is a misprint of *C. sanguineolentum*, a mistake made by Teijsmann & Binnendijk in 1866.

Schlechter (1910) described *C. pulchellum*, from Sarawak, as a relative of *C. pubescens* Lindley (= *C. bicolor* Lindley) because of their similarity in flower size and structure, but did not compare it with *C. chloranthum*, or any of its later synonyms. The description closely resembles that of *C. chloranthum*. Although the type has been destroyed, his illustration of *C. pulchellum* published in 1934 is unmistakably of *C. chloranthum*. *C. pulchellum* is therefore included here in the synonymy of *C. chloranthum*.

C. chloranthum is distributed from W Malaysia and Sumatra (J.J. Smith, 1933) to Java, Sarawak and Sabah. Although no reference has been found to specimens collected in the rest of Borneo, it is likely that this species also occurs there. It appears to be uncommon throughout its range, although it may be more abundant in localised areas. It is a tropical forest species, growing epiphytically in moist shade.

In Sabah it is found at altitudes of 500–1000 m (1640–3280 ft), and flowers less well when cultivated at sea level (Du Puy & Lamb, 1984). In Java, it can be found at about 800 m (2625 ft) (Comber, 1980) and J.J. Smith (1933) has recorded it from 250 m (820 ft) in Sumatra. It probably flowers throughout the year in the tropics at these relatively low elevations where climatic seasonality is not pronounced.

11. C. madidum *Lindley* in Bot. Reg. 26: misc. 9-10 (1840); Hawkes in Australian Orchid Rev. 26: 135 (1961); Dockrill in Australian Plants 3: 293, 328 (1966) & Australian Indig. Orchids 1: 634-6, + fig. (1969); Clements, Prelim. Checkl. Australian Orchid.: 61-4 (1982); Seth & Cribb in Arditti (ed.), Orchid Biol., Rev. Persp. 3: 299 (1984). Type: cult. *Rollinsons* (holotype K!).

C. iridifolium Cunn. in Bot. Reg. 25: misc. 34 (1839); Rupp in Proc. Linn. Soc. New South Wales 59: 94 (1934), in *loc. cit.* 62: 300-1 (1937); & Guide Orchids New South Wales: 128 (1943); *non* Roxb., Hort. Bengal: 63 (1814) & Fl. Indica 3: 458 (1832) = *Oberonia iridifolia*; *non* Sw. ex Steud. Nom., ed. 2, 1: 460 (1840) = *Oncidium iridifolium*. Type: Australia, Brisbane R., *Cunningham* (holotype BM!).

C. albuciflorum F. Muell., Fragm. Phyt. Aust 1: 188 (1859); Benth., Fl. Australia 6: 303 (1873); Bailey, Queensland Fl. 5: 1547 (1902); Rupp, Guide Orchids New South Wales: 48 (1930). Type:

Australia, Queensland, Moreton Day, *Hill* s.n. (holotype MEL!).

C. leai Rendle in J. Bot. 36: 221-2 (1898); Rupp in Proc. Linn. Soc. New South Wales 62: 300-1 (1937); Tierney in Amer. Orchid Soc. Bull. 26:168 (1957). Type: Australia, Queensland, Sonata, *Lea* (holotype BM!).

C. queeneanum Klinge in Acta Hort. Petrop. 17: 137-8, t.2 (1899); Rupp in Proc. Linn. Soc. New South Wales 59: 93 (1934) & in l.c. 62: 299 (1937). Type: Australia, *Persich* (holotype LE).

C. leroyi St Cloud in North Queensland Nat. 24 (112): 3-5, + fig. (1955). Type: Australia, North Queensland, Emmagen Creek, *Le Roy*, cult. Cairns (holotype QRS).

C. madidum var. *leroyi* (St Cloud) Menninger in Amer. Orchid Soc. Bull. 30: 870-1 (1961); Dockrill in Austral. Pl. 3: 293, 328, + fig. (1966) & Austral. Indig. Orchids 1: 634-6, + fig. (1969), **syn. nov.**

A medium to large, perennial, epiphytic *herb. Pseudobulbs* very large and conspicuous, up to 10–16 × 4–6 cm, ellipsoidal, bilaterally flattened, covered by the persistent, sheathing leaf bases, and by 4–5 cataphylls up to 10 cm long, the longest like true leaves, with an abscission zone near the apex, and a very short lamina, becoming scarious and eventually fibrous with age. *Leaves* 6–8(9), distichous, up to 50–80(90) × 2.3–3.4(4) cm, linear-elliptic, acute to cuspidate, sometimes slightly oblique at the apex, erect, flexible, mid-green, channelled at the base, articulated very close to the pseudobulb to a broadly sheathing base 1.5–9 cm long, with a membranous margin up to 4 mm broad. *Scape* (30)40–80 cm long, pendulous, produced at the base of the pseudobulb in the axil of the upper cataphylls, which are often ruptured to allow the scape to develop; peduncle usually 11–25 cm long, horizontal to pendulous, covered in the basal half by about 6-7 overlapping, keeled, cymbiform, acute sheaths up to 6–7 cm long, which are somewhat spreading apically with a short cylindrical base, peduncle exposed in the apical half, but with some short, sterile bracts up to 1 cm long; rhachis (20)30–56 cm long, with (17)22–60 flowers; bracts 1–4 mm long, triangular, acute. *Flowers* 2.6–2.8 cm across, although sometimes not opening fully; slightly sweetly scented; rhachis, pedicel and ovary purple-brown, occasionally apple-green; sepals and petals straw-yellow stained with pale brown, especially on the backs of the sepals, producing an olive-green effect inside, to clear yellow or yellow-green; lip primrose-yellow with a broad deep yellow to red-brown stripe from the base of the lip into the base of the mid-lobe, bordered with brownish-red and with two deep red-brown blotches at the base of the mid-lobe; column clear yellow or greenish, paler at the base; anther-cap pale yellow. *Pedicel and ovary* slender, (10)16–33 mm long. *Dorsal sepal* 11–17 × (4)6–9 mm, elliptic to obovate, rounded to obtuse, weakly mucronate, erect; lateral sepals similar, spreading. *Petals* narrower than the sepals, 10.5–15 × (4)5–7 mm, elliptic, obtuse, weakly mucronate, slightly oblique, usually porrect, usually almost parallel with the column, somewhat spreading at the tips. *Lip* 10.5–13.5 × 5–6 mm when flattened, 3-lobed, minutely pubescent; side-lobes small, erect, almost as long as the column, apex acute, porrect; mid-lobe 5.5–7 × (3.6)4–5.5 mm, narrowly obovate, truncate, weakly mucronate, porrect, margin entire; callus a shallow, viscid, secretory depression extending from the base of the lip well on to the base of the mid-lobe. *Column* short, 7.5–8 mm long, weakly winged towards the apex; pollinia 2, ellipsoidal, cleft, on a very small, crescent-shaped viscidium about 0.7 mm across, extending into two short, hair-like processes at the lower corners. *Capsule* about 4 cm long, orbicular-ellipsoidal, with an apical beak.

ILLUSTRATIONS. Plate 9; photographs 68, 69; figure 21.4.

DISTRIBUTION. Eastern Australia, from northern Cape York to northern New South Wales (map 5C); sea level to 1300 m (4265 ft).

HABITAT. In the damper, tropical forests on the eastern side of the coastal ranges in areas of high rainfall, often in clumps of the epiphytic staghorn or elkhorn fern (*Platycerium* sp.), or in rotting wood in hollows of trees and branches; flowering August to October in tropical regions, and until December in the south. Flowering about March in cultivation in north-temperate regions of the world.

C. madidum, one of the three endemic Australian species of *Cymbidium*, was described by Lindley in

1840, from a specimen cultivated by Messrs Rollinsons which was mistakenly said to have originated in the East Indies.

It may attain greater dimensions in the wild than those given here in the description, which have been taken from cultivated specimens. The pseudobulbs are as large as any in the genus. The leaves are articulated very close to the pseudobulb so that when the leaf eventually falls, the pseudobulb does not have the prominent tips of the leaf bases projecting beyond it. This character appears to be typical of section *Austrocymbidium*. The tip of the leaf base is, however, sharply spiny when the lamina falls. *C. madidum* has a pendulous scape which is very long and has been reported to reach over 120 cm (47 in) (Leaney, 1966), with numerous, well-spaced, yellowish or greenish flowers.

The colour variant most commonly found in the wild has olive-green flowers, with brownish staining on the backs of the sepals and a paler yellow lip, but selected clear-coloured variants are commonly found in cultivation. Those are apple-green to bright yellow in colour. A yellow-green variant is also found occasionally which lacks the typical dark red pigmentation on the disc of the lip (M. Clements, pers. comm.). Rupp (1937) states that there is an unusual variant which is clear, pale green in colour with a yellowish lip, found in the region of Cairns. This has straighter, more rigid racemes, more densely spaced flowers and narrower, more widely spreading floral segments. He also states that 'the two are so nearly identical that it would be unwise to separate them', and that opinion is followed here.

The callus ridges normally found in *Cymbidium* species are absent in *C. madidum* and in *C. suave*, and are replaced by a tongue-shaped, shiny, secretory depression which extends from the base of the lip, into the base of the mid-lobe. This most unusual and easily observed character immediately separates them from all other species of *Cymbidium*. *C. suave*, however, is easily differentiated as it has smaller leaves with an obtuse, strongly oblique apex, and no pseudobulb, the stem growing in an indeterminate fashion, and flowering for several seasons on the same growth. The stem eventually becomes long and has fresh leaves only towards the apex, the lower part being covered by the fibrous remains of the leaf bases. The scape is a lot shorter, usually only about 15–20 cm (6–8 in) long, and is arching rather than pendulous, and the flowers are more densely spaced. The flowers are rather similar in these species, but the mid-lobe of the lip of *C. suave* is ovate and obtuse, not obovate and truncated, and the flowers are rather smaller than those of *C. madidum*.

C. madidum has had an unusually convoluted taxonomic history. Although Lindley gave it this name in 1840, it was not until 1961 that it was correctly applied. Until then the name commonly used for this species was *C. albuciflorum*, a name given by F. Mueller (1859) and used by both Bentham (1873) and Bailey (1902). Rupp (1930) also used this name, but in 1934 he resurrected the name *C. iridifolium*, described by Cunningham in 1839. Rolfe had previously noted, as early as 1889 on the type sheet of *C. madidum*, that it was the same species as *C. albuciflorum*, and that the latter was therefore a later synonym. Rupp (1934) concluded that because *C. iridifolium* was published before *C. albuciflorum* and *C. madidum*, it should be given priority and accepted as the correct name for this species. He therefore used that name for this species, with *C. albuciflorum* and *C. madidum* as synonyms (Macpherson & Rupp, 1935; Rupp, 1937, 1943).

Rupp had noted in his 1934 article that Roxburgh (1814, 1834) had used the name *C. iridifolium* for a different species which had subsequently been transferred to the genus *Oberonia*, by Lindley, but apparently did not realise that Cunningham's name was therefore a later homonym, illegitimate according to the International Code. It was not until 1961 that A.D. Hawkes rectified this mistake and correctly applied the name *C. madidum* Lindley, more than 120 years after its original publication.

Three further synonyms have been included in *C. madidum* by Dockrill (1966). *C. leai* was described by Rendle (1898), and was said to be close to *C. canaliculatum*, but could be distinguished from this by its broader, blunter sepals, and by its 10–11 mm (0.4 in) long lip, with a 5 mm (0.2 in) long mid-lobe which was almost as broad at the tip, and with a broad, tongue-like depression on the disc which passed into the base of the mid-lobe between shallow, lateral crests, and had a button-like tubercle at the base. This unusual callus structure is similar to that of *C. madidum* and *C. suave*, except for the lateral crests and the tubercle, which are not present in either of these species. The rest of the description and the type specimen agree with *C. madidum*, except that the leaf apex was described as obtuse, suggesting *C. suave*. Rupp (1937) was sent scapes of a plant which appeared to fit the description of *C. leai*. He described the colour as 'brownish, with darker blotches on the perianth', and the lip 'appeared to agree precisely with *C. leai*'. During the following two years, further scapes were sent from the same plant, but the flowers were identical to those of *C. madidum*, and none had the peculiar lip characters of *C. leai*. The flower colour, with blotching on the tepals, and the presence of some rudimentary callus ridges on the lip suggest that *C. leai* might be a hybrid of *C. madidum* or *C. suave* with *C. canaliculatum*, but Dockrill (1966) included *C. leai* in *C. madidum*, and this treatment is followed here until further evidence is available. Tierney (1957) states that '*C. leai* has not been recorded since its discovery in the Mulgrave Ranges when mining was active at Goldsborough. Sugar growing has since destroyed much of this range, and may have destroyed this species.'

Similarly, very little is known about *C. queenianum*, which was described by Klinge in 1899, without any locality or other collection data. There does not appear to be any outstanding difference between this and some of the darker-flowered variants of *C. madidum* with olive-green and brownish flowers. We have therefore followed Dockrill (1966) and included it as a synonym of *C. madidum*.

Dockrill (1966, 1969) recognised *C. leroyi* St Cloud (1955) as distinct at varietal rank. E. Menninger (1961) first reduced it to varietal rank but did not include any explanation of her move. *C. leroyi* was described by Dockrill as having a lip with a pointed, cymbiform mid-lobe and a flowering time during November and December, slightly later than would be expected in northern Queensland where it was collected, but not outside of the normal flowering period in more southerly localities. The unusual lip can be understood as a failure of the lip to expand fully when the flower opened, possibly due to environmental factors. Indeed, when specimens have been removed from the wild and are cultivated, they appear to be indistinguishable from the typical variety. This variety is not, therefore, accepted as distinct. Dockrill gave its distribution as 'apparently confined to the area between the Barron River and the Endeavour River', in Queensland, but further field work into this variant is required to establish its geographical and morphological range.

C. madidum prefers a more humid habitat than the other Australian species and is found in eastern Australia from near the Clarence River in northern New South Wales, northwards probably to the tip of Cape York, growing in tropical rainforest or in adjacent more open forest which has a high rainfall. It occurs at altitudes from sea level up to about 1300 m (4265 ft), preferring the coastal plains and the eastern slopes of the coastal range which have a high rainfall. It is an epiphyte, often reported growing in the bases of staghorn or elkhorn ferns (*Platycerium* sp.). It also grows commonly on rotting trees or in hollows in branches or trunks of trees such as *Eucalyptus* or *Melaleuca*, usually in conjunction with rotting wood, and may eventually form huge clumps. It occasionally grows on the rough, fibrous covering on the trunks of palm trees, and has been observed growing as a terrestrial on roadside cuttings on the Atherton Plateau, and near Cape Tribulation in northern Queensland (M. Clements, pers. comm.). It usually occurs in positions where it has full sun for at least part of the day, but will tolerate even quite heavy shade.

It usually flowers from August to October in the more northerly, tropical regions, and until December in its more southerly localities. The unusually late flowering, from October to December, of a population found between the Barron and Bloomfield Rivers in northern Queensland has already been discussed (see var. *leroyi*).

12. C. suave *R.Br.* Prodr.: 331 (1810); Benth., Fl. Australia 6: 303 (1873); Bailey, Queensland Fl. 5: 1548 (1902); Rupp in Proc. Linn. Soc. New South Wales 62: 299-302, f.2-3 (1937); Tierney in Amer. Orchid Soc. Bull. 26: 168-9 (1957); Leaney in Australian Pl. 3: 291, + fig. (1966); Dockrill in Australian Pl. 3: 293-4, 330-1, + fig. (1966) & Australian Indig. Orchids 1: 638-9, + fig. (1969); Clements, Prelim. Check. Australian Orchid.: 61-4 (1982); Seth & Cribb in Arditti (ed.), Orchid Biol., Rev. Persp. 3: 299 (1984). Type: Australia, *R. Brown* (holotype BM!).

C. gomphocarpum Fitzg. in J. Bot. 21: 203-4 (1883); Rupp in Australian Orchid Rev. 4: 65-6, + fig. (1939). Type: Australia, *Fitzgerald* (holotype BM!).

A medium-sized, perennial, epiphtyic *herb*. *Pseudobulbs* not strongly inflated, apparent only in young plants, developing in older specimens into an elongated stem up to about 50 cm long, which is covered by the fibrous remains of the sheathing leaf bases. Each shoot will grow and flower for many years before a new growth is produced near the base, the shoot extending only a few centimetres in length each year. Cataphylls present only in young plants. *Leaves* 6–11, carried apically on the shoot, distichous, up to 30–60 × 0.8–1.4 cm, ligulate, somewhat tapering, obtuse to subacute, oblique at the apex, rather thin and grass-like in texture, arching, eventually deciduous, articulated to a sheathing base 4–7 cm long with a narrow membranous margin. *Scape* usually 15–24(35) cm long, arching or pendulous, produced in the axils of the leaf bases just below the current leaves, often more than one per stem and often persistent on the stem for several years; peduncle 5–10 cm long, suberect at the base, arching above, covered in the basal half by 7–8 keeled, cymbiform, acute, somewhat spreading sheaths up to 3.5 cm long, exposed in the apical half, with a few small, sterile bracts; rhachis about 10–20 cm long with (10)20–40(50) very closely spaced flowers; bracts very small, 1–2 mm long, triangular, subacute. *Flowers* 1.5–2.5 cm across; strongly sweet-scented; rhachis often stained purplish, pedicel and ovary green; sepals and petals light green, sometimes yellow-green or olive-green, occasionally with pale reddish blotches; lip bright yellow to greenish, side-lobes stained orange-brown, or rarely green; disc dark red-brown in front, often paler behind; column pale yellow-green. *Pedicel and ovary* slender, (8)12–18 mm long. *Dorsal sepal* 12–15 × 5–6 mm, oblong-elliptic to obovate, obtuse, mucronate, erect or somewhat porrect; lateral sepals similar, often slightly broader than the dorsal sepal. *Petals* shorter than the sepals, 10.5–13 × 4–5.5 mm, oblong-elliptic, slightly oblique, weakly porrect to almost parallel with the column. *Lip* 9.5–11 × 5–6 mm, almost entire to strongly 3-lobed, minutely papillose; side-lobes often weakly differentiated, erect, with rounded to acute apices; mid-lobe about 4 × 4 mm, ovate, obtuse to subacute, porrect, of thicker texture than the rest of the lip, margin entire; callus a raised, ligulate ridge which is glabrous and shiny in front, with a shallow depression in front which extends into the base of the mid-lobe. *Column* 6–8 mm long, winged towards the apex; pollinia 2, 1.3 mm long, ellipsoidal, deeply cleft, on a minute (0.5 mm), broadly crescent-shaped viscidium extending into two, short, thread-like appendages at the tips. *Capsule* about 3 × 2 cm, orbicular-ellipsoidal, on a 10–18 mm pedicel and with a 6 mm long apical beak formed by the persistent column.

ILLUSTRATIONS. Photographs 70, 71; figure 21.5.

DISTRIBUTION. Eastern Australia, from southern New South Wales to northern Queensland (map 5E); sea level to 1200 m (3940 ft).

HABITAT. In damp, open woodland, usually near the coast, often on *Eucalyptus* trees where it grows in hollows and rotting wood left by fallen branches, or on *Melaleuca* trees; flowering August-October in the tropics, and until December (January) in the more temperate part of its range.

C. suave was described in 1810 by Robert Brown. It has an unusual growth habit which is very distinct from any other species in subgenus *Cymbidium*, except for the newly described *C. elongatum* from Sabah. Instead of producing a new pseudobulb annually, the same stem continues to grow and flower indeterminately for many years, eventually reaching 50 cm (20 in) or more in length.

The scape is usually about 15–25 cm (6–10 in) long, with many closely spaced flowers, but Rupp (1937) suggests that plants from northern Queensland have shorter scapes, averaging only about 13 cm (5 in) in length, while those from southern New South Wales may reach 30 cm (12 in) or more. Usually the flower colour varies between bright green and yellow-green, often tinged with brown on the backs of the sepals. The colour of the flowers of plants from northern Queensland is paler, while Rupp (1937) noted that he had seen a plant with dull green flowers with indistinct red blotches at Illawarra, New South Wales.

The shape of the lip is very variable. R. Brown (1810) in his type description gave the lip shape as 'indiviso'. Bentham (1873) and Bailey (1902) described the variation in lip shape as undivided or obscurely sinuately 3-lobed. Rupp (1937) extended this even more to include a variant from coastal New South Wales which had a lip which was as strongly 3-lobed as any of the other Australian species. Although it is much less common and widespread than the entire or obscurely lobed variants, he was able to illustrate a complete range of variation from entire to strongly lobed lips.

Dockrill (1966, 1969) cites *C. gomphocarpum*, described by Fitzgerald in 1883, as the only synonym of *C. suave*. No locality information was given in the original description, and until Rupp discovered an unpublished plate drawn by Fitzgerald in the Herbarium of the Royal Botanic Gardens, Sydney, little information was available about the species. The description and plate, subsequently published by Rupp (1939), are both undoubtedly of *C. suave*, differing somewhat from the typical species only in minor details. Fitzgerald stated that it differed from *C. suave* in having a club-shaped or almost terete capsule, but his plate shows that it is simply immature. Rupp stated that the lip also differed from *C. suave*, and that the scape was erect rather than arching. The plate does show a suberect scape, but it appears to be supported by the leaves, and older scapes are shown as arching. The lip is shown as strongly 3-lobed, and the mid-lobe appears to be truncated. The side-lobes and the base of the mid-lobe are finely dark-spotted. Although less common than plants with an entire or weakly lobed lip, specimens with a strongly 3-lobed lip are not outside of the range of variation in *C. suave* (see Rupp, 1937). Furthermore, the lip has not been flattened, and the scape has flowers with a lip shape similar to *C. suave*. The coloured markings on the lip therefore appear to be the only unusual character. This seems insufficient basis on which to recognise it as a distinct taxon. The spotting may, perhaps, be indicative of hybridisation with *C. canaliculatum*, a possibility reinforced by the rare colour variants of *C. suave* discussed previously (see Rupp, 1937) which are dull green with faint red-brown blotches on the tepals which could also be explained by natural hybridisation. Tierney (1957) claimed that *C. gomphocarpum* had been re-collected at Harvey's Creek on the Atherton Tableland, but this record has not been confirmed. Therefore *C. gomphocarpum* is treated here as a synonym of *C. suave*.

Leaney (1966) includes an interesting note on the vegetative growth of this species. He states that *C. suave* can send rhizomes considerable distances inside the rotting trees in which it grows. These eventually reappear further down the trunk from the main plant, and form a new young plant. He notes that careful splitting of the trunk has shown connections between apparently separate plants on the same tree, and suggests that the distance between connected plants may be as much as 2.5 m (8.2 ft).

C. suave has the most restricted distribution of the Australian species. It is found from southern New South Wales, 14 km (9 miles) from the border with Victoria, northwards along the east coast to southern Cape York. It has the most southerly distribution of all the *Cymbidium* species, at about 36°S, almost equivalent to the most northerly latitudes at which the genus occurs, as represented by *C. goeringii* in Japan and Korea. *C. suave* is found from sea level up to about 1200 m (3940 ft), but does not occur on the drier western side of the Great Dividing Range.

This species prefers a drier habitat than *C. madidum*, and only occasionally occurs in cleared areas in the rain forest. Neither will it tolerate the very dry conditions in which *C. canaliculatum* thrives, although it is more cold tolerant than either of these species. It will also tolerate having its growths burnt in bush fires, perhaps because its roots are well protected.

C. suave is found in two main types of habitat, both of which provide some moisture and protection for the roots. In damp, open hardwood forest, particularly on *Eucalyptus*, it grows in hollows in branches and trunks of trees, forming a huge root system which may penetrate deep into the damp, rotting core of the tree, the roots sometimes extending 10 m (33 ft) from the plant. It is also found on tree stumps and even on fence posts made from *Eucalyptus*, if they have started to rot.

Its other common habitat is on *Melaleuca* trees which grow along water courses or on swamp margins. The bark of this tree peels off like sheets of paper, and the roots of the orchid penetrate beneath the top layers, eventually forming an extensive network around the tree. When plants are taken from the wild into cultivation, so much of this root system is lost that the plants will often die. Rupp (1937) also reported this species growing on tree fern trunks.

It requires an open position where it receives a lot of sunlight. It flowers during August to October in northern Australia, and until December or January in its temperate, more southerly localities.

13. C. elongatum *J.J. Wood, Du Puy & P.S. Shim* **sp. nov.**, ab omnibus speciebus Malesicis caulibus maturis elongatis reclinatis 30–130 cm vel ultra longis apice foliosis caulium basibus vaginatibus omnino celatis. *Folia* disticha, late linearia, ensiformia vel ligulata et leviter curvata. *Labium* subintegrum, callo e costis duobus levibus sistente. *Pollinia* duo, fissa. Habitus vegetativus *Agrostophyllum* Bl. vel *Dipodium* R.Br. in mentem vocat. Typus: Sabah, *Collenette* A47 (holotype BM! isotype K!).

A medium-sized, perennial, terrestrial or epiphytic *herb*. *Pseudobulbs* absent; stem elongated, 30–130 cm or more long, growing and flowering indeterminately, erect when young, becoming reclinate as it lengthens, the base of the shoot covered by the scarious or fibrous remains of the leaf bases, through which new roots emerge in the basal portion. *Leaves* 4–9, distichous, 10–19(37) × 1.2–2.3(3.5) cm, broadly linear, linear-elliptic or ligulate, conduplicate, somewhat curved, obtuse and mucronate to acuminate, slightly oblique, somewhat cucullate at the apex, coriaceous, eventually deciduous, articulated to distichous, overlapping, sheathing, persistent bases (3.5)6–7 cm long, with scarious margins 1 mm broad. *Scape* 9–28 cm long, suberect, produced in the upper leaf axils; peduncle 8–23 cm long, suberect, covered in the basal portion by about 6 cymbiform, acute, overlapping sheaths up to 8 cm long; rhachis 1–8 cm long, with 1-5 (perhaps more) distant flowers; bracts 4–7 mm long, ovate to triangular, acute, scarious. *Flowers* about 4 cm in diameter; slightly scented; rhachis, pedicel and ovary apple-green; sepals and petals purplish-red outside, olive-green to cream and sometimes stained with red-brown inside; lip pale yellow-green or green to cream, sometimes suffused with pink, usually spotted and blotched with red; callus ridges yellow; column greenish-yellow to red-brown, flushed with purple at the base. *Pedicel and ovary* slender, 2–4.3 cm long. *Dorsal sepal* 26–36 × 10–12 mm, oblong-elliptic, acute, erect; lateral sepals similar, slightly oblique, spreading. *Petals* narrower than the sepals, 22–31 × 6–8 mm, narrowly elliptic, acute, slightly porrect. *Lip* 20–24 × 11–13 cm, obscurely 3-lobed, minutely papillose; side-lobes erect, weakly differentiated; mid-lobe 12–14 × 10–12 mm, ovate, acute, porrect, entire; callus ridges 2, inflated towards the apex, somewhat convergent, from near the base of the lip to the base of the mid-lobe. *Column* 15–16 mm long, with a very short foot, curved, strongly winged towards the apex; pollinia 2, 1.8 mm long, triangular-ellipsoidal, deeply cleft, on a broadly triangular viscidium extending into two short, thread-like appendages at the lower corners.

ILLUSTRATIONS. Photographs 76, 77; figure 22.1.

DISTRIBUTION. Sabah, Sarawak (map 5F); 1200–1750 m (3940–5740 ft) in Sabah, 2100–2300 m (6890–7545 ft) in Sarawak.

HABITAT. Terrestrial in marshy areas in open, scrubby woodland of stunted trees, often rooted at the base of *Leptospermum* or amongst rattans, sedges, Ericaceae and *Begonia* on sandstone or ultrabasic serpentine rock, or occasionally epiphytic (Sarawak); flowering during September-October, but also in March and June so this is perhaps not strictly seasonal.

C. elongatum is one of the most unusual species in the genus, having a monopodial habit, with indeterminately growing stems which tend to lean on to and scramble over the surrounding vegetation as they elongate.

The distinctive vegetative habit of *C. elongatum*, and its few-flowered scape with greenish flowers, distinguish it from all other species of *Cymbidium* in Malesia. The habit is unusual in *Cymbidium*, and is reminiscent of certain climbing species of *Dipodium*, such as *D. pictum* (Lindley) Reichb.f. However, the flowers of *Dipodium* species differ markedly in having distinct dark spots on the reverse of the tepals, a hairy lip and bilobulate stipes on the viscidium. This latter character similarly prevents the inclusion of *C. elongatum* in *Grammatophyllum*.

The two pollinia and free lip place this species in subgenus *Cymbidium*, and in its indeterminate growth habit and leaf shape it closely resembles *C. suave* alongside which it has been placed in section *Austrocymbidium*. Further affinities with this section can be found in the relatively broad, oblong-elliptic

Map 6: The distribution of the species in section *Floribundum*:
A *C. floribundum*
B *C. suavissimum*
and in section *Himantophyllum*: **C** *C. dayanum*
and in section *Bigibbarium*: **D** *C. devonianum*

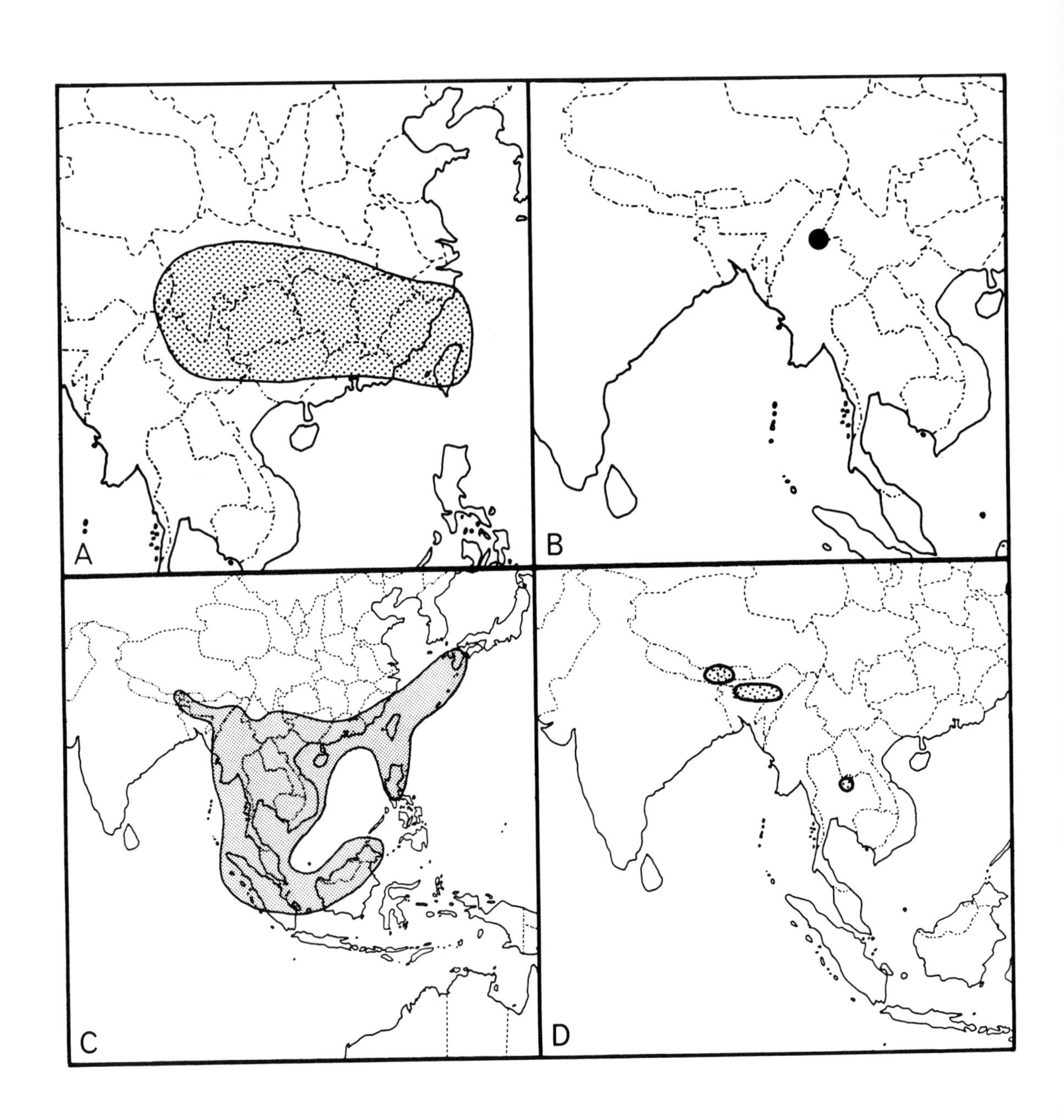

sepals, the obscurely 3-lobed lip, the broad wings at the apex of the column and the ellipsoidal pollinia. It is, however, easily distinguished from the Australian *C. suave* by its shorter, coriaceous, more distant leaves, its much fewer-flowered scape, its larger flowers and the presence of two distinct callus ridges on the lip.

Section **Floribundum** *Seth & Cribb* in Arditti (ed.), Orchid Biol., Rev. Persp. 3: 298 (1984). Type: *C. floribundum* Lindley.

Cymbidium section *Suavissimum* Seth & Cribb, l.c.: 297 (1984). Type: *C. suavissimum* Sander ex Curtis.

The two species in this section are almost identical florally, but have highly distinct vegetative appearances. These two species were placed separately in two sections established by Seth & Cribb (1984). Section *Suavissimum* was differentiated on the basis of its much broader, longer and less leathery leaves. This is not sufficient basis for its separation at sectional level, the leaves being very similar in their anatomy. The similarity of the flowers has also led to the fusion of these two sections in the present revision. Schlechter (1924) included these species in section *Jensoa* on the mistaken suggestion of the presence of four pollinia.

Section *Floribundum* is characterised by having numerous (15–50), closely spaced, red-brown flowers on a short, suberect scape. The sepals and petals are obtuse at the apex. The lip has broad side-lobes which are rounded at the apex and a broadly ovate mid-lobe. This section is closely allied to section *Austrocymbidium*, both sections having many closely spaced flowers on a relatively short scape, and obtuse sepals and petals which have a greenish background colour. It is distributed in southern China and Burma, much further north than section *Austrocymbidium* which does not occur north of W Malaysia, and it can be further distinguished by its strongly red-flushing lip with broader side-lobes, its red-brown flower colour and its triangular pollinia.

14. C. floribundum *Lindley*, Gen. Sp. Orchid. Pl.: 162 (1833); Wu & Chen in Acta Phytotax. Sin. 18: 301 (1980); Seth & Cribb in Arditti (ed.), Orchid Biol., Rev. Persp. 3: 297-8 (1984). Lectotype: Icon. in R. Hort. Soc. London.

C. pumilum Rolfe in Kew Bull.: 130 (1907); Makino & Nemoto, Fl. Japan, 1628-32 (1931); Nagano & Nagano in Amer. Orchid Soc. Bull. 24: 739-40, + fig. (1955); Lin, Native Orchids of Taiwan 2: 123-5, t.51-2, + fig. (1977); Liu & Su, Fl. Taiwan 5: 947 (1978). Type: Yunnan, Tsekou, *Monbeig* (lectotype K!, isolectotypes P, K!, chosen here).

C. illiberale Hayata in Icon. Pl. Formosa 4: 78 (1914); Mark, Ho & Fowlie in Orchid Dig. 50: 16 (1986). Type: Taiwan (Formosa), cult. Taihoku, *B. Hayata* (holotype TI).

C. pumilum Rolfe f. *virescens* Makino in Iinuma, Somoku-Dzusetsu 18: 1185 (1912), **syn. nov.**

C. floribundum Lindley var. *pumilum* (Rolfe) Wu & Chen, in Acta Phytotax. Sin. 18: 301 (1980).

A perennial, lithophytic or epiphytic, often clump-forming *herb. Pseudobulbs* small, up to 3.3 cm long, 2.2 cm in diameter, ovoid, slightly bilaterally compressed, covered by persistent sheathing leaf bases with a 1 mm wide scarious margin, and surrounded by about 5 cataphylls which become scarious, and eventually fibrous with age. *Leaves* 5–6, up to 20–55 × 0.8–1.5(2.0) cm, the shortest merging with the cataphylls, linear-elliptic, arching, acute, the apex usually slightly oblique, articulated 1.5–6 cm from the pseudobulb. *Scape* usually 15–25(40) cm long, robust, suberect, arching upwards from the base of the pseudobulb, with (6)15–30(45) closely spaced flowers; peduncle about 10(-15) cm long, covered in 6–8 sheaths; sheaths up to 6 cm long, becoming scarious, cylindrical in the basal half, expanded and cymbose in the upper half, acute; bracts short, 2–6(17) mm long, triangular, acute. *Flowers* 3–4 cm across, red-brown or occasionally green, not scented; rhachis apple-green; pedicel and ovary often stained with red-brown; sepals and petals strongly flushed red-brown, with a narrow, yellow or green margin; lip white, mottled purple-red on the side-lobes and blotched with purple-red on the mid-lobe (occasionally with pink markings instead), yellow at the base, becoming bright red on pollination or with age; callus ridges bright yellow; column yellow-green, flushed red-brown above, pale green below, dark red-purple at the base; anther-cap yellow. *Pedicel and ovary* 1.5–3.3 cm long. *Dorsal sepal* suberect, 1.8–2.1 × 0.5–0.8 cm, oblong-elliptic, obtuse, margins slightly recurved; lateral sepals similar, spreading or slightly porrect. *Petals* 1.6–2.0 × 0.5–0.8 cm, elliptic, obtuse to subacute, slightly curved, somewhat spreading. *Lip* about 1.8 cm long, 3-lobed; side-lobes up to 7 mm broad, erect, rounded or obtuse, minutely papillose; mid-lobe 0.7 × 0.8 cm, broadly ovate, subacute, weakly recurved, minutely papillose, the margin sometimes weakly undulating; callus of 2 parallel ridges which tend to converge at their apices, with a shallow channel between them. *Column* 1.2–1.5 cm long, curved, broadening into two narrow wings near the apex, minutely papillose below; pollinia 2, triangular, cleft, on a small rectangular viscidium. *Capsule* about 3 cm long, cylindrical, tapering at each end, apex with a short beak.

ILLUSTRATIONS. Plate 10, photographs 78, 79; figure 22.4.

DISTRIBUTION. S China, Taiwan (map 6A); (800)1500–2800 m ((2625) 4920–9185 ft).

HABITAT. On rocks in shaded gorges, often in pine forest, or in full sun, but also recorded as an epiphyte. Flowering (March) April-June.

C. floribundum was described by Lindley (1833) in a footnote appended to his description of *C. sinense.* The short Latin diagnosis was made from a painting by a Chinese artist, in the possession of the London Horticultural Society. Lindley distinguished *C. floribundum* from *C. sinense* by its smaller, more numerous flowers, its obtuse sepals and the red lip with a yellow centre. Although its leaf is not unlike *C. sinense*, it is not closely related. The two cleft pollinia, the lack of fusion of the lip and column base, the leaf anatomy, leaf-tip and seed shape place this species in section *Cymbidium.* It has recently gained importance as one of the parents of the modern miniature hybrids.

C. floribundum is allied to *C. suavissimum*, but the latter can be easily distinguished by its broader leaves and the small auricles at the base of the column (see discussion of *C. suavissimum* for further details).

The name *C. floribundum* was unfortunately ignored by later authors and the species has become well known in cultivation under the later synonym *C. pumilum. C. pumilum* was described by Rolfe in 1907, based on a specimen imported from Japan and cultivated by Barr, and a second collected by the French missionary Monbeig in Yunnan (China). This latter specimen has been selected here as the lectotype. The description of *C. pumilum* is far more detailed than that of *C. floribundum*, but there can be little doubt that they are the same species and that *C. pumilum* is therefore a later synonym. The strongly red-coloured lip in the painting of *C. floribundum* is unusual in that the species usually has a spotted lip, but the lip does turn deep red, obscuring the spots, when the rostellum is disturbed following pollination, or when the flower ages.

C. illiberale was described in 1914, by Hayata, from a plant cultivated in Taiwan. He differentiated it from *C. pumilum* (= *C. floribundum*) by the 'light reddish green petals and sepals, and by the lips which are light red with red maculatum blotches on the front lobe, and numerous minute red spots on the side-lobes'. These colour patterns are typical of *C. floribundum*, and it appears that the two taxa are not distinct. Lin (1977) states that specimens collected in Taiwan cannot be distinguished from those collected on mainland China.

A variant lacking the red-brown pigment in the flowers, which are consequently green with a white lip, is commonly cultivated as var. *album*, and has been named f. *virescens* by Makino (1912). The lip of this colour variant also turns red on pollination, or as the flower ages. Specimens with pale pink markings on the lip are also known.

The Japanese name 'Kinryohen' means 'Golden-margined' (Nagano & Nagano, 1955). This probably refers to the narrow yellowish margin on the sepals and petals, but may also refer to the highly prized specimens with variegated leaves which are cultivated in Japan, although the Japanese name for the latter is 'Jitsugetsu'. There are many named variants maintained in cultivation in Japan and China, where *C. floribundum* has been grown for several centuries.

C. floribundum is not native to Japan, although it may now be found naturalised in the warmer regions (Nagano & Nagano, 1955). In Taiwan and China it grows in the southern provinces. Specimens have been seen from Taiwan, Guangdong, Guizhou, Xizang, Zhejiang, Yunnan and Sichuan, and Wu & Chen (1980) extend this list to include Fujian, Guangxi, Hubei and Jiangxi. Plants from Taiwan and eastern China usually have broader leaves (1.5–2 cm, 0.6–0.8 in), and flower earlier, in March and April.

Wu & Chen (1980) recognise two varieties differentiated on the shape of the side-lobes of the lip: var. *floribundum* with broad, rounded side-lobes, and var. *pumilum* with narrower, more acute side-lobes. This difference is not apparent in the specimens we have seen, but further study may uphold this distinction.

C. floribundum is a montane plant, usually found between 1500 and 2800 m (4920 and 9185 ft) altitude, although it has been recorded from as low as 800 m (2625 ft) in Taiwan (Lin, 1977). It is normally lithophytic in mountain gorges, in shaded situations, often in pine forest, although it may grow epiphytically, and has been observed on *Liquidambar formosana* (Hamamelidaceae) in eastern China. It has also occasionally been recorded as growing on limestone. It has a high tolerance of dry conditions, growing in open situations, such as on cliffs, where it can form extensive clumps.

15. C. suavissimum *Sander ex C. Curtis* in Gard. Chron. 84: 137, 157, *f8.* 67 (1928); E. Cooper in Orchid Rev. 40: 296 (1932); C. Curtis in Orchid Rev. 42: 292, 299, + fig. (1934); R. Doig in Amer. Orchid Soc. Bull. 9: 86-7, + fig. (1940); B.H. Ghose in Amer. Orchid Soc. Bull. 29: 824 (1960); P. Taylor in Bot. Mag. 180: n.s. t.671 (1974). Lectotype: Icon. in Gard. Chron., *loc. cit.*: fig. 67 (1928), chosen here.

A large perennial *herb. Pseudobulbs* about 6 cm long, 3 cm in diameter, ovoid, lightly bilaterally compressed, covered by persistent, sheathing leaf bases wih a membranous margin up to 3 mm wide, and surrounded by about 5 purple cataphylls which become scarious and eventually fibrous with age. *Leaves* 5–7, up to 70 × 3–3.8 cm, the shortest merging with the cataphylls, linear-elliptic, arching, acute, the apex usually slightly oblique; articulated 2–6 cm from the pseudobulb. *Scape* usually about 50 cm long, robust, suberect, arching upwards from the base of the pseudobulb, with about 50 closely spaced flowers; peduncle about 15 cm long, covered by about 7 sheaths; sheaths up to 7 cm long, inflated, cymbose, acute. *Flower* about 3.5 cm across, red-brown; sweetly fruit-scented; sepals and petals green or yellow at the margins, strongly flushed red-brown in the centre; lip white, mottled purple-red on the side-lobes and blotched with purple-red on the mid-lobe, yellow at the base, turning bright red on pollination, callus ridges bright yellow; column yellow-green, speckled red-brown above, pale green below, with a dark purple-red blotch at the base; anther-cap yellow. *Pedicel and ovary* 3–4.3 cm long. *Dorsal sepal* suberect, 2–2.5 × 0.6–0.7 cm, oblong-elliptic, obtuse, margins recurved; lateral sepals similar, spreading. *Petals* 1.8–2 × 0.6–0.7 cm, elliptic, obtuse to subacute, somewhat spreading or slightly porrect and covering the column. *Lip* about 1.6 cm long, 3-lobed; side-lobes 6 mm broad, erect, rounded, apex obtusely angled, minutely hairy; mid-lobe 0.6 × 0.7 cm, broadly elliptic, rounded, minutely acuminate, recurved, minutely papillose, sometimes with a weakly undulating margin; callus of 2 parallel ridges, which tend to converge at their apices and are not strongly raised but have a shallow channel between them. *Column* 1.3–1.4 cm long, curved, broadening into 2 narrow wings near the apex, and with 2 small (1–2 mm), incurved auricles on the margins at the base, minutely papillose; pollinia 2, triangular, cleft, on a small rectangular viscidium.

ILLUSTRATIONS. Plate 11; photographs 80, 81; figure 22.5.

DISTRIBUTION. N Burma (map 6B); about 800–1000 m (2625–3280 ft).

HABITAT. Not known but probably evergreen, tropical, lower montane forest; flowering period July-August.

C. suavissimum was first described in the *Gardeners' Chronicle* (1928) by its editor C.H. Curtis, after it had been exhibited at the Royal Horticultural Society by Messrs Sander, in August of that year. The name refers to the sweet-smelling flowers (Latin: *suaveolens*). No type specimen was preserved, so the figure given with the original description is taken as the type. A specimen at Kew (spirit collection no. 12858), taken from a plant bought from Sander in 1933, may be from the type plant. This species has been collected only once, and it is thought that all of the plants now in cultivation are derived from the original collection. It is not known for certain where these plants originally came from, but Ghose (1960) states that Kohn collected the plants near Bhamo in northern Burma close to the border with Yunnan (China), and shipped them to him in Darjeeling. Sander probably obtained plants from this source. Ghose was expecting a consignment of *C. tracyanum*, and grew the plants under similar conditions to *C. tracyanum* and *C. lowianum* he had acquired from the same region. Unfortunately they were less hardy and many plants were lost during the winter. Therefore, it seems probable that *C. suavissimum* was collected in the valleys around Bhamo at lower elevations than *C. tracyanum* or *C. lowianum*.

Vegetatively the plant appears similar to the large-flowered species in section *Iridorchis*, but the flowers are very different, lacking the fusion at the base of the lip and column, and very much smaller. The flower is almost identical to *C. floribundum*, albeit slightly larger, and the scape has a similar structure, but is longer, more robust, and has more flowers. Table 10.4 gives the characters by which these two species can be separated. These characters are almost all concerned with the larger size and stronger growth of *C. suavissimum*, but, in the manner of growth, the species are very similar. The flowers themselves are almost identical.

C. suavissimum appears to be a distinct species, but the possibility of a hybrid origin must be considered, especially as it appears to have such a restricted distribution. The leaves and pseudobulbs strongly resemble those found in the sympatric species in sect. *Iridorchis*; *C. lowianum*, *C. tracyanum* and *C. iridioides*. Either of the last two is the likeliest candidate as one of the putative parents, as both of these have scented flowers and the red-brown colour, whilst the other parent would be *C. floribundum*. The flowering time of *C. suavissimum* is intermediate between those of the possible parents. Its flowers and inflorescences strongly resemble those of *C. floribundum*. It is difficult to imagine a hybrid flower which is so similar to one parent and almost unaffected by the other. However, the small auricles at the base of the column in *C. suavissimum* are unique in the genus, and may be the rudimentary expression of the lip and column fusion found in *C. tracyanum* and *C. iridioides*. Furthermore, *C.* Pumilow, the primary hybrid between *C. floribundum* and *C. lowianum* (closely related to *C. iridiodes* and *C. tracyanum*), is similar in many respects to *C. suavissimum*.

TABLE 10.4: A comparison of the characters used to distinguish between *C. floribundum* and *C. suavissimum*

Character	***C. suavissimum***	***C. floribundum***
Pseudobulb size	up to 6 × 3 cm	up to 3.3 × 2 cm
Cataphyll colour	purple	green
Leaf length	up to 70 cm	up to 50 cm
Leaf width	3–3.8 cm	1.5(2.0) cm
Scape length	about 50 cm	15–25(40) cm
Number of flowers	about 50	15–30(40)
Ovary length	3–4.3 cm	1.5–3.3 cm
Scent	fruity	not scented
Dorsal sepal length	2–2.5 cm	1.8–2.1 cm
Petal length	1.8–2.2 cm	1.6–2.0 cm
Auricle at the column base	present	absent
Flowering period	July-August	(March)April-June

FIGURE 22

1. ***C. elongatum*** (Kew spirit no. 47629/29462)
   a Perianth, × 1
   b Lip and column, × 1
   c Pollinarium, × 3
   d Pollinium (reverse), × 3
   e Pollinium (side), × 3
   f Flower, × 0.8
   g Flowering plant, × 0.1
2. ***C. borneense*** (Kew spirit no. 48015)
   a Perianth, × 1
   b Lip and column, × 1
   c Pollinarium, × 3
   d,e Pollinium (1 pair), × 3
3. ***C. dayanum*** (Kew spirit no. 47962)
   a Perianth, × 1
   b Lip and column, × 1
   c Pollinarium, × 3
   d Pollinium (reverse), × 3
4. ***C. floribundum*** (Kew spirit no. 44942)
   a Perianth, × 1
   b Lip and column, × 1
   c Pollinarium, × 3
   d Pollinium (reverse), × 3
5. ***C. suavissimum*** (Kew spirit no. 47448)
   a Perianth, × 1
   b Lip and column, × 1
   c Pollinarium, × 3
   d Pollinium (reverse), × 3
6. ***C. devonianum*** (Kew spirit no. 47905)
   a Perianth, × 1
   b Lip and column, × 1
   c Pollinarium, × 3
   d Pollinium (reverse), × 3

1a
1b
1c
1d
1e
1f
1g
2a
2b
2c
2d
2e
3a
3b
3c
3d
4a
4b
4c
4d
5a
5b
5c
5d
6a
6b
6c
6d

Section **Bigibbarium** *Schltr.* in Fedde, Repert. 20: 105 (1924); P. Hunt in Kew Bull. 24: 94 (1970); Seth & Cribb in Arditti (ed.), Orchid Biol., Rev. Persp. 3: 299 (1984). Type: *C. devonianum* Paxt. (lectotype chosen by P. Hunt, 1970).

Schlechter (1924) established this section which contains a single, highly distinctive species characterised by its unusual leaves which are narrowed to a slender petiole from an unusually broad, elliptic lamina. The petals are rhombic in shape, and the almost entire lip has two small swellings replacing the callus ridges, and two large deep purple spots at the base. The callus ridges are reduced to two small swellings at the base of the mid-lobe. There is a short column-foot.

The two cleft, triangular pollinia and the pendulous scape suggest an affinity with section *Cymbidium*. The leaf margin is acuminate in transverse section suggesting an affinity with subgenus *Cyperorchis*, but the small flowers and the lack of a fusion between the base of the lip and the base of the column precludes its inclusion there.

16. C. devonianum *Paxton*, Mag. Bot. 10: 97-8, + fig. (1843); Hook. f., Fl. Brit. India 6: 10 (1891); Veitch, Man. Orchid. Pl. 9: 13-14 (1893); King & Pantling in Ann. Roy. Bot. Gard. Calcutta 8: 190-1, t.253 (1898); Summerhayes in Bot. Mag. 156: t.9327 (1933); Hara, Stearn & Williams, Enum. Flow. Pl. Nepal 1: 37 (1978); Easton in Orchid Rev. 91: 135-8, f.125 (1983); Seidenfaden in Opera Bot. 72: 83-6, f.45 (1983); Du Puy in Orchid Rev. 91: 369 (1983) & in Kew Mag. 1: 78 (1984); Seth & Cribb in Arditti (ed.), Orchid Biol., Rev. Persp. 3: 299-300 (1984). Type: Icon. in *Paxton*, loc. cit. (1843).

C. sikkimense Hook.f., Fl. Brit. India 6: 9 (1891) & Icon. Plant. 12: t.2117 (1894); Schlechter in Fedde, Repert., Beih. 20: 105 (1924). Type: Sikkim, Lachen Valley, *J.D. Hooker* s.n. (holotype K!).

A medium to small, perennial, lithophytic or epiphytic *herb*. *Pseudobulbs* represented by a swelling of the base of the shoot, about 3 × 2 cm, covered by 5–6 scarious cataphylls and the sheathing, persistent leaf bases; new growths strongly bilaterally flattened, about 2–3 cm broad, composed of the folded, flattened and keeled, distichous, purple cataphylls up to 10 cm long, from which the 2-4 true leaves emerge. *Leaves* suberect, 20–49 cm long including the slender, channelled petiole; lamina 17–30 × 3.5–6.2 cm, elliptic, somewhat coriaceous, smooth, with a prominent mid-vein, obtuse to subacute, oblique, shortly mucronate, margin entire; articulated 5–15 cm from the base. *Scape* (15)24–44 cm long, pendulous, with about 15-35 closely spaced flowers; peduncle 7–12 cm long, horizontal to pendulous, covered by 6–7 cymbose, acute, slightly spreading, purple sheaths up to 6 cm long; bracts 2–5 mm long, triangular, acute. *Flowers* 2.5–3.5 cm across, usually orientated towards one side of the scape; not scented; rhachis, pedicel and ovary purple to greenish; sepals and petals pale yellow to dull green, lightly to heavily mottled with purple-brown; lip purple, the side-lobes and disc cream with maroon mottling, the mid-lobe maroon with two large, deep purple spots at the base; column greenish, speckled red-brown above, and with a dark maroon spot at the base ventrally; anther-cap cream. *Pedicel and ovary* (8)12–24 cm long. *Dorsal sepal* 20–29 × 7–10 mm, elliptic, obtuse, erect, with somewhat recurved margins; lateral sepals similar, slightly oblique, spreading, somewhat pendulous. *Petals* 16–22 × 6–10 mm, subrhombic, subacute to acute, slightly oblique, somewhat spreading. *Lip* broad, 14–16(20) × 10–15 mm, rhombic, minutely papillose, weakly 3-lobed, attached to a very short column-foot; side-lobes obscure; mid-lobe about 8 × 8 mm, triangular-ovate, decurved to weakly recurved, obtuse to mucronate; callus reduced to two small swellings at the base of the mid-lobe. *Column* 11–14 mm long, narrow at the base, broadening into a strongly winged apex; pollinia 2, 1.5 mm long, triangular, deeply cleft, on a triangular viscidium. *Capsule* 3.5 × 2 cm, oblong-ellipsoidal, with a short stalk, and an apical beak about 1 cm long formed by the persistent column.

ILLUSTRATIONS. Plate 12; photographs 82, 83; figure 22.6.

DISTRIBUTION. Nepal, N E India (Sikkim, Meghalaya), Bhutan, N Thailand (map 6D); 1450–2200 m (4760–7220 ft).

HABITAT. On mossy rocks and moss-covered trees where humus and leaf litter has accumulated, in broken shade; flowering April-June.

C. devonianum is a beautiful and distinctive species, both in its foliage and its flowers, and it has been used successfully in the breeding of miniature hybrids. Its leaves are not unlike those of *C. lancifolium*, but are thicker in texture, oblique or sharply mucronate at the apex, and always have an entire margin. Moreover, *C. lancifolium* has a totally different growth habit, with cigar-shaped pseudobulbs, an erect scape and fewer, very dissimilar, white flowers. The scape of *C. devonianum* is pendulous and usually bears 15–35 closely spaced, purplish flowers.

Although this is not a highly variable species, there is some variation in the flower colour, especially in the amount of red-brown on the sepals and petals. The type had very pale tepals and a lip with a pale margin. Variation is also found in the number of leaves produced per pseudobulb.

Several characters of *C. devonianum* are unusual, if not unique in the genus, and it is difficult to say to which species it is most closely related. The two pollinia and the pendulous scape place it alongside the species in subgenus *Cymbidium*, but the differences from the other species are so great that it has been placed in a separate monotypic section.

C. devonianum first flowered in Britain in 1843, at Chatsworth, in the collection of the Duke of Devonshire, a great orchid enthusiast and leading patron of horticulture of the day. It had been collected in 1837 by Gibson in the Khasia Hills in northern India. Paxton (1843) described and figured this specimen in his *Magazine of Botany*. As with other species named by Paxton, no type specimen was preserved, so the illustration must serve as the type. Gamble and Pantling later collected it in Sikkim and Darjeeling, extending the known range to the extent to which it was, until recently, understood.

J.D. Hooker collected an unusual flowering specimen, which lacked mature leaves, in May 1849, in the Lachen Valley in Sikkim at about 1700 m (5580 ft). He named and described *C. sikkimense* from this material in 1891, indicating that he considered it to be a deciduous species which produced flowers which were later followed by the leaves. The type specimen consists of a flowering specimen of *C. devonianum* which lacks leaf laminas, which have broken off at the abscission zone towards the apex of the false petiole. An examination of the specimen shows that the growths are not immature, but rather that they are composed of several cataphylls which conceal the bases of 2–3 mature leaves. No new growths are visible, but this is normal in *C. devonianum* which tends not to produce new growths until flowering has finished. The illustration by Hooker (1894) shows flowers that are identical with those of *C. devonianum*. As the specimen was collected in the same region as *C. devonianum*, at similar altitudes and flowering at a similar time, and as the plant appears identical to this species in all characters except its lack of leaves, it must be assumed that this is in fact a specimen of *C. devonianum* which has, probably for some environmental reason, lost its leaves. Furthermore, this variant has never been found again, all other collections possessing leaves on the newer growths. Hooker also noted on the type specimen that the flowers and the scape appeared to belong to *C. devonianum*.

C. devonianum has been collected most often in Sikkim, Darjeeling and Meghalaya. Hara, Stearn & Williams (1978) include eastern Nepal and Bhutan in this distribution. This restricted range, and the apparent rarity of this species in the wild, made the discovery of *C. devonianum* in northern Thailand especially exciting (Du Puy, 1983, 1984; Seidenfaden, 1983). There had hitherto been several questionable reports of this species in Laos, Cambodia and Vietnam by Guillaumin (1932, 1960, 1961a, 1961b). In 1932, he identified a leafless specimen collected by Contest-Lacour from Cambodia as *C. sikkimense*, but this has been shown by Seidenfaden (1983) to be *C. macrorhizon*. A leafless specimen (*d'Orleans* 395) from Laos and also mentioned by Guillaumin has now been identified by Seidenfaden as *C. aloifolium*. The three later references identify various specimens from near Dalat, Annam (Vietnam) as *C. devonianum*. Guillaumin (1961b) claims that one of these specimens, collected by Tixier, corresponds exactly with the King & Pantling figure of *C. devonianum*. An examination of the flowers shows that this specimen, as represented in the Paris herbarium (flowers only), lacks the rhomboid petal and lip shape and strongly winged column apex of *C. devonianum*. Moreover, the callus is of two well-defined, sigmoid ridges, the lip has well-defined side-lobes with obtuse apices and forms a sac towards the base, all features strongly suggestive of *C. bicolor* Lindley. Seidenfaden (1983) suggests that the flowers are close to *C. devonianum* as they have a distinct, short column-foot but this feature is also found in *C. bicolor*. Therefore, all of the references to *C. devonianum* and *C. sikkimense* by Guillaumin can be rejected.

In northern Thailand, several specimens of *C. devonianum* have been found growing in *Rhododendron lyi/Lyonia ovalifolia/Lithocarpus/Agapetes saxicola* dominated scrubland, at 1450–1500 m (4760–4920 ft) altitude (Du Puy 1983, 1984). They were usually found growing on moss- and lichen-covered sandstone rocks in the broken shade cast by taller (3 m (9.8 ft)) shrubs, and in taller stands of vegetation surrounding the numerous creek beds. A layer of leaf litter and moss covered the rocks and the orchid's roots spread widely through this moisture-retentive substrate, the flower spike often lying horizontally on the mossy surface. Some specimens were also seen growing epiphytically on old, shaded, moss-covered branches, others in very dry and exposed situations below small shrubs on almost bare rock. Although the plants were seen in flower in early April, during the dry season, the substrate retained enough moisture to maintain humidity around the roots of the orchid. New growth takes place during the rainy season from June to November. The pollination of this species seemed to be very efficient, and almost all of the flowers were seen to have swollen ovaries. However, attempts to observe the pollinator were unsuccessful.

The first specimens of *C. devonianum* collected by Gibson were growing epiphytically on the trunks of decayed trees and in the forks of the branches of old trees where some humus had collected. J.D. Hooker, in his *Himalayan Journals* (2: 294-5, 1854) mentioned that it was growing lithophytically with other orchids on the top of Kollong Rock (Meghalaya). This is a red granite outcrop covered on top by mosses, lichens, club-mosses and ferns, amongst which, he reported, the orchids grew and flowered freely, even though they were exposed to the elements, including occasional frosts. *C. devonianum* is sometimes found high up in the crevices of rocky cliffs in northern India. However, C. Bailes (pers. comm.) recently observed populations of this species in the Teesta Valley, in Sikkim (N India), where it commonly occurs as an epiphyte on mossy branches.

C. devonianum is found at altitudes of 1500–2200 m (4920–7220 ft) in northern India, and was collected at 1450 m (4760 ft) in northern Thailand. It usually flowers during May and June in northern India, and was in flower in early April in Thailand. Some out-of-season flowering has been observed in cultivated specimens.

Subgenus Cyperorchis

Cymbidium *subgenus* **Cyperorchis** *(Bl.) Seth & Cribb* in Arditti (ed.), Orchid Biol. Rev. Persp. 3: 300 (1984); Seidenfaden in Opera Bot. 72: 86 (1983). Type: *Cyperorchis elegans* (Lindley) Bl. *(C. elegans* Lindley).

Cyperorchis Bl. in Rumphia 4: 47 (1848), Mus. Bot. 1: 48 (1849) and Orchid Arch. Ind. 1: 92-3 (1858); Schltr. in Fedde, Repert. 20: 105 (1924).

Iridorchis Bl., Orchid. Arch. Ind. 1: 90-2, t.26 (1858). Type: *Iridorchis gigantea* (Wall. ex Lindley) Bl. (*C. iridioides* D. Don).

Arethusantha Finet in Bull. Soc. Bot. France 44: 178-80, pl.5 (1897). Type: *Arethusantha bletioides* Finet (*C. elegans* Lindley).

The taxonomic status of *Cyperorchis* is discussed in the introduction to the genus. *Cyperorchis* in the broad sense of Schlechter (1924) is given subgeneric status, following the treatment by Seth & Cribb (1984).

The subgenus contains five sections which are all characterised by having narrow, somewhat hyaline, incurved leaf margins which appear acuminate in transverse section (photograph 11). Section *Bigibbarium* in subgenus *Cymbidium* shares this characteristic. The flowers are relatively large, and the dorsal sepal is somewhat porrect, tending to cover the column. The base of the lip is fused to the base of the column for about 2–6 mm. The pollinia are paired and are deeply cleft behind, but are variable in shape

Map 7: The distribution of the species in section *Iridorchis*:
A *C. tracyanum*
B *C. iridioides*
C *C. erythraeum*
D *C. wilsonii*
E *C. hookerianum*
F *C. lowianum*

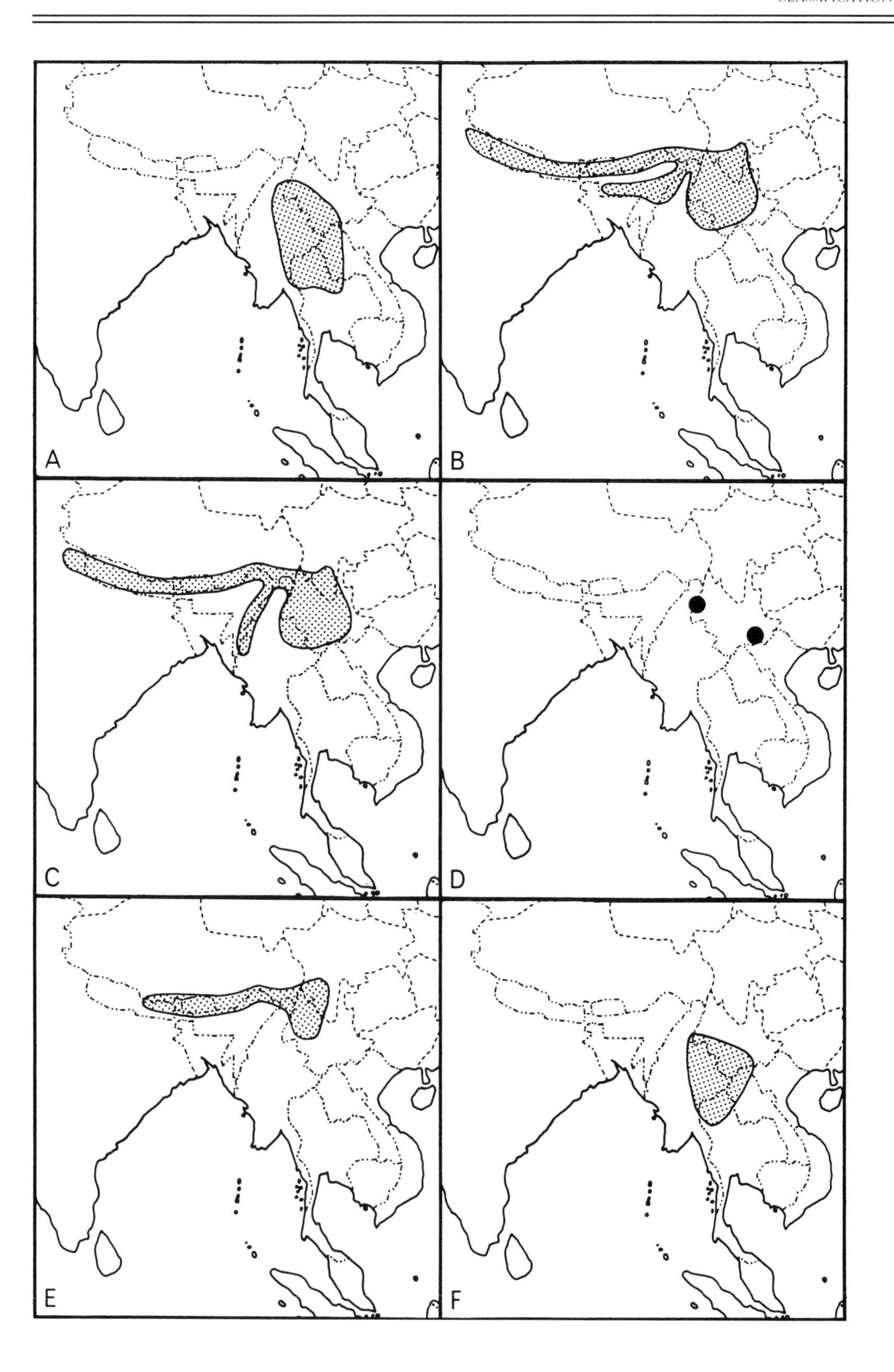
A
B
C
D
E
F

from triangular to quadrangular or clavate, the shape often being characteristic for the sections (figure 14). These characters do not merit the separation of this taxon as a distinct genus.

Section **Iridorchis** *(Bl.)P. Hunt* in Kew Bull. 24: 94 (1970); Seth & Cribb in Arditti (ed.), Orchid Biol., Rev. Persp. 3: 300 (1984). Type: *Iridorchis gigantea* (Wall. ex Lindley) Bl. (*C. iridioides* D. Don).

Iridorchis Bl., Orchid. Arch. Ind. 1: 90-2, t.26 (1858), *non* 'Iridorkis' Thouars (1809).

Cyperorchis section *Iridorchis* (Bl.)Schltr. in Fedde, Repert. 20: 107 (1924).

Iridorchis Bl. was reduced to sectional status within the genus *Cyperorchis* by Schlechter (1924), and was later transferred to the genus *Cymbidium* by P. Hunt (1970). Seth & Cribb (1984) placed it in subgenus *Cyperorchis*. *C. sanderae* is here included in section *Iridorchis*.

The species in this section, along with *C. eburneum* and *C. erythrostylum*, have formed the basis for the breeding of the large-flowered modern hybrids. The flowers are large (about 7–15 cm in diameter), and open widely except for the dorsal sepal which is usually porrect and covers the column. The two cleft pollinia are triangular on a triangular viscidium which has long, thread-like appendages at the lower corners (figure 14). The vegetative plant is robust, with long, acute leaves and large, bilaterally compressed pseudobulbs which are produced annually and flower from the base.

17. C. tracyanum *L. Castle* in J. Hort., ser. 3, 21: 513 (1890); O'Brien in Gard. Chron., ser. 3, 8: 702, 718 (1890) & in Gard. Chron., ser. 3, 9: 137, f.34 (1891); Rolfe in Orchid Rev. 19: 39-40 (1911); Kingdon-Ward in Gard. Chron., ser. 3, 108; 155 (1940); Summerhayes in Bot. Mag. 166: t.56 (1949); Seidenfaden in Opera Bot. 72: 86, f.46, t.5d (1983); Easton in Orchid Rev. 91: 225-8 (1983); Du Puy in Orchid Rev. 91: 370 (1983); Du Puy, Ford-Lloyd & Cribb in Kew Bull. 40: 421-34 (1984). Type: cult. *Tracy* (holotype not located).

Cyperorchis traceyana (L. Castle) Schltr. in Fedde, Repert. 20: 108 (1924) [sphalm. for *tracyana*].

A perennial, epiphytic or lithophytic *herb*. *Roots* thick, fleshy, often with erect aerial roots present around the base of the plant. *Pseudobulbs* 5–15 cm long, 2–6 cm in diameter, elongate-ovate, bilaterally flattened, 5-11-leaved. *Leaves* up to 95 × 4 cm, linear-elliptic, acute, mid-green, articulated 6–13 cm from the pseudobulb to a yellow-green, broadly sheathing base. *Scape* up to 130 cm long, suberect, arching; peduncle covered in sheaths up to 20 cm long; rhachis robust, yellow-green, bearing 10-20 flowers; bracts triangular, 3 mm long. *Flowers* up to 15 cm across; scent strong, sweet; rhachis, pedicel and ovary green; tepals yellowish-green to olive-green, strongly stained with irregular, dull, red-brown veins and spots; lip pale yellow to cream, side-lobes veined dark red-brown, mid-lobe marked with scattered spots and vertical dashes of red-brown, with a central reddish stripe almost to the front of the callus; callus cream, spotted with red along the crests; column yellowish with many red-brown dashes on the ventral surface, and red-brown shading around the base and apex on the dorsal surface; anther-cap cream. *Pedicel and ovary* 2.5–7.5 cm long. *Dorsal sepal* 5.5–8 × 1.4–1.9 cm, narrowly obovate, acute, concave, porrect; lateral sepals similar, slightly asymmetric and twisted. *Petals* 5.4–7.2 × 0.8–1.4 cm, ligulate, usually strongly falcate, spreading or reflexed, giving the flower a drooping appearance. *Lip* 3-lobed, fused to the column for the basal 4–5.5 mm; side-lobes 1.1–1.6 cm broad, triangular, acute and porrect, margins ciliate (cilia over 2 mm long), indumentum otherwise short with the hairs mainly pigmented red-brown and confined to the veins; mid-lobe 1.8–2.1 × 1.8–3.0 cm, elliptic, mucronate, strongly recurved, with scattered hairs on the upper surface, except in the centre where there are three rows of long cilia continuing from the callus ridges to the centre of the mid-lobe, margin finely erose, very strongly undulate and partially reflexed; callus ridges 2, tapering towards the base and dilated towards the apex, densely covered in long cilia, with a third line of cilia between them. *Column* 3.4–4.4 cm long, winged to the base, with short hairs present ventrally towards the base, and tufts of short hairs at the apex on either side of the anther; pollinia 2, 3.2–4.0 mm long, triangular. *Capsule* up to 15 × 6 cm, fusiform-ellipsoidal, stalked, with a short (2 cm) apical beak.

ILLUSTRATIONS. Plate 13, photograph 84; figure 23.1.

DISTRIBUTION. China (S Yunnan), E & N Burma, N Thailand (map 7A); 1200–1900 m (3940–6235 ft).

HABITAT. On damp rocks or on trees in damp, shaded forest, often near or overhanging streams; flowering in September-January.

C. tracyanum was first noticed in December 1890, when it flowered in the collection of Mr Tracy in Twickenham, having been imported three years previously with *C. lowianum*, which it closely resembles vegetatively. It was exhibited at the Royal Horticultural Society, and was described in the *Journal of Horticulture* by Castle (1890). A few days later O'Brien (1890) described the plant again in the *Gardeners' Chronicle* where it was illustrated early in the following year. The plant was sold by auction, and no type specimen appears to have been preserved.

In 1895 a second plant appeared, again in a collection of *C. lowianum*, and a year later a third. This latter was said to have been collected in Upper Burma, giving the first clue to the origin of the species. Soon afterwards, several more plants appeared, imported with *C. lowianum* or occasionally *C. hookerianum*. In about 1900, several specimens, collected by Kerr near Chieng Mai in N Thailand, flowered in cultivation. Kingdon-Ward (1940) described its habitat in N Burma as 'in wet evergreen hill forests' and 'growing in the fork of a tree overhanging a stream in a deep gulley'. It grows in broken shade, in a constantly humid atmosphere, well protected from the occasional frosts. The plants grow in similar situations in northern Thailand (Du Puy, 1983). *C. tracyanum* has characteristic erect, aerial roots which may be an adaptation to a moist environment.

C. tracyanum is similar to *C. erythraeum* in flower colour and shape, but can be readily distinguished by its much larger flower size (up to 15 cm (6 in) diameter), its broader leaves and the long cilia present on the callus, mid-lobe and margins of the side-lobes. *C. iridioides* is similar in colour, and is the only other species in the section which has long hairs in the centre and base of the mid-lobe (see discussion of *C. iridioides*).

C. hookerianum has flowers of a similar size and with ciliate side-lobe margins, so that herbarium specimens of the two species are often confused. *C. hookerianum*, however, has clear apple-green sepals and petals, red-brown spotted side-lobes, and submarginal spots and blotches on the mid-lobe. *C. tracyanum* can also be distinguished by its strongly falcate petals and its longer, denser indumentum. It has very strongly hairy callus ridges, lines of cilia in the centre and base of the mid-lobe, and lines of short hairs confined to the veins on the side-lobes (Du Puy, Ford-Lloyd & Cribb, 1984).

18. C. iridioides *D. Don*, Prodr. Fl. Nepal.: 36 (1824); Hook.f., Fl. Brit. India 6: 14 (1890); Wu & Chen in Acta Phytotax. Sin. 18: 302 (1980); Du Puy, Ford-Lloyd & Cribb in Kew Bull. 40: 421-34 (1984). Type: Nepal, *Wallich* (holotype BM!).

C. giganteum Wall. ex Lindley, Gen. Sp. Orchid. Pl.: 163 (1833) & in J. Proc. Linn. Soc. Bot. 3: 29 (1859); Paxton in Paxton's Mag. Bot. 12: 241 (1845); Griffith, Notulae 3: 341 (1851); Hook. in Bot. Mag. 81: t.4844 (1855); Hook.f., Fl. Brit. India 6: 12-13 (1891); King & Pantling in Ann. Roy. Bot. Gard. Calcutta 8: 191 (1878); Duthie in Ann. Roy. Bot. Gard. Calcutta 9(2): 137-8 (1906); Schltr. in Fedde, Repert., Beih. 4: 267 (1919); *non* Sw. in Schrad. J. Bot. 2: 224 (1800) = *Cyrtopera gigantea*. Type: Nepal, *Wallich* 7355 (lectotype K!, chosen here).

Cyperorchis gigantea (Wall. ex Lindley) Schltr. in Fedde, Repert. 20: 107 (1924).

Iridorchis gigantea (Wall. ex Lindley)Bl., Coll. Orchid Arch. Ind. Jap. 90: t.26 (1858).

A perennial, epiphytic or lithophytic *herb*. *Pseudobulbs* 5–17 cm long, 2–6 cm in diameter, elongate-ovate, bilaterally flattened, with about 10 leaves. *Leaves* up to 90 cm or more long, 2.0–4.2 cm broad, linear-elliptic, acute, mid-green, articulated 6–11 cm from the pseudobulb to a yellow-green, broadly sheathing base. *Scape* 45–85 cm long, suberect to horizontal; peduncle stiff, covered in scarious sheaths up to 11 cm long; rhachis 25–50 cm long, robust, tapering above, yellowish-green, bearing 7-20 flowers; bracts triangular, up to 2.5 mm long. *Flowers* up to 10 cm across, scented; rhachis, pedicel and ovary green;

sepals and petals yellowish-green heavily stained with irregular veins and spots of red- or ginger-brown, with a narrow cream margin; lip yellowish, side-lobes dark red-veined, mid-lobe yellow at the base, marked with a broad submarginal band of confluent deep red spots and blotches; callus ridges yellowish, spotted maroon in front becoming muddy red behind; column yellowish-green, streaked red-brown below. *Pedicel and ovary* 2.2–4.2 cm long. *Dorsal sepal* 4.5–4.7 × 1.2–1.8 cm, narrowly obovate, acute, concave, porrect; lateral sepals similar, slightly asymmetric and twisted forward giving the flower a half-open appearance. *Petals* 4.4–4.8 × 0.7–1.0 cm. ligulate, slightly curved, spreading. *Lip* 3-lobed, fused to the column base for 4–5 mm; side-lobes 1.0–1.2 cm broad, triangular, slightly rounded at the apex, porrect, margin fringed with short hairs, indumentum of short hairs evenly distributed; mid-lobe 1.2–1.6 × 1.4–1.8 cm, ovate, mucronate, strongly recurved, sparsely hairy except in the centre where two or three lines of long hairs extend from the callus to beyond the centre of the mid-lobe; margin erose and strongly undulating, fringed with short hairs; callus ridges 2, short, reaching half way down the disc, dilated at the apex, tapering off rapidly below, covered in long hairs. *Column* 2.5–2.9 cm long, winged but narrowing at the base, giving a more slender appearance than in *C. tracyanum*, short hairs present ventrally near the base; pollinia 2.1–2.5 mm long, triangular. *Capsule* about 6–8 × 3–4 cm, fusiform-ellipsoidal, stalked, with a persistent column forming a short (1.5–2 cm), apical beak.

ILLUSTRATIONS. Plate 14; photograph 87; figure 23.2.

DISTRIBUTION. Nepal, N India (Kumaon, Assam, Sikkim, Meghalaya), Bhutan, Burma & S W China (Yunnan, Sichuan, Xizang) (map 7B); 1200–2200 m (3940–7220 ft).

HABITAT. On trees in dense forest, especially in hollows containing decomposing vegetable matter; flowering August to December.

C. iridioides, first collected in Nepal by Wallich in 1821, was described by David Don in 1825 from the material in the Wallich herbarium. In 1832, Lindley published the name *C. giganteum*, also based on a specimen in the Wallich herbarium. He noted the previous description by Don, but questioned to which species the name referred. Don did not cite a collection number, and it was generally assumed that the name *C. iridioides* was applicable to a *Coelogyne*, and therefore the name *C. giganteum* came into general usage. Don's description is brief, but the only characters which do not fit the present species are the short leaves only 30 cm (12 in) long, and the flower colour ('albi'). The Wallich specimens do indeed have rather short leaves, and agree with Don's description. They also lack colour notes, and it is likely that Don did not know the colour of the living flowers. There is no reason why *C. iridioides* D. Don should not be accepted as the valid name for this species. Furthermore, Lindley's *C. giganteum* is a later homonym of *C. giganteum* Sw. (1800).

C. iridioides is one of the less showy species in the section, with dull, brownish sepals and petals which often don't open fully, hiding the relatively small lip. It is similar in colouring, size and distribution to *C. erythraeum* but the latter has much narrower floral

FIGURE 23

1. ***C. tracyanum*** (Kew spirit no. 45390)
   a Perianth, × 0.7
   b Lip and column, × 0.7
   c Pollinarium, × 3
   d Pollinium (reverse), × 3
2. ***C. iridioides*** (Kew spirit no. 45961)
   a Perianth, × 0.7
   b Lip and column, × 0.7
   c Pollinarium, × 3
   d Pollinium, × 3
3. ***C. erythraeum*** (a–e, Sikkim, *Pantling* 8; f–g, Yunnan, *Henry* 11371)
   a Perianth, × 0.7
   b Lip and column, × 0.7
   c Pollinarium (reverse), × 3
   d Pollinium, × 3
   e Flowering plant, × 0.1
   f Lip and column, × 0.7
   g Lip (flattened), × 0.7
4. ***C. hookerianum*** (Kew spirit no. 47872)
   a Perianth, × 0.7
   b Lip and column, × 0.7
   c Pollinarium, × 3
5. ***C. wilsonii*** (Kew spirit no. 37623)
   a Perianth, × 0.7
   b Lip and column, × 0.7
   c Pollinarium, × 3
   d Pollinium (reverse), × 3
   e Flower, × 0.7

1a
1b
1c
1d
2a
2b
2c
2d
3a
3b
3c
3d
3e
3f
3g
4a
4b
4c
4d
5a
5b
5c
5d
5e
CES

segments, giving a spidery appearance, and the leaves are much narrower. The lip is also distinctive, lacking the undulating margin and central lines of cilia present in *C. iridioides.*

C. tracyanum also has a similar colouring, and has the lines of cilia on the mid-lobe, but it can be distinguished from *C. iridioides* by the larger size of its flower and especially of the mid-lobe, its strongly falcate petals, the hairs confined to the veins on the side-lobes and the fringe of long cilia on the margins of the side-lobes of the lip. *C. iridioides* is also allied to *C. wilsonii* (Du Puy, Ford-Lloyd & Cribb, 1984; see *C. wilsonii* for a discussion of their affinity and distinguishing characters).

In Kumaon, Nepal and Sikkim, the distribution of *C. iridioides* coincides with those of *C. hookerianum* and *C. erythraeum*, both of which, however, usually occur at higher altitudes. It also occurs to the south, in Meghalaya, a region from which the latter two species are absent. The ranges of these three species also overlap in Yunnan, but *C. iridioides* extends further south. It has been accidentally imported with collections of *C. lowianum*, a species endemic to the tropical forests of Thailand, Burma and S Yunnan.

C. iridioides grows at altitudes of 1200–2200 m (3940–7220 ft), the altitude rising in the more southerly localities. It has usually been reported growing epiphytically on mossy trees in damp, shaded forest, the strongest specimens growing on rotting wood, or in tree hollows which have collected humus and leaf litter. Generally, the habitat of this species is frost-free, with some winter rain, but in regions where night frosts occur the winters are dry, as in Meghalaya.

19. C. erythraeum *Lindley* in J. Proc. Linn. Soc. Bot. 3: 30 (1859); Hara, Stearn & Williams, Enum. Flow. Pl. Nepal 1: 37-8 (1979); Wu & Chen in Acta Phytotax. Sin. 18: 303 (1980), as '*erythraecum*'; Hara in Taxon 34: 690-1 (1985). Type: Sikkim, *J.D. Hooker 229* (holotype K!).

C. longifolium sensu Lindley, Gen. Sp. Orchid. Pl.: 163 (1833); Hook.f., Fl. Brit. India 6: 13 (1891); King & Pantling in Ann. Roy. Bot. Gard. Calcutta 8: 191, t.254 (1898); Duthie in Ann. Roy. Bot. Gard. Calcutta 9(2): 137 (1906); Tuyama in Hara, Fl. East. Himal.; 430 (1966) & ibid., 2nd edn: 183 (1971); Cribb & Du Puy in Kew Bull. 38: 65-7 (1983); Du Puy, Ford-Lloyd & Cribb in Kew Bull. 40: 421-34 (1984), *non* D. Don (1825).

C. hennisianum Schltr. in Orchis 12: 46 (1918). Type: Burma, *Hennis* (holotype B†), **syn. nov.**

Cyperorchis longifolia (D. Don) Schltr. in Fedde, Repert. 20: 108 (1924).

Cyperorchis hennisiana (Schltr.)Schltr. in Fedde, Repert. 20: 107 (1924), **syn. nov.**

A perennial, epiphytic or lithophytic *herb. Pseudobulbs* up to 5 cm long, 5 cm in diameter, ovoid, bilaterally flattened, with 5-9 distichous leaves. *Leaves* narrow, up to 90 × 1.6 cm, tapering gradually from the middle to a fine point, base broadly sheathing, narrowing to the abscission zone 3–6 cm from the base. *Scape* slender, 25–75 cm long, suberect to horizontal, arching; peduncle covered in scarious sheaths up to 9.5 cm long; rhachis slender, bearing 5–14 flowers; bracts triangular, up to 4 mm long. *Flowers* small in the section, up to 8 cm across; scented; leaf bases and abscission zone purplish; rhachis, pedicel and ovary green; petal and sepals greenish, heavily spotted and irregularly striped red-brown; lip yellowish to white, side-lobes veined dark red-brown, mid-lobe sparsely spotted red or red-brown, with a central stripe; callus cream or white; column pale yellow, shading to red-brown towards the tip, with red-brown dashes below; anther-cap cream. *Pedicel and ovary* 1.1–3.7 cm long. *Sepals* narrowly obovate, acute; dorsal sepal 3.7–4.1(5.3) × 0.75–1.1(1.25) cm, concave, porrect. *Petals* 3.6–4.3(4.9) × 0.5–0.8 cm, very narrow, ligulate, acute, falcate, spreading. *Lip* 3-lobed, fused to the base of the column for 2–4 mm; side-lobes, 0.7–1.2 cm broad, erect, acute and forward pointing or truncate, densely covered in short hairs at the front, papillose towards the base, margin sometimes fringed with short hairs; mid-lobe 0.9–1.1 × 1.0–1.4 cm, cordate to reniform, acute, not strongly recurved, papillose, sometimes with scattered short hairs, margin flat or weakly undulating, callus tapering to the base of the lip, slightly swollen at the tip, densely covered with short hairs, the indumentum not extending on to the mid-lobe. *Column* 2.3–2.9(3.3) cm long, sparsely hairy below, with wings narrowing strongly towards the base; pollinia obliquely triangular, 2.1–2.7 mm long.

Capsule about 5–6 × 3 cm, fusiform-ellipsoidal, stalked, with a short (1.5 cm) apical beak.

ILLUSTRATIONS. Photographs 88, 89; figure 23.3.

DISTRIBUTION. Nepal, N India (Kumaon, Sikkim, Assam), Bhutan, Burma and China (Yunnan, Sichuan, Xizang) (map 7C); 1000–2800 m (3280–9185 ft).

HABITAT. On trees, rocks and steep banks in open forest; flowering August-November.

This species is best known under the name *C. longifolium*. However, Hara *et al.* (1978) and Hara (1985) have convincingly demonstrated that the type specimen of *C. longifolium*, collected by Nathaniel Wallich in Nepal, is referable to the species commonly known as *C. elegans*. Lindley (1833) misapplied both of these names in his *Genera & Species of Orchidaceous Plants*, and they have been incorrectly used since then. The correct name for this species is, therefore, *C. erythraeum*, described by Lindley in 1859 from a collection made by J.D. Hooker in Sikkim.

Schlechter (1918) also described this species as *C. hennisianum*, distinguishing it from *C. erythraeum* by its very narrow leaves (3–5 mm (0.12–0.2 in) broad) and its fewer flowered (4–7), more slender, shorter, (about 40 cm, 16 in) spike. However, the other features mentioned are typical of *C. erythraeum*. The type specimen has been destroyed, but other specimens from Burma agree well with *C. erythraeum*. It seems likely that the type specimen of *C. hennisianum* was a slender variant of *C. erythraeum* and should not be maintained as a separate taxon.

The flowers of *C. erythraeum* are heavily pigmented with brown on the sepals and petals, although a yellow-flowered variant, lacking red and brown pigment, has also been collected. They resemble those of *C. tracyanum* and *C. iridioides*. The petals are falcate as in *C. tracyanum*, but are much narrower, and the lip is much smaller and lacks the strongly undulating margin and cilia present in *C. tracyanum* and *C. iridioides*. Vegetatively, *C. erythraeum* is distinct in that its pseudobulbs are much smaller, and the leaves narrower than the other species in this section. Its flowers are borne on a slender rhachis, and are smaller than in the other species.

C. erythraeum has been found to be variable in the shape of the lip (Du Puy *et al.*, 1984). One variant, with a lip with truncated side-lobes and a cordate mid-lobe, is found in W China, while the other from Nepal, N India and Bhutan has acute, porrect side-lobes and a reniform mid-lobe. Both of these vary greatly in the density of the indumentum on the lip and in the breadth of the leaf, although this latter is usually about a centimetre wide.

These two major variants occur at opposite ends of the distribution of this species. This variation reflects its large east-west range. *C. erythraeum* grows in the forests of the Himalaya of northern India from Kumaon, through Nepal, Sikkim and Bhutan to Burma and Yunnan and Sichuan, the western provinces of China. It has not been recorded in Meghalaya. It is usually epiphytic in moist forest, but is also found growing on rocks, and there is one record of it on an overhanging grassy bank. Some plants may survive as terrestrials if they fall from overhanging branches. It appears to tolerate cool conditions, and in the Dafla Hills of N E Assam it has been collected in forest which is regularly subjected to snow in December.

20. C. hookerianum *Reichb.f.* in Gard. Chron.: 7 (1866); Bateman in Bot. Mag. 92: t.5574 (1866); Du Puy, Ford-Lloyd & Cribb in Kew Bull. 40: 421-34 (1984). Type: cult. *Veitch* (holotype W).

C. grandiflorum Griffith, Notulae 3: 342 (1851), Itiner. Not. 145 (1851) & Icon. Pl. Asiat. 3: t.321 (1851); Hook.f., Fl. Brit. India 6: 12 (1891); King & Pantling in Ann. Roy. Bot. Gard. Calcutta 8: 192, t.256 (1898); Schltr. in Fedde, Repert., Beih. 4: 268 (1919); Easton in Orchid Rev. 91: 60-1 (1983); *non* Sw. (1799) = *Pogonia grandiflora*. Type: Bhutan, *Griffith* 698 (holotype CAL).

C. giganteum Wallich ex Lindley var. *hookerianum* (Reichb.f.) Bois., Orchid.: 119 (1893), seen in Lindenia 9:13 (1893), **syn. nov.**

C. grandiflorum Griffith var. *punctatum* Cogn. in J. Orch. 4: 76 (1893); L. Linden in Lindenia 9: 13-14, t.389 (1893). Type: cult. *Linden* (holotype not located, but see the above reference for an illustration of the type).

Cyperorchis grandiflora (Griffith)Schltr. in Fedde, Repert. 20: 107 (1924).

C. hookerianum Reichb.f. var. *hookerianum*. Y.S. Wu & S.C. Chen in Acta Phytotax. Sin. 18: 302 (1980), **syn. nov.**

A perennial, epiphytic or lithophytic *herb*. *Pseudobulbs* 3–6 cm long, 1.5–3.5 cm in diameter, elongate-ovoid, bilaterally flattened, 4–8 leaved. *Leaves* up to 80 × 1.4–2.1 cm, linear-elliptic, acute, articulated 4–10 cm from the pseudobulb to a broadly sheathing, strongly yellow- and green- striated base. *Scape* up to 70 cm long; peduncle suberect, loosely covered in scarious sheaths up to 12 cm long; rhachis slender, arching to pendant, slightly flattened, bearing 6-15 flowers; bracts triangular, up to 4 mm long. *Flowers* up to 14 cm in diameter, with a strong fresh scent; rhachis, pedicel and ovary green; sepals and petals clear apple-green with some deep red spots towards the base, occasionally lightly shaded with red-brown; lip cream-coloured, becoming greenish at the margin, flushing strong purplish-pink after pollination, side-lobes spotted deep maroon, mid-lobe with a submarginal ring of red-brown blotches and spots, and a broken central line of confluent red blotches, base of lip bright yellow with maroon spots; callus ridges cream with reddish spots; column cream, spotted maroon below, cream becoming green at the apex above, with a fine maroon line at the apex; anther-cap cream. *Pedicel and ovary* 3.5–6.0 cm long. *Dorsal sepal* 5.6–6.0 × 1.7–1.9 cm, narrowly obovate, acute, porrect; lateral sepals similar, spreading. *Petals* 5.3–5.6 × 1.2–1.4 cm, ligulate or very narrowly obovate, slightly curved, spreading. *Lip* 3-lobed, fused to the column base for 4.5–5 mm; side-lobes 1.1–1.6 cm broad, triangular, acute and porrect, margins ciliate (cilia over 1 mm long), papillose or shortly hairy especially towards the apex; mid-lobe 1.7–2.0 × 2.7–2.9 cm, broadly ovate-cordate, mucronate, recurved, papillose, sometimes with scattered hairs, never with the central lines of hairs present in *C. tracyanum* or *C. iridioides*, margin erose and strongly undulate; callus ridges 2, tapering to the cavity at the base of the lip, strongly hairy along the top. *Column* 3.3–4.0 cm long, winged, with a few short hairs or large papillae ventrally towards the base; pollinia 3.2–4.0 mm long, triangular. *Capsule* up to 13 × 4 cm, fusiform-ellipsoidal, stalked with a 2.5 cm apical beak.

ILLUSTRATIONS. Plate 15; photograph 85; figure 23.4.

DISTRIBUTION. E Nepal, Sikkim, Bhutan, Assam, S W China (map 7E); 1500–2600 m (4920–8530 ft).

HABITAT. On trees or steep banks in dense, damp forest; flowering January-March.

Griffith described this species in 1851 as *C. grandiflorum*, based on a specimen he had collected in Bhutan. This name had, unfortunately, previously been used by Swartz (1799) for a distinct species now placed in *Pogonia*. The next available and legitimate name for the species is *C. hookerianum*, based on material collected by Lobb in the early 1850s. It flowered soon after its introduction, in the nursery of Messrs Veitch, but not again until 1866, when Reichenbach described it. A flower from the same plant was figured in the *Botanical Magazine* (Bateman, 1866).

Var. *punctatum*, described by Cogniaux, based on a colour variant which has a few red-brown spots at the base of the sepals and petals, and more numerous spots on the mid-lobe of the lip, is otherwise identical to *C. hookerianum*. Two further varieties ascribed to *C. hookerianum*, var. *kalawense* T.A. Colyear and var. *lowianum* (Reichb.f)Y.S. Wu & S.C. Chen, are discussed under *C. lowianum*.

C. hookerianum is similar in flower size and lip shape to *C. tracyanum* (see discussion of *C. tracyanum* for distinguishing characters), and is closely allied to *C. lowianum*, which also has clear green sepals and petals. The spots on the side-lobes of the lip are absent in *C. lowianum*, and the mid-lobe spots are replaced by a large, red, V-shaped patch. *C. hookerianum* can also be distinguished by its acute, porrect side-lobe apices, its larger, more strongly recurved, ovate mid-lobe with a strongly undulating margin and more evenly distributed indumentum, and its callus ridges which taper towards the base. *C. hookerianum* is also allied to *C. wilsonii* (see *C. wilsonii* for a discussion of their affinity and distinguishing characters).

C. hookerianum is found in the Himalaya of northern India from eastern Nepal, through to China where it occurs in N Yunnan, S W Sichuan and S E

Xizang, growing epiphytically in damp, shady forest or on steep banks or rocks, often where thick moss cover occurs. It is found at higher altitudes than *C. iridioides* in regions where the distributions of these two species overlap. The tendency in cultivated specimens to drop their flower buds, or even to fail to form flower spikes, if the growing conditions are too warm, reflects the cooler climatic conditions which this species prefers.

21. C. wilsonii *(Rolfe ex Cook)Rolfe* in Orchid Rev. 12: 97 (1904); Anon. in Gard. Chron., ser. 3, 35:157, f.66 (1904); Schltr. in Fedde, Repert., Beih. 4: 272 (1919); Wu & Chen in Acta Phytotax. Sin. 18: 303 (1980); Du Puy, Ford-Lloyd & Cribb in Kew Bull. 40: 421-34 (1984). Type: China, Yunnan, cult. Veitch, *Wilson* Cym. Sp. 2 (holotype K!).

C. giganteum Wallich ex Lindley var. *wilsonii* Rolfe ex Cook in The Garden 65: 158, 189 + fig. (1904).
Cyperorchis wilsonii (Rolfe) Schltr. in Fedde, Repert. 20: 108 (1924).
C. giganteum Wallich ex Lindley cv. Wilsonii (Rolfe)P. Taylor & P. Woods in Curtis's Bot. Mag. 181: n.s. t.704 (1976).

A perennial, epiphytic *herb. Pseudobulbs* up to 6 cm long, 3 cm in diameter, elongate-ovate, bilaterally flattened, about 7-leaved. *Leaves* up to 90 × 2.5 cm, linear-elliptic, acute, articulated 6–11 cm above the pseudobulb, to a broadly sheathing base. *Scape* 25–70 cm long, suberect to horizontal, arching; peduncle loosely covered by sheaths up to 11 cm long; rhachis slender, pendulous at the tip, bearing 5–15 flowers; bracts triangular. *Flowers* 9–10 cm across; scented; rhachis, pedicel and ovary green; sepals and petals green with some pale brown shading over the veins, and distinct red-brown speckles on the veins in the basal half; lip cream, flushing purplish pink on pollination, side-lobes veined dark red-brown, becoming broken and spotted towards the margins and the tips, mid-lobe with a submarginal ring of confluent red-brown blotches, and a reddish median line to the front of the callus ridges; callus cream, spotted red-brown along the crests; column pale yellow-green with a conspicuous dark maroon apex above, cream with spots and dashes of red-brown below; anther-cap pale yellow. *Pedicel and ovary* 2.2–4.2 cm long. *Dorsal sepal* 4.4–5.7 × 1.2–1.9 cm, narrowly obovate, acute, porrect; lateral sepals similar, spreading. *Petals* 4.0–5.3 × 0.7–1.3 cm, ligulate or very narrowly obovate, curved, acute, spreading. *Lip* 3-lobed, fused to the column base for 3.5–5 mm; side-lobes 1.2–1.4 cm broad, triangular, acute and porrect, shortly hairy, margin fringed with short hairs; mid-lobe 1.5–1.8 × 1.5–2.1 cm, ovate, tapering to a fine point, recurved, papillose with scattered short hairs, margin erose, strongly undulating; callus ridges 2, tapering to the base and dilated at the apex, densely hairy. *Column* 2.7–3.2 cm long, broadly winged at the tip, narrowly at the base, with sparse short hairs and inflated papillae present ventrally towards the base; pollinia 2, 2.4–2.65 mm long, triangular.

ILLUSTRATIONS. Photographs 90, 91; figure 23.5.

DISTRIBUTION. China (S Yunnan) (map 7D); about 2400 m (7875 ft).

HABITAT. On tall trees in deeply shaded forest; flowering February to April.

This species was discovered by E.H. Wilson, in 1901, near Mengzi in south-eastern Yunnan. He sent plants to the nurseries of Messrs Veitch, who flowered it in February 1904, and exhibited it at the Royal Horticultural Society as *C. giganteum* (= *C. iridioides*) var. *wilsonii* on the advice of Rolfe. It was subsequently described under that name by Cook (1904) in the February edition of *The Garden.* The Orchid Committee would not accept this name, but awarded it under the name *C. wilsonii.* Rolfe subsequently validated this name in the *Orchid Review,* March 1904.

Since then, there has been continuing debate about the status of *C. wilsonii,* and Taylor & Woods (1976) recently treated it as a cultivar of *C. iridioides.* There have been very few collections of *C. wilsonii.* A single living specimen, probably from Wilson's original collection, has survived at the Botanic Garden Edinburgh, and some pieces from this plant, and selfed seedlings, are being grown at Kew. An herbarium specimen collected from Hpimaw Fort on the border of western Yunnan and Burma, which closely resembles the type and is attributable to *C. wilsonii*, is preserved at Kew (R.A. 840). Its rarity, and its superficial similarity to *C. iridioides* especially in lip shape and colour, lends strength to the argument that *C. wilsonii* is a variant of *C. iridioides*. However, numerical taxonomic analyses (Du Puy *et al.*, 1984) have indicated a more complex relationship. Analyses of the species in section *Iridorchis* have shown *C. wilsonii* to be very closely linked to *C. hookerianum*, rather than to *C. iridioides* as might have been expected. Secondly, scatter diagrams of these three species show *C. wilsonii* to be intermediate between *C. hookerianum* and *C. iridioides*. It therefore appears probable that hybridisation has been involved in the origin of *C. wilsonii*. The characters in table 10.5 illustrate this intermediacy.

C. wilsonii occurs within the range of *C. iridioides*, but at slightly higher altitudes. *C. hookerianum* occurs further to the north, but at similar altitudes to *C. wilsonii*. *C. hookerianum* and *C. wilsonii* flower in spring, *C. iridioides* in autumn. *C. hookerianum* and *C. iridioides* are therefore normally prevented from hybridising by both spatial and temporal isolation. Records indicate that the spatial isolation may be less distinct towards the southern extreme of the range of *C. hookerianum*, and occasional out-of-season flowering can occur in either species. A combination of these factors might allow occasional hybridisation between the two species.

C. wilsonii was originally collected in south-eastern Yunnan, distant from the known range of *C. hookerianum*. The original hybrid must have been able to disperse and establish itself as a reproductively

TABLE 10.5: A comparison of *C. hookerianum*, *C. iridioides* and *C. wilsonii*, showing the intermediacy of *C. wilsonii*

Character	*C. hookerianum*	*C. wilsonii*	*C. iridioides*
Leaf size	up to 80 × 2.1 cm	up to 70 × 2.5 cm	up to 90 × 4.0 cm
Rhachis	slender	slender	robust
Number of flowers	6–15	5–15	7–20
Flower size	up to 12 cm	9–10 cm	up to 10 cm
Dorsal sepal size	5.6–6.0 × 1.7–1.9 cm	4.4–5.7 × 1.2–1.9 cm	4.5–4.7 × 1.2–1.8 cm
Petal and sepal colour	green, with some red-brown spots near the base	green, spotted red-brown over the veins to the middle	green to yellow, heavily veined and blotched red-brown
Side-lobe shape	apex acute	apex slightly rounded	apex slightly rounded
Side-lobe colour	spotted red-brown along the veins	veined red-brown, spotted towards the margins	veined red-brown
Side-lobe margin	ciliate (hairs 1 mm + long)	fringed with short hairs	fringed with short hairs
Mid-lobe shape	broadly ovate-cordate (B = L × 1.5)	ovate (B = L)	ovate (B = L)
Mid-lobe indumentum	papillose with scattered hairs	papillose with scattered hairs, glabrous at the base and centre	hairy, with lines of long hairs extending from the callus ridges to the centre
Callus length	to the base of the lip	to the base of the lip	not to the base of the lip
Callus indumentum	densely short-hairy	densely short-hairy	densely long-hairy
Pollinium length	3.2–4.0 mm	2.4–2.7 mm	2.1–2.5 mm
Flowering period	January-March	February-April	August-December
Altitude	1500–2600 m	2300 m	1200–2000 m

Note: B = breadth; L = length.

viable population, occupying the high altitude niche of *C. hookerianum*, but further south than that species. It is therefore suggested that *C. wilsonii* is a distinct species, originating as a hybrid between *C. hookerianum* and *C. iridioides.*

C. wilsonii can be distinguished from *C. iridioides* by its green sepals and petals, its lack of long cilia in the centre and base of the mid-lobe, its narrower leaves and spring flowering season. *C. hookerianum* differs in its spotted side-lobes, more acute side-lobes, ciliate side-lobe margins, broader and narrower leaves, narrower, more strongly falcate sepals and petals, and larger flowers.

22. C. lowianum *(Reichb.f.)Reichb.f.* in Gard. Chron. n.s. 11: 332, 404, t.56 (1879); Easton in Orchid Rev. 91: 66-9 (1983); Seidenfaden in Opera Bot. 72: 86, t.47 (1983); Du Puy, Ford-Lloyd & Cribb in Kew Bull. 40: 421-34 (1985). Type: Burma, *Boxall* cult. *Low* (holotype W).

C. giganteum Wallich ex Lindley var. *lowianum* Reichb.f. in Gard. Chron. n.s., 7: 685 (1877); Hooker f., Fl. Brit. India 6: 13 (1891); Grant, Orchids of Burma: 228 (1895).

C. lowianum (Reichb.f.)Reichb.f. var. *concolor* Rolfe in Gard. Chron. ser. 3, 10: 187 (1891). Type: cult. *Eastwood* (holotype K!). **syn. nov.**

C. lowianum (Reichb.f.) Reichb.f. var. *superbissimum* L.Linden, Lindenia 9: 19, 20, t.392 (1893). Type: Burma, cult. *Linden* (holotype not located, but see the above reference for an illustration of the type), **syn. nov.**

C. lowianum (Reichb.f.)Reichb.f. var. *flaveolum* L. Linden, Lindenia 12: 91-2, t.572 (1896). Type: cult. *Linden* (holotype not located, but see the above reference for an illustration of the type), **syn. nov.**

C. lowianum (Reichb.f.)Reichb.f. var. *viride* Warner & Williams, Orchid Album 11: t.527 (1897). Type: cult. *Smee* (holotype not located but see the above reference for an illustration of the type).

Cyperorchis lowiana (Reichb.f.)Schltr. in Fedde, Rep. 20: 108 (1924).

C. hookerianum Reichb.f. var. *lowianum* (Reichb.f.)Y.S. Wu & S.C. Chen in Acta Phytotax. Sin. 18: 303 (1980).

A perennial, epiphytic or lithophytic *herb. Pseudobulbs* up to 13 cm long, 5 cm in diameter, bilaterally flattened, about 9-leaved. *Leaves* up to 90 × 3.5 cm, linear-elliptic, acute, articulated 6–10 cm from the pseudobulb to a yellow-green, broadly sheathing base. *Scape* up to 100 cm or more long, suberect to horizontal arching; peduncle covered by sheaths up to 10 cm long; rhachis slender at the tip, bearing 12-30(40) flowers; bracts triangular, 3 mm long. *Flowers* up to 10 cm in diameter; not scented; rhachis, pedicel and ovary green; sepals and petals bright apple-green to yellowish, sometimes lightly shaded with red-brown; lip yellowish to white, side-lobes not marked, the mid-lobe with a deep red, broad, V-shaped, submarginal patch, extending from near the tip to the angles between the mid-lobe and side-lobes, and a central red line to the callus tips; the base of the lip is bright yellow or orange, spotted with red-brown (*C. lowianum* var. *iansonii* has light to heavy red-brown shading on the sepals and petals, and a light brown V-shaped mark on the mid-lobe); callus cream or white, occasionally with a few red spots; column green or yellowish above, tipped with red-brown, spotted red-brown below towards the base; anther-cap bream or white. *Pedicel and ovary* 3–5 cm long. *Dorsal sepal* 4.8–5.7 × 1.6–1.8 cm, narrowly obovate, acute, concave, porrect; lateral sepals similar, spreading. *Petals* 4.8–5.3 × 1.1–1.3 cm, ligulate to narrowly obovate, acute, slightly curved, spreading. *Lip* 3-lobed, fused to the column base for 3–4.5 mm; side-lobes 1.4–1.6 cm broad, triangular, apex right-angled, slightly rounded, margin not distinctly fringed, indumentum of short velvety hairs especially at the front; mid-lobe 1.4–1.8 × 1.5–2.1 cm, cordate, mucronate, porrect to vertical (angled down at the base), with the indumentum in two zones comprising short, silky hairs in the centre and base, and very dense velvety hairs on the V-shaped, coloured region (var. *iansonii* has longer hairs on the lip and callus than var. *lowianum*), margin of the mid-lobe erose, minutely undulating; callus ridges 2, not reaching the base of the lip, spreading behind, dilated at the apex, finely hairy. *Column* 3.0–3.5 cm long, winged, with large papillae or short hairs present ventrally towards the base; pollinia 3–3.5 mm long, triangular. *Capsule* about 6–8 × 3–4 cm, fusiform-ellipsoidal, stalked, with a 1.5–2 cm apical beak.

1c
2c
2d
1d.
2b
2a
1a
3d
1b
3b
3a
3c
4b
4d
3e
3f
5b
5c
4a
5a
4c
5d

Figure 24

1. ***C. sanderae*** (Kew spirit no. 46238)
   - a Perianth, × 0.7
   - b Lip and column, × 0.7
   - c Pollinarium, × 3
   - d Pollinium (reverse), × 3
2. ***C. insigne*** (Kew spirit no. 48010/22593)
   - a Perianth, × 0.7
   - b Lip and column, × 0.7
   - c Pollinarium, × 3
   - d Pollinium (reverse), × 3
3. ***C. schroederi*** (Kew spirit no. 14989/14553)
   - a Perianth, × 0.7
   - b Lip and column, × 0.7
   - c Pollinarium, × 3
   - d Pollinium (reverse), × 3
   - e Flower, × 0.7
   - f Flowering plant, × 0.1
4. ***C. lowianum*** var. **lowianum** (Kew spirit no. 45530/14543)
   - a Perianth, × 0.7
   - b Lip and column, × 0.7
   - c Pollinarium, × 3
   - d Pollinium (reverse), × 3
5. ***C. lowianum*** var. **iansonii** (Kew spirit no. 37656)
   - a Perianth, × 0.7
   - b Lip and column, × 0.7
   - c Pollinarium, × 3
   - d Pollinium (reverse), × 3

ILLUSTRATION. Plate 16; photographs 92-95, figures 24.4, 24.5.

DISTRIBUTION. N and E Burma, China (southern Yunnan), N Thailand (map 7F); 1200–2400 m (3940–7875 ft).

HABITAT. On trees in damp, shaded evergreen or mixed forest; flowering February-June.

C. lowianum was first collected in 1877, in Burma, by Boxall, who sent plants to Messrs Low. Reichenbach (1877) described it as a variety of *C. giganteum* (= *C. iridioides*) from a dried specimen, but noted at the time that it was possibly a distinct species, and when the imported plants flowered, in 1879, he raised it to species level.

Wu & Chen (1980) consider *C. lowianum* to be a variety of *C. hookerianum*. Recent numerical taxonomic analyses (Du Puy *et al.*, 1984) have demonstrated conclusively that these two taxa are distinct at specific rank. (See *C. hookerianum* for a discussion of their distinguishing characters.)

There are several variants of this species which have been recognised as distinct by some authors. Var. *superbissimum* L. Linden is a fine colour form with clear green sepals and petals, and a strong red colour on the mid-lobe of the lip. Var. *concolor* Rolfe (photograph 95) lacks red pigment in the flowers, giving clear green sepals and petals, and the red mark on the mid-lobe is replaced by yellow. Var. *viride* R. Warner & H. Williams and var. *flaveolum* L. Linden are the same. Var. *concolor* does not differ otherwise from the red-pigmented variants and does not form distinct populations. The most appropriate taxonomic treatment for this variant would appear to be as a cultivar (photograph 95).

C. lowianum occurs in N Thailand, N E and E Burma and S Yunnan. Seidenfaden (1983) includes the N W Himalaya in this range. Collections of this species occasionally include specimens of *C. iridioides* which has a more northerly distribution. The range of *C. lowianum* largely coincides with that of *C. tracyanum*, but it usually grows at higher altitudes. All of the species in section *Iridorchis* from northern Thailand, *C. insigne*, *C. lowianum* and *C. tracyanum*, have been under great collection pressure and are now rare and confined to a few of the more inaccessible mountains.

C. lowianum is variable in its flower size and shape, in its flower colour and in the indumentum on the lip. A distinctive variant is recognised at varietal rank, as indicated in the following key.

Key to the varieties of *C. lowianum*

Scape arching, with a slender, pendulous rhachis; flowers well spaced; sepals and petals light green, sometimes slightly veined and shaded red-brown; lip with a dark red (or rarely yellow) V-shaped mark on the mid-lobe, and shortly pubescent on the inner surface .... *var.* **lowianum**

Scape suberect, with a stiff, suberect rhachis; flowers closely spaced; sepals and petals lightly to strongly shaded and veined red-brown; lip with a light chestnut-brown V-shaped mark on the mid-lobe, and with a longer indumentum on the inner surface and especially on the callus .... *var.* **iansonii**

var. **lowianum**

The typical variety is highly distinctive and easily recognised by the short indumentum on its lip, as well as by its lip shape and markings. It is one of the most beautiful of the *Cymbidium* species, with its striking, large, apple-green flowers carried gracefully on a slender, arching scape. The mid-lobe is marked with a single, bold, V-shaped patch of maroon (occasionally yellow) which covers the entire apical and submarginal region. Only *C. schroederi* has a similar mid-lobe marking (see *C. schroederi* for a discussion of the relationship between these species and their distinguishing characters). It has frequently been used in hybridisation, the strong lip marking being apparent in many modern hybrids.

ILLUSTRATIONS. Plate 16; photographs 92, 93; figure 24.4.

DISTRIBUTION. Burma, China, N Thailand.

var. **iansonii** *(Rolfe) Cribb & Du Puy.* in Kew Bull. 40: 432 (1984). Type: Burma, cult. *Low* (holotype K!).

C. mandaianum Gower, cited in Lindenia 12:91 (1896); Anon, in Orchid Rev. 20: 167 (1912). Type: cult. *Manda* (holotype not located).

C. × *iansonii* Rolfe in Orch. Rev. 8: 191, 209, f.34 (1900); Crosby in Amer. Orchid Soc. Bull. 21: 257-9 (1952).

C. grandiflorum Griffith var. *kalawensis* T.A. Colyear in Orch. Rev. 42: 248 (1934) & 43: 165, 167 (1935). Type: Burma, *Colyear* (holotype CAL).

The flowers are slightly larger than in var. *lowianum*, and the lip and callus have longer hairs. The petals and sepals are weakly to strongly marked with irregular red-brown lips as in *C. tracyanum*, and the markings on the mid-lobe of the lip is light brown.

ILLUSTRATIONS. Photograph 94; figure 24.5.

DISTRIBUTION. Burma.

Var. *iansonii* first flowered in the collection of Messrs Low in 1900, having originally been collected in northern Burma near Bhamo. Rolfe described it in 1900, considering it to be a hybrid between *C. lowianum* and *C. tracyanum*.

Map 8: The distribution of the species in section *Iridorchis* (cont):

	A	*C. insigne*
	B	*C. sanderae*
	C	*C. schroederi*
and in section *Annamaea*:	**D**	*C. erythrostylum*
and in section *Eburnea*:	**E**	*C. eburneum*
	F	*C. parishii*

In 1912, a second plant was exhibited, under the name *C. mandaianum*, which Rolfe could not distinguish from the original specimen (Anon., 1912). In 1934, a specimen collected in Burma, near Kalaw, was described by Colyear as *C. grandiflorum* var. *kalawensis.* In 1935 it was figured in the Orchid Review. A plant of similar appearance is now in cultivation at Kew. These plants are intermediate between var. *lowianum* and the original plant of var. *iansonii.* The flowers are slightly larger than other specimens of *C. lowianum*, and the lip has a longer pubescence. The mid-lobe has a light brown, V-shaped mark, but the sepals and petals are much less strongly marked with brown than in the type collection. This colour variation shows that a range of intermediates exists between the two varieties.

In a numerical taxonomic analysis of sect. *Iridorchis* (Du Puy *et al.*, 1984), the specimens of *C. iansonii* linked very closely with those of *C. lowianum*, although some distinction was apparent. No evidence could be found to suggest that hybridisation with any other species in the section was involved. *C. iansonii* has therefore been formally reduced to varietal rank within *C. lowianum.* Its distribution appears to be in northern and eastern Burma.

23. C. schroederi *Rolfe* in Gard. Chron. ser. 3, 37: 243 (1905), in Orchid Rev. 14: 39 (1906) & 25: 101 (1917); Summerh. in Bot. Mag. 163: t.9637 (1942); Crosby in Amer. Orchid Soc. Bull. 28: 584-6 (1952); Du Puy, Ford-Lloyd & Cribb in Kew Bull. 40: 421-34 (1984). Type: Vietnam, cult. *Schroeder* (holotype not located).

Cyperorchis schroederi (Rolfe)Schltr. in Fedde, Repert. 20:108 (1924).

A perennial, epiphytic *herb. Pseudobulbs* up to 15 cm long, 4 cm in diameter, bilaterally flattened, about 6-leaved. *Leaves* up to 60 × 2.5 cm, linear-elliptic, acute, articulated to a broadly sheathing base. *Scape*

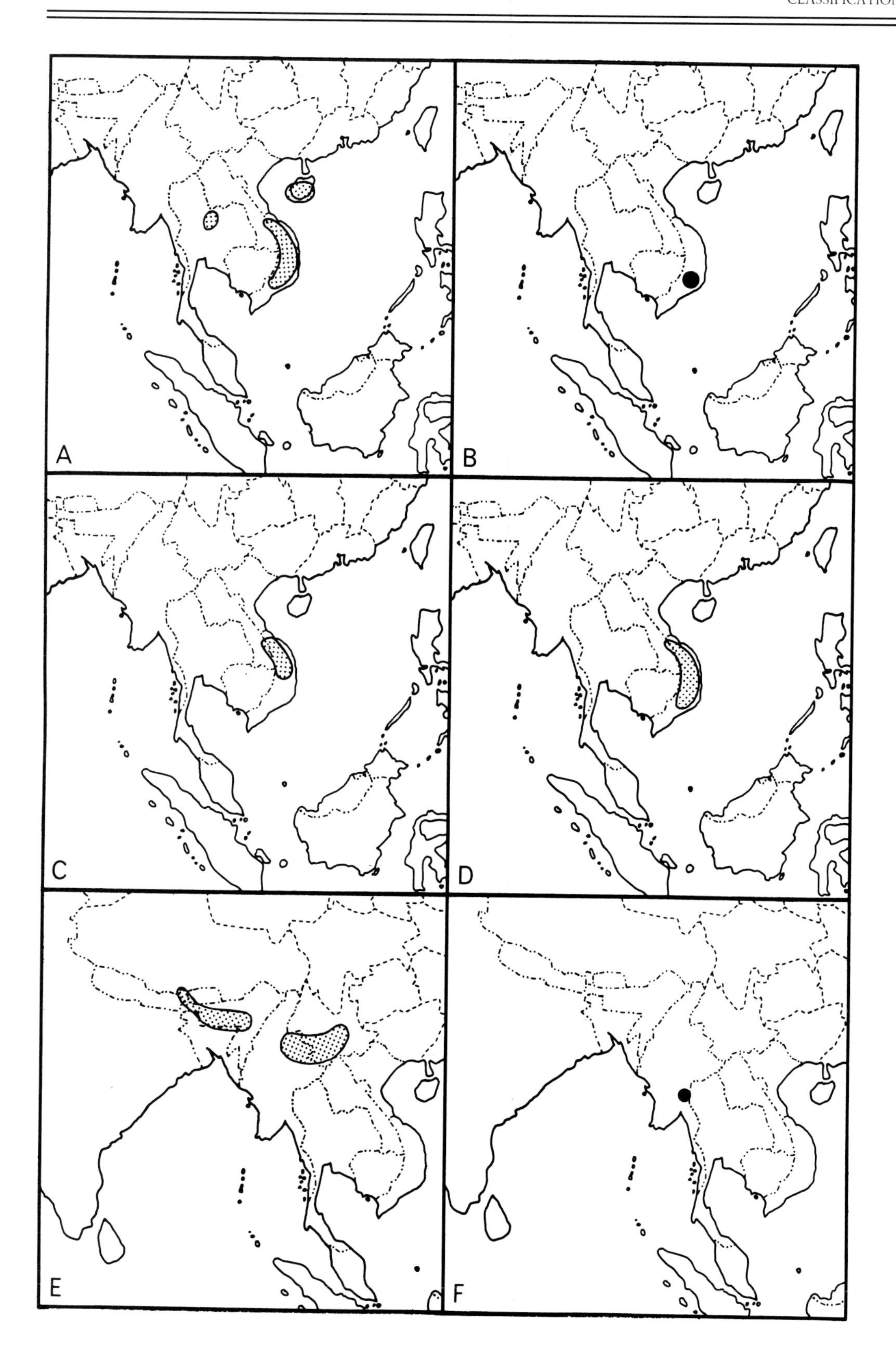
A
B
C
D
E
F

44–65 cm long, suberect, arching; peduncle covered in sheaths up to 14 cm long; rhachis robust at the base, becoming more slender towards the apex, bearing 14–23 flowers; bracts triangular, 3 mm long. *Flowers* 8–9 cm across; not scented; rhachis, pedicel and ovary bright green; sepals and petals green to yellowish-green, with dull, irregular, light brown veins and spots; lip pale yellow, side-lobes veined red-brown, mid-lobe marked with a large, red-brown, submarginal, V-shaped patch and median line to the front of the callus; callus whitish, sparsely spotted red-brown; column greenish-yellow shading to dark purple around the anther above, cream with red dashes below; anther-cap cream. *Pedicel and ovary* 2.0–4.5 cm long. *Dorsal sepal* 4.5–4.9 × 1.4–1.6 cm, narrowly obovate, acute, porrect; lateral sepals similar, spreading. *Petals* 4.2–4.5 × 0.8–1.2 cm, narrowly obovate, acute, slightly curved, spreading. *Lip* 3-lobed, fused to the column base for 3 mm; side-lobes 1.1–1.2 cm broad, triangular, densely hairy with long hairs towards the apex, apex right-angled to slightly rounded, margin fringed with short hairs; mid-lobe 1.4–1.6 × 1.5 cm, ovate-cordate, mucronate, slightly recurved, with silky hairs in the centre and towards the base, and a densely short-haired region corresponding to the V-shaped colour patch; callus ridges 2, tapering to the base of the lip, densely hairy, with a small glabrous patch in front of the callus ridges over the mid-vein on the mid-lobe. *Column* 2.8 cm long, winged, narrowing slightly towards the base, with inflated papillae ventrally towards the base.

ILLUSTRATION. Figure 24.3.

DISTRIBUTION. Central Vietnam (map 8C); altitude not known.

HABITAT. Not known: flowering March-June.

C. schroederi is similar to *C. lowianum* in habit, but its flowers are slightly smaller. Their colour combination of heavily red-brown marked sepals and petals, red-brown veined side-lobes, and brownish, submarginal, V-shaped patch on the mid-lobe, distinguish *C. schroederi* from the other species in this section. It can also be recognised by its hairy lip, its abruptly terminating side-lobe tips, and its callus ridges which are long, tapering towards the base.

This species was first collected at Trung Phan in central Vietnam (formerly Annam), and was imported by Messrs Sander. It flowered in 1905 in the collection of Schroeder, and Rolfe (1905) based his description of *C. schroederi* on this material. Unfortunately, the type specimen has not been located. It may have been placed in Rolfe's collection of spirit-preserved material, which has not survived. Rolfe compared the plant with *C. iridioides* and *C. lowianum*, concluding that it was most closely related to the former, while Summerhayes (1942) came to the opposite conclusion. Table 10.6 illustrates the intermediacy of this species.

Numerical taxonomic analyses (Du Puy *et al.*, 1984) have produced scatter diagrams which further confirm the intermediacy of *C. schroederi* between *C. iridioides* and *C. lowianum*. Dendrograms based on morphological similarity have shown *C. schroederi* to be closer to *C. lowianum*. The geographical distribution of *C. schroederi* is well to the south of either of the other species. However, *C. lowianum* and *C. iridioides* have been collected from the same locality in Yunnan. This suggests that *C. schroederi* could have originated from the natural hybridisation of these two species, and probably following some introgression with *C. lowianum* has stabilised to form a distinct species occupying a distribution to the south of either of the proposed progenitors.

Seidenfaden (1983), in his account of the Cymbidieae in Thailand, tentatively refers one specimen to *C. schroederi*. This specimen is, however, a small-flowered *C. tracyanum*, having narrow, curved petals, a very hairy lip with the indumentum on the side-lobes confined to the veins, lines of cilia extending from the callus well into the mid-lobe and acute, porrect side-lobe tips.

A natural hybrid with the sympatric *C. insigne* has been recorded and named as *C.* × *cooperi* Rolfe (1914). This cross has also been made artificially and named *C.* J. Davis.

24. C. insigne *Rolfe* in Gard. Chron. ser. 3, 35: 387 (1904) & in Bot. Mag. 136: t.8312 (1910); Wu & Chen in Acta Phytotax. Sin. 18: 303 (1980); Seidenfaden in Opera Bot. 72: 90, f.50 (1983); Easton in Orchid Rev. 91: 28-32 (1983); Du Puy in Orchid Rev. 91: 366–71, t.374 (1983) & in Kew Mag. 1: 75-84 (1984); Du Puy, Ford-Lloyd & Cribb in Kew Bull. 40: 421-34 (1984); Kjellsson, Rasmussen & Du Puy in J. Tropical Ecology 1: 289-302 (1985). Type: Vietnam, *Bronckart* 43 (holotype K!).

C. sanderi O'Brien in Gard. Chron., ser. 3, 37: 115, t.49 (1905). Type: Vietnam, cult. Sander, *Micholitz* no. 1 (holotype K!).

C. insigne Rolfe var. *sanderi* (O'Brien) Hort. in J. Hort., ser. 3, 58: 415, + fig. (1909), **syn. nov.**

Cyperorchis insignis (Rolfe)Schltr. in Fedde, Repert. 20: 108 (1924).

C. insigne Rolfe var. *album* Hort. ex Gard. Chron., ser. 3, 61: 101, f.35 (1917). Type: cult. *Armstrong & Brown* (holotype not located, but see the above reference for an illustration of the type), **syn. nov.**

TABLE 10.6: A comparison of *C. lowianum*, *C. iridioides* and *C. schroederi*, showing the intermediacy of *C. schroederi*

Character	***C. lowianum***	***C. schroederi***	***C. iridioides***
Spike length	up to 100 cm	44–65 cm	45–85 cm
Spike strength	slender	slender	more robust
Number of flowers	12–30(40)	14–23	7–20
Flower shape	petals spreading, dorsal sepal porrect. Flower open	petals spreading, dorsal sepal porrect. Flower open	petals not fully spreading, dorsal sepal porrect. Flower not opening fully
Petal and sepal colour	clear green	greenish, stained with red-brown	greenish, heavily stained with red-brown
Length of fusion of lip and column	3–4.5 mm	3 mm	4–5 mm
Lip indumentum	short	long	long
Lip margin	fringe absent	fringed with short hairs	fringed with short hairs
Side-lobe tip shape	right angled, not porrect	right angled, not porrect	acute, porrect
Side-lobe markings	no markings	veined red-brown	veined red-brown
Mid-lobe shape	cordate	ovate-cordate	ovate
Mid-lobe recurvature	not recurved	slightly recurved	strongly recurved
Mid-lobe margin	minutely undulating	weakly undulating	strongly undulating
Mid-lobe indumentum	in two distinct regions, corresponding with V-shaped markings	in two distinct regions corresponding with V-shaped markings	shortly hairy, not in two regions. Lines of long hairs present from the base to the centre
Mid-lobe markings	submarginal V-shaped patch of deep maroon	submarginal V-shaped patch of red-brown	yellowish in the centre, with a submarginal ring of confluent red-brown spots and blotches
Callus shape	spreading towards the base	tapering towards the base	tapering towards the base
Callus indumentum	short hairs	long hairs	long hairs
Column length	3.0–3.5 cm	2.8 cm	2.5–2.9 cm
Column colour	spotted red-brown below towards the base	streaked red-brown below	streaked red-brown below
Scent	none noticeable	none noticeable	sweet-scented
Flowering	February-June	March-June	August-December

A perennial, terrestrial *herb*. *Pseudobulbs* up to 8 cm long, 5 cm in diameter, ovoid, lightly bilaterally flattened, with 6–10 leaves. *Leaves* up to 100 × 0.7–1.8 cm, narrowly linear-elliptic tapering to an acute apex, articulated to a broadly sheathing base, 5–15 cm from the pseudobulb. *Scape* 100–150 cm long; peduncle erect, very long (up to 120 cm), robust, covered in about 13 scarious sheaths up to 23(33) cm long; rhachis 17–40 cm long, arching, becoming slender towards the tip, with up to 27 closely spaced flowers; bracts triangular, 3–15 mm long. *Flowers* 7–9 cm in diameter; not scented; rhachis, pedicel and ovary stained purple; sepals and petals white or pale pink, sometimes with some red spots at the base and over the mid-vein; lip white, becoming deep pink on pollination, side-lobes usually strongly veined and spotted maroon-red, mid-lobe yellow at the base and in the centre, usually veined and spotted maroon-red, callus bright yellow at the tips, paler behind; column pale to deep purple-pink above, white, usually streaked with red below; anther-cap cream. *Pedicel and ovary* (1.5)2.3–4.5(6.5) cm long. *Dorsal sepal* 4.2–5.6 × 1.5–2.0 cm, obovate, acute, concave, erect or porrect; lateral sepals similar, spreading, slightly drooping. *Petals* 4.0–5.2 × 1.4–1.8 cm, narrowly obovate, acute, spreading. *Lip* 3-lobed, fused to the base of the column for 2.0–4.0 mm; side-lobes 1.2–2.2 cm broad, lightly clasping the column, papillose or minutely hairy, broadly rounded, margin not fringed; mid-lobe 1.4–1.7 × 1.8–2.4 cm, triangular to subcircular, acute, weakly recurved, papillose, with a basal and central patch of short, dense hairs, margin entire and undulating; callus ridges 2, strongly inflated at the apex, tapering to the base, densely hairy. *Column* 3.0–3.2 cm long, winged to the base, with some minute basal hairs ventrally; pollinia triangular to sub-quandrangular, 2.0–2.5 mm across. *Capsule* 4–5 cm long, round to oblong-fusiform, with a short apical beak formed from the persistent column.

ILLUSTRATIONS. Plate 17, photographs 12, 13, 96–100; figure 24.2.

DISTRIBUTION. Vietnam, China (Hainan), N Thailand (map 8A); 1000–1700 m (3280–5580 ft).

HABITAT. In sandy soil in low, open woodland; flowering February-May(November).

C. insigne is easily distinguished from the other species in sect. *Iridorchis* by its white or pink flowers with broad petals, its broad side-lobes with rounded apices, its very tall, erect flower spike which has flowers only in the apical region, its longer bracts, its more rounded pseudobulbs (in transverse section) and its smaller, almost spherical capsules.

This species was first collected by Bronckart in 1901 in Annam, now Vietnam. He sent dried flowers and an inaccurate watercolour sketch to Kew (via Mr Schneider), from which Rolfe (1904) described the species as having rosy-lilac flowers. Plants exhibited by Bronckart in 1906 had paler flowers. Specimens collected by Micholitz in 1903, on the Lang Bian Plateau in South Vietnam, were exhibited by Sander and described by O'Brien (1905) as *C. sanderi*. Subsequent comparison with Bronckart's specimens has shown the two collections to be conspecific. *C. insigne* has been awarded about ten times by the Royal Horticultural Society, and many unpublished cultivar names are therefore in circulation.

This species varies especially in the size, colour and shape of the mid-lobe of the lip. Specimens from Vietnam have a large, rounded mid-lobe, spotted with maroon-red. Specimens from northern Thailand have smaller, triangular mid-lobes, more strongly veined and spotted maroon-red, or sometimes entirely lacking the red markings on the lip.

In early April 1983, a population recently discovered by Dr G. Seidenfaden, in northern Thailand, was studied by the author (Du Puy, 1983, 1984). Two colour variants were found there in almost equal proportions. One of these had strong red spots and veins on the lip, whilst in the other any trace of this pigmentation was lacking, the lip being pure white or pinkish, with a pale yellow patch on the centre of the mid-lobe and the apical region of the callus. The vegetation of the area was dominated by the white-flowered *Rhododendron lyi*, and another large-flowered white orchid, *Dendrobium infundibulum*, which was also very common. All three were found to be pollinated by the same species of bumble-bee, *Bombus eximius*, the pollinaria of the two orchid species being placed on different regions of the thorax of the bee. The apparent absence of a food reward or other attractant in the orchid flowers, and their similar colouring, suggested that they were both mimicking the *Rhododendron* flower (see Kjellson, Rasmussen & Du Puy, 1985 and Chapter 5).

C. insigne is the only truly terrestrial orchid in sect. *Iridorchis* (photograph 96). In Thailand it grows in shallow, very sandy soil. The region has many outcrops and boulders of sandstone which serve to break the vegetation cover, providing a diversity of habitats. *C. insigne* was found growing at about 1500 m (4920 ft) altitude, in the shade of low bushes, its long flowering spike pushing through the twigs so that the flowers were held clear of the foliage of the bush. This meant that the flowers were at a similar height, and close to the *Rhododendron* flowers they were mimicking (see also Du Puy 1983, 1984 for more details of the habitat). Bronckart also stated that *C. insigne* was a terrestrial growing in sandy soil along ravines. Micholitz (in *The Garden* 66: 141, 1904) found it growing on steep banks among thick grass, usually in a stiff clay type of soil.

There is a natural hybrid between *C. insigne* and *C. schroederi*, *C.* × *cooperi*.

25. C. sanderae *(Rolfe)Cribb & Du Puy*, **comb. et stat. nov.**. Type: Vietnam, cult. *Sander, Micholitz* (holotype K!).

C. parishii Reichb.f. var. *sanderae* Rolfe in Gard. Chron., ser. 3, 35: 338-9, f.146 (1904) & in Orchid Rev. 12: 163-4, 279 (1904); Micholitz in The Garden 66: 141 (1904); Menninger in Amer. Orchid Soc. Bull. 34: 892-7 (1965); D.E. & D.R. Wimber in Amer. Orchid Soc. Bull. 37: 572-6 (1968), **syn nov.**

A perennial, epiphytic *herb. Pseudobulbs* up to 6 cm long, 4 cm in diameter, ovoid, lightly bilaterally flattened, produced annually, with about 10 leaves. *Leaves* up to 50 × 2.5 cm, linear-elliptic, tapering to an entire, acute apex; articulated to a broadly sheathing base about 9 cm from the pseudobulb. *Scape* 30–50 cm long; peduncle erect to suberect, robust, covered in numerous cymbiform sheaths up to 19 cm long; rhachis about 10 cm long, suberect, becoming slender towards the tip, with 3-15 flowers; bracts 5–12 mm long, triangular. *Flowers* 8 cm across; scented; rhachis green, pedicel and ovary lightly stained with purple; petals and sepals white, usually flushed pink on the reverse with a few purple spots at the base of the petals; lip cream, usually heavily marked with maroon, side-lobes usually strongly blotched and stained with maroon except towards the margins, mid-lobe yellow in the centre and at the base, with a 2 mm wide cream margin and usually marked with a submarginal band of confluent strong maroon blotches and random spots, callus bright orange-yellow at the front; column cream above, usually streaked and stained maroon below and flushed yellow towards the base; anther-cap cream. *Pedicel and ovary* 2.5–4.3 cm long. *Dorsal sepal* 4.5–4.6 (5.7) × 1.4(2) cm, narrowly obovate to elliptic, acute, concave, erect or porrect; lateral sepals similar, spreading, slightly curved. *Petals* 4.4–4.5(5.4) × 1.1–1.2(1.6) cm, narrowly obovate to elliptic, acute, spreading. *Lip* 3-lobed, fused to the base of the column for 4–5 mm; side-lobes 1.2–1.4(1.6) cm broad, lightly clasping the column, minutely hairy, apex broadly rounded to obtuse, margin not fringed; mid-lobe 1.2–1.5(1.9) × 1.1–1.5(2.4) cm, ovate, rounded, obtuse or mucronate, weakly recurved, papillose with a dense basal and central patch of short hairs, margin entire and weakly undulating; callus ridges 2, strongly inflated at the apices, tapering to near the base of the lip, shortly hairy. *Column* 3.2–3.5 cm, narrowly winged, with some minute hairs ventrally towards the base; pollinia subquadrangular, about 2.5 mm long.

ILLUSTRATIONS. Plate 18; photograph 101; figure 24.1.

DISTRIBUTION. Vietnam (map 8B); 1400–1500 m (4595–4920 ft).

HABITAT. On trees, frequently in association with a *Polypodium* fern; flowering January-March(May).

This species was collected in Vietnam (Annam) on the Lang Bian Plateau, from the same locality as *C. insigne*, by Micholitz, in 1904. The specimens of *C. insigne* which he sent to Sander's nursery were given the name *C. sanderi*, while those of this species were named after Sander's wife as *C. sanderae*. This latter name was not validly published, and Rolfe subsequently published it at varietal rank within *C. parishii*.

That decision was based on the rather similar flower colour of the two species, but further study has shown that the two are quite distinct, and that *C. sanderae* is more closely related to *C. insigne* than to any of the species in section *Eburnea*. The International Rules of Botanical Nomenclature allow *C. sanderae* to be used, despite *C. sanderi* having been used for a distinct taxon, since they are not homonyms (R.K. Brummitt, pers. comm.).

C. sanderae has a very different growth habit from *C. parishii*. It has ovoid, well-developed pseudobulbs which are produced annually, rather than fusiform pseudobulbs which grow and flower for several seasons before producing new growths; its leaves are acute, rather than unequally bilobed or forked at the apex; the flower spike is produced from the base of the pseudobulb, not from the leaf axils towards the centre or apex of the pseudobulb; the spike is more robust, and produces more (up to 15) flowers. The flower is also distinct, opening more fully, and with elliptic rather than oblong sepals and petals. The lip is not elongated, and has a relatively large mid-lobe and rounded side-lobes which are not acute at the apex and has callus ridges which extend nearer to the base of the mid-lobe.

All of these characters are shared by *C. insigne*, indicating that *C. sanderae* should be included in section *Iridorchis*, with *C. insigne*. Vegetatively these two species are very similar. Their pseudobulbs are of a similar size, and are ovoid with little bilateral flattening. The leaves are similar in shape, including the acute apices, but the leaves of *C. sanderae* are broader. Both species produce new pseudobulbs annually, and the flower spike grows from the base of the pseudobulb. *C. sanderae* has a much shorter flower spike with fewer flowers, but is otherwise similar. The flowers of these two species are similar in shape, except that the mid-lobe of *C. sanderae* is rounded at the apex, not acute, and the markings on the lip are usually stronger. *C. insigne* is a terrestrial species, but Micholitz (1904) reported that *C. sanderae* usually grew epiphytically, in clumps of *Polypodium* fern, and that the two species could be separated in the field.

C. sanderae is very uncommon in cultivation, and was believed to have been lost until, in 1961, Mrs E.D. Menninger uncovered a single specimen in the nursery of Armacost and Royston, California. This plant was flowered in 1963 (Menninger, 1965). The named cultivar 'Emma Menninger' is a tetraploid plant converted from the diploid by Dr D. Wimber. The measurements in brackets in the description usually refer to this plant. The increase in the size of the flower, especially in the width of the sepals and petals, and the size of the lip can be directly attributed to the increased ploidy level (D.E. & D.R. Wimber, 1968).

The type specimen of *C. sanderae* is supplemented by two further specimens imported by Sander from Lang Bian. One is similar to the type, and a note is attached saying 'only 2 or 3 plants flowered with dark spotted lip. All the others turned out an inferior form which was named var. *Ballianum* — without the purple markings on the lip — and sepals and petals waxy white.' The second sheet has leaves and flower spikes of this white-flowered variant. This latter is very similar in appearance, but dissection of the flowers reveals a similarity to *C. mastersii* in the longer lip shape and the narrower callus ridges. The variation in the flower colour and lip shape suggest that the plants are part of a hybrid swarm, with *C. mastersii* as one parent, and either *C. insigne* or *C. sanderae* as the other. The possibility therefore exists that *C. sanderae* might be a natural hybrid. However it would be difficult to account for the very heavy marking on the lip and the leaves which are broader than either of the proposed parents. Furthermore, *C. mastersii* has not been collected in this region. Colour variation of this nature (presence or absence of red spots on the lip) is encountered in many related species including *C. insigne*, *C. roseum*, *C. mastersii* and *C. eburneum*, and might therefore also be expected in *C. sanderae*.

Sander's *Orchid Guide* (1927), mentions *C.* × *Ballianum* from two separate sources. The first importation was from Burma, and was a natural hybrid between *C. eburneum* and *C. mastersii*. The use of this name for the collections made in Annam is therefore inadmissible, as they are certainly not hybrids of that parentage. Sander notes that 'the typical (Burmese) form has flowers slightly larger and of a more pure white than the Annamese varieties'.

Section **Eburnea** *Seth & Cribb* in Arditti (ed.), Orchid Biol., Rev. Persp. 3: 304 (1984). Type: *C. eburneum* Lindley.

When Seth & Cribb (1984) established this section they distinguished it from section *Cyperorchis* on the basis of its more widely opening flowers and its quadrangular pollinia. The two species included were *C. eburneum* and *C. parishii* to which *C. mastersii* and *C. roseum* are added here.

Section *Eburnea* is characterised by its growth habit, where the pseudobulbs are somewhat slender and fusiform, and grow and flower indeterminately for two to many years. The leaf apex tends to be acutely bilobed with a small mucro in the sinus. The few-flowered scape is produced from the axils of the leaves, not from the base of the pseudobulb. In common with section *Cyperorchis* the flowers do not usually open fully, the hypochile is relatively long, and the rostellum is somewhat beaked. The quadrangular pollinia are placed on a rectangular viscidium with two long, hair-like processes from the lower corners (figure 14).

26. C. eburneum *Lindley* in Bot. Reg. 33: t.67 (1847); Paxton, Mag. Bot. 15: 145-6 + plate (1849); Hook. in Bot. Mag 85: t.5126 (1859); Mueller in Walp. Ann. 6: 625 (1864); O'Brien in Gard. Chron., n.s., 17: 496, f.78 (1882) & in *op. cit.* 22: 77, f.17 (1884); Hook.f., Fl. Brit. India 6: 11-12 (1891); King & Pantling in Ann. Roy. Bot. Gard. Calcutta 8: 196, t.262 (1898); Rolfe in J. Linn. Soc. Bot. 36: 29 (1903): Hara, Stearn & Williams, Enum. Flow. Pl. Nepal 1: 37 (1978); Wu & Chen in Acta Phytotax. Sin. 18: 303 (1980); Seth & Cribb in Arditti (ed.), Orchid Biol., Rev. Persp. 3: 304 (1984). Type: Meghalaya (Khasia Hills), cult. *Loddiges* (holotype K!).

C. syringodorum Griffith, Notulae 3: 338 (1851). Type: N India, Khasia Hills, Myrung, *Griffith* 228 (holotype K!).

C. eburneum Lindley var. *dayi* Jennings, Orchids: t.16 (1875). Type: Icon. in loc. cit. (1875), **syn. nov.**

C. eburneum Lindley var. *williamsianum* Reichb.f. in Gard. Chron., ser. 2, 15: 530 (1881). Type: cult. *Williams* (holotype W), **syn. nov.**

C. eburneum Lindley var. *philbrickianum* Reichb.f. in Gard. Chron., ser. 2, 25: 585 (1886). Type: cult. *Philbrick* (holotype W), **syn nov.**

Cyperorchis eburnea (Lindley)Schltr. in Fedde, Repert. 20:107 (1924).

A perennial, epiphytic *herb. Pseudobulbs* about 10 cm long, 3 cm in diameter, ovoid to fusiform, bilaterally flattened, not produced annually but growing in an indeterminate fashion for about three years before a new growth is produced, often covered in persistent leaf bases and bearing about 7 fresh leaves, each pseudobulb producing about 15–17 distichous leaves in total. *Leaves* up to 60 × 1.3(2.0) cm, narrowly ligulate, acute, with a finely unequally bilobed apex with a minute mucro in the sinus formed as an extension of the mid-vein; articulated to a broad, sheathing base 5–10 cm from the pseudobulb. *Scape* about 25(36) cm long, from within the axils of the leaves; peduncle erect or suberect, covered in about 8–12 sheaths up to 15 cm long; upper sheath cymbiform, middle sheaths mostly cylindrical below, inflated, cymbiform above; rhachis short, with 1–2(3) flowers; bracts 0.4–2.0 cm long, triangular, acute. *Flower* large, 8–12 cm across, not opening fully; sweetly lilac-scented; rhachis, pedicel and ovary bright green; petals and sepals white or faintly pink; lip white with a bright yellow central and basal patch on the mid-lobe and bright yellow callus, occasionally with some pale purple-pink spots on the mid-lobe; column white or flushed pale pink, sometimes spotted pink ventrally and with a small yellow patch at the base; anther-cap white. *Pedicel and ovary* 3.2–4.1 cm long. *Dorsal sepal* 5.6–7.6 × 1.8–2.9 cm, narrowly oblong-elliptic, acute, concave, porrect; lateral sepals similar, not fully spreading. *Petals* 5.0–7.3 × 1.5–2.2 cm, narrowly spathulate, slightly curved, acute, tips recurved and spreading. *Lip* 3-lobed, elongated, fused to the base of the column for 4–6 mm; side-lobes 1.2–1.8 cm broad, clasping the column, papillose or minutely hairy, apex broadly rounded, margin not fringed; mid-lobe small 1.6–2.0 × 1.4–1.7 cm, ovate-triangular, rounded or mucronate, porrect or weakly recurved, minutely hairy with a dense central and basal patch of short hairs, margin entire and undulating; callus long, with three slightly raised ridges on a broad, glabrous disc behind, becoming confluent and strongly inflated and forming a cuneate apex terminating well behind the junction of the mid- and side-lobes, papillose or minutely hairy. *Column* (3.4)4.1–4.6 cm long, narrowly winged to the base, curved down near the apex, almost glabrous; pollinia quadrangular, about 4 mm long, on a rectangular viscidium with hair-like processes from the lower corners. *Capsule* about 7–10 cm long, ellipsoidal, with a short, apical beak.

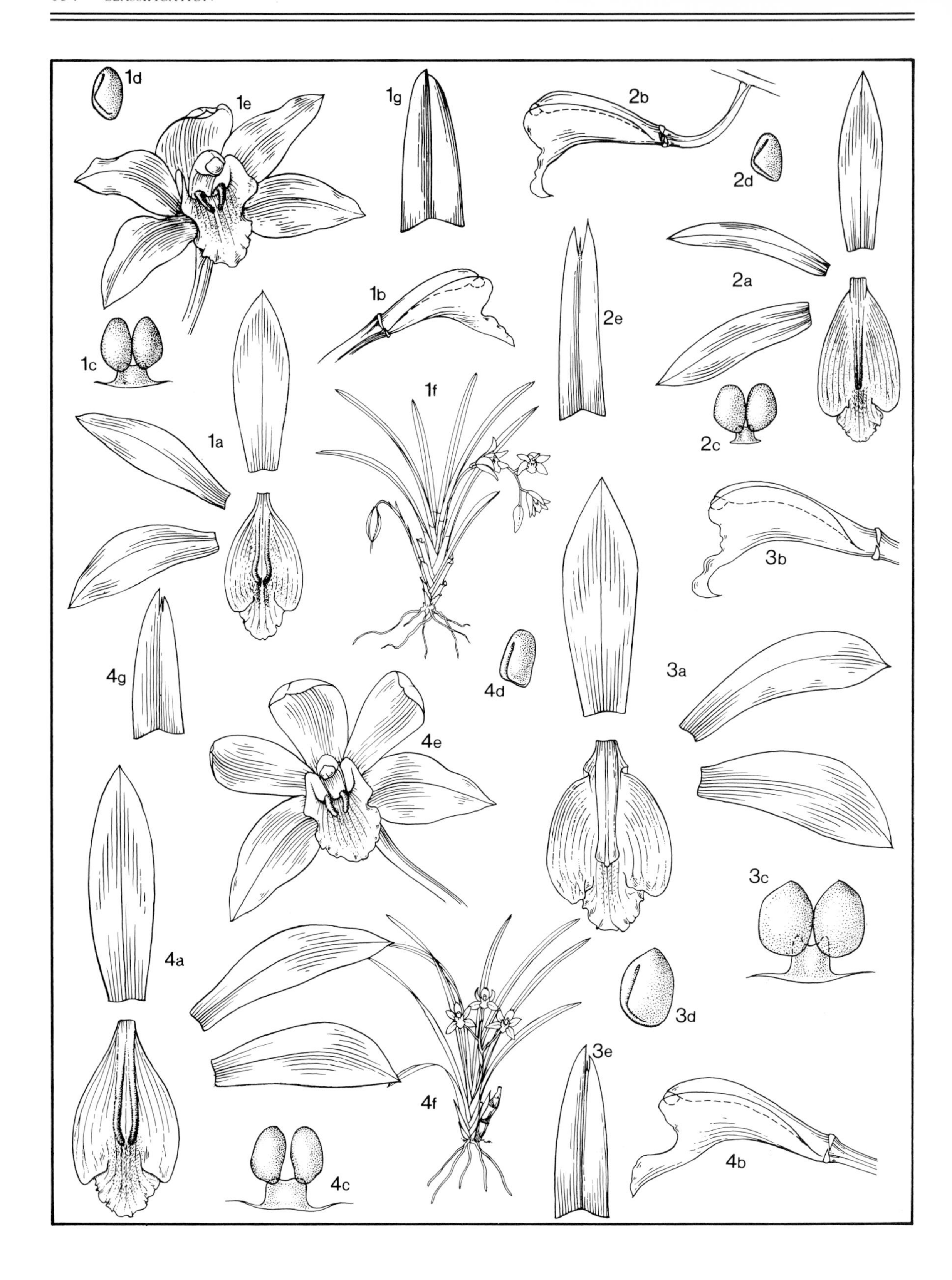
1d
1e
1g
2b
2d
2a
1b
2e
1c
1f
1a
2c
3b
4g
4d
3a
4e
3c
4a
3d
3e
4f
4b
4c

FIGURE 25

1. ***C. roseum*** (Malaya, *Segerbeck* 2089, + photo)
   - a Perianth, × 0.7
   - b Lip and column, × 0.7
   - c Pollinarium, × 3
   - d Pollinium (reverse), × 3
   - e Flower, × 0.7
   - f Flowering plant, × 0.1
   - g Leaf tip, × 0.5
2. ***C. mastersii*** (Kew spirit no. 14986/47814)
   - a Perianth, × 0.7
   - b Lip and column, × 0.7
   - c Pollinarium, × 3
   - d Pollinium (reverse), × 3
   - e Leaf tip, × 0.5
3. ***C. eburneum*** (Kew spirit no. 29007/33346)
   - a Perianth, × 0.7
   - b Lip and column, × 0.7
   - c Pollinarium, × 3
   - d Pollinium (reverse), × 3
   - e Leaf tip, × 0.5
4. ***C. parishii*** (Burma, *Parish* 56)
   - a Perianth, × 0.7
   - b Lip and column, × 0.7
   - c Pollinarium, × 3
   - d Pollinium (reverse), × 3
   - e Flower, × 0.7
   - f Flowering plant, × 0.1
   - g Leaf tip, × 0.5

ILLUSTRATIONS. Plate 19, photographs 3, 102, 103; figure 25.3.

DISTRIBUTION. N India (Sikkim, Assam, Khasia Hills), Nepal, N Burma, China (S Yunnan, ?Hainan) (map 8E); 300–1700 m (985–5580 ft).

HABITAT. On trees in warm, damp forest, in shade; flowering March-May (November-January in cultivation).

This species has short inflorescences which usually carry a single, large, white flower, with a single, broad, yellow callus ridge. It is amongst the most beautiful of the *Cymbidium* species, and has been used extensively in hybridisation. Its strong fragrance is usually not inherited when it is hybridised.

C. eburneum was described by Lindley in 1847, from a specimen cultivated by Loddiges' nursery and said to be from the 'East Indies'. When Griffith (1851) described *C. syringodorum*, the lilac-scented *Cymbidium*, he gave the first precise locality information as the Khasia Hills in Assam. Lindley (1858) reduced this name to synonymy in *C. eburneum*. Later, Clarke and then King & Pantling collected it from Darjeeling and Sikkim. It is likely that the distribution extends along the Himalaya to include Nepal (Hara *et al.*, 1978) and Bhutan. Forrest collected a cultivated specimen of *C. eburneum* (wrongly identified as *C. hookerianum*) in western Yunnan where it is undoubtedly native. In 1939 a specimen from Burma, but without an exact locality, was sent to Kew, extending the known distribution of this species.

C. eburneum can easily be distinguished from the other large, white-flowered species in this section by its characteristic callus ridges which are fused into one single, inflated, wedge-shaped structure at the apex. *C. mastersii* shares the tendency to grow indeterminately for several years before producing a new growth, the characteristic leaf tip shape, the elongated lip, and has similar colouring. Its flowers are much smaller and more numerous, the petals are much narrower, and it lacks the typical callus structure of *C. eburneum*. *C. insigne* lacks the complex leaf tip structure, has a much longer and more robust flower spike bearing many more flowers and has two separate, hairy callus ridges. The relationship between *C. eburneum* and *C. parishii* is discussed under the latter species.

This species was often well cultivated in the warm, humid stove greenhouses of the late nineteenth century, and large specimen plants were exhibited bearing upwards of 25 flowers. It was used as one of the parents of the first artificial *Cymbidium* hybrid, *C.* × Eburneo-lowianum in 1889. The popularity of *C. eburneum* led to the naming of several cultivars. Reichenbach described two varieties based on variations in colour; the pure white-flowered var. *philbrickianum* and the pink-tinged var. *williamsianum*, both with a yellow callus and mid-lobe patch. This species includes both white and pale pink-tinged variants, and some pale purple-pink spots may also occur on the mid-lobe. This latter has been called var. *dayi* (Jennings, 1875).

A natural hybrid, *C.* × *ballianum* between *C. eburneum* and *C. mastersii*, has been recorded by Rolfe (*Orchid Rev.* 12: 85, 1904), originally imported from Burma. Specimens of a distinct species from Annam (Vietnam) have also been given this name (see under *C. sanderae*).

27. C. parishii *Reichb.f.* in Trans. Linn. Soc. 30: 144 (1874), in Gard. Chron., n.s., 1:338, 566 (1874) and n.s., 10: 74 (1878); R. Warner & B.S. Williams, Orchid Album 1: t.25 (1882); Reichb.f., Xenia Orchidacea 3: 55-6, t.224 (1883); Parish in Mason (ed.) Burma, its people and productions 2: 171-2 & 197 (1883); L. Linden, Orchid. Exot.: 683 (1894) & Lindenia 15: 93, t.717 (1900); P.J. Cribb in Orchid Rev. 83:332-3 (1975). Type: Burma, *Parish* 56 (holotype K!).

C. eburneum Lindley var. *parishii* (Reichb.f.)Hook,f., Fl. Brit. India. 6: 12 (1891); Veitch, Man. Orchid. Pl. 9: 15-16 (1893); Grant, Orch. Burma: 227 (1895).
Cyperorchis parishii (Reichb.f.)Schltr. in Fedde, Repert. 20: 108 (1924), **syn. nov.**

A perennial, probably epiphytic *herb*, resembling *C. eburneum*. *Pseudobulbs* about 11.5 × 4 cm, fusiform, not produced annually but growing in an indeterminate fashion for several seasons, often covered in persistent, distichous leaf bases, with 11–14 apical leaves. *Leaves* 38–53 × 1.8–3.0 cm, ligulate, acute, with a finely, unequally bilobed apex with a short mucro in the sinus, articulated to a broadly sheathing base. *Scape* about 25 cm long, from within the axils of the leaves; peduncle covered with several sheaths up to 15 cm long; upper sheath cymbiform, central sheaths cylindrical below, expanded and cymbiform above; rhachis short with 2-3 flowers; bracts triangular, to 4 mm long. *Flower* slightly smaller than that of *C. eburneum*, not opening fully; scented; rhachis, pedicel and ovary green; petals and sepals white; lip white with a yellow central and basal patch on the mid-lobe and orange-yellow callus ridges, with strong, interrupted streaks of purple on the side-lobes, and a few purple spots on the mid-lobe except on the broad, white submarginal area; column white, yellowish and red-spotted below towards the base; anther-cap white. *Pedicel and ovary* 3–4 cm long. *Dorsal sepal* 5.9 × 1.5 cm, narrowly oblong-elliptic, subacute, concave, suberect to porrect; lateral sepals similar but shorter, 5.0 × 1.6 cm, slightly curved. *Petals* 5.7 × 1.3 cm, oblong-spathulate, subacute, slightly curved, not fully spreading but reflexed and spreading at the tips. *Lip* 3-lobed, elongated, fused to the base of the column for about 3 mm; side-lobes 1.5 cm broad, weakly clasping the column, minutely hairy, acute, porrect, margins not fringed; mid-lobe 1.6 × 1.7 cm. circular, rounded, mucronate, slightly recurved, with a dense central patch of short hairs extending from the front of the callus ridges, margin undulating; callus long, with three slightly raised ridges on a glabrous disc, the outer two becoming strongly inflated and shortly hairy towards the apex, converging but not confluent, terminating well behind the junction of the mid- and side-lobes. *Column* 3.7 cm long, narrowly winged, curved down near the apex, almost glabrous.

ILLUSTRATIONS. Photograph 106; figure 25.4.

DISTRIBUTION. Burma (Tenasserim; map 8F); 1500 m (4920 ft).

HABITAT. Montane forest; flowering June-July.

C. parishii was discovered in Burma, near Moulmein, on the border with Thailand in 1859 by the Rev. Charles Parish, who collected a number of plants which were lost in transit. In 1867 he sent two further plants to Messrs Low, and a dried flower to Hooker at Kew from the plant he had cultivated and painted. Hooker considered it to be a variety of *C. eburneum*, but did not publish this name until 1891, in his *Flora of British India.* Meanwhile, in 1872, Reichenbach examined Parish's dried material and named it *C. parishii*, publishing the name in 1874. Coincidentally, the two plants with Messrs Low were sold, one to John Day, the orchid enthusiast and artist, and another to the collection of Mr Leech. Both of these plants flowered in June 1878, those grown by Mr Swan, gardener for Mr Leech, being the earliest. Leech's plant was subsequently sold to B.S. Williams for 100 guineas. Thus, 19 years after its discovery, living material was at last available to Reichenbach, enabling him more fully to describe the species, and to justify his claim that it was distinct from *C. eburneum* because the flower was smaller and the callus extended nearer to the base of the mid-lobe and lacked the middle velvet line of *C. eburneum*. He claimed that the viscidium had two long, spreading, hair-like appendages, which were said to be lacking in *C. eburneum*. Further investigation has shown the presence of these appendages in *C. eburneum* also.

Since then, *C. parishii* has been lost to cultivation. Very little material has been preserved. The type specimen, *Parish* 56, consists of a 2-flowered scape and a leaf, and is the only specimen at Kew. Two unpublished paintings by Day and by Parish, both preserved at Kew, and Reichenbach's painting in *Xenia Orchidacea*, supply extra information about the colouring of the flower and, in particular, the habit of the plant. The paintings in R. Warner and B.S. Williams' *Orchid Album*, and L. Linden's *Lindenia* are both rather stylised, and depict an immature growth and old pseudobulb, and therefore do not contribute much to the information available about this species.

Little is known about the exact locality or habitat of this species. Parish, on a note with the type specimen, says he found it 'on the ascent of Nat-taung, near Toungoo (about 18°50′N, 96°50′E), at 5000 ft', in the same region as *Dendrobium crassinode*. In 1874, Reichenbach wrote about *C. parishii* at the request of Low, probably as a note had been received from his collector Mr Boxall, which Day quotes as saying 'I found 50 plants of this on the top of Moulle Tongue, one of the highest mountains in Burma, with *Dendrobium jamesonianum* and *Coelogyne reichenbachii*.' No records remain of the results of this collection.

C. parishii is closely related to both *C. eburneum* and *C. roseum*. Its tendency to grow indeterminately for more than one year, producing flowers over several seasons from the axils of the leaves towards the top of the growth, is very similar to both of the above species, and is typical of species in section *Eburneum*, the most extreme case of this being found in *C. mastersii*. The unequally bilobed, often split leaf tips, sometimes with a small needle-like extension in the fork, are also typical of this section.

C. parishii is similar to *C. eburneum*, differing in the characters listed in Table 10.7.

The differences between *C. parishii* and *C. roseum* are discussed under the latter species. *C. parishii* var. *sanderae* Rolfe is treated here as a distinct species, *C. sanderae*.

Table 10.7: A comparison of the morphological differences between *C. parishii* and *C. eburneum*

Character	*C. parishii*	*C. eburneum*
Leaf width	1.8–3.0 cm	up to 1.3(2.0) cm
Number of flowers	2–3	1–2(3)
Dorsal sepal size (as a measure of flower size)	5.9 × 1.5 cm	5.6–7.6 × 1.8–2.9 cm
Lip colour	very strong purple markings	no purple markings, or weakly spotted
Side-lobe apex shape	acute and porrect	broadly rounded
Mid-lobe shape	circular	ovate-triangular
Callus structure	two ridges, converging	one single cuneate ridge
Flowering time	June–July	March–May (November–January)

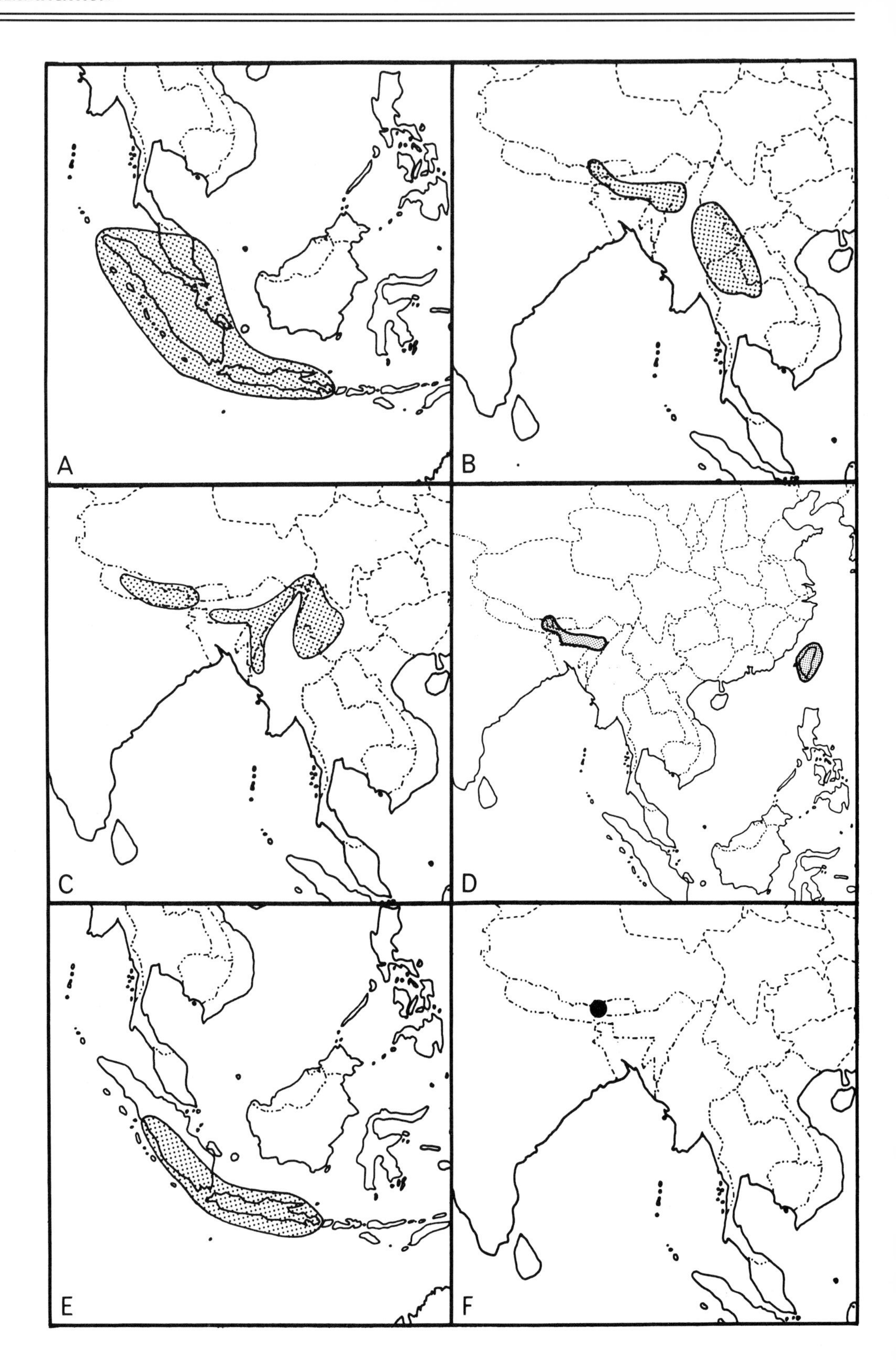
A
B
C
D
E
F

28. C. roseum *J.J. Smith*, Orchid. Java 475 (1905) & in Bull. Jard. Bot. Buitenzorg, ser. 3, 6: t.11 (1924); Holttum, Fl. Malaya 1: 514-15 (1953); Backer & Bakhuizen, Fl. Java 3: 394 (1968); J.B. Comber in Orchid Dig. 44: 165-7, + figs. (1980). Types: Java, Malabar, *Bosscha*!; Wanaredja, Desa Godong, *Ader*, Tijikorai, *Kessler*, Goentoer, *Raciborski*; Slamat, J.J. Smith (syntypes BO!).

Cyperorchis rosea (J.J. Smith)Schltr. in Fedde, Repert. 20:107 (1924); J.J. Smith in Bull. Jard. Bot. Buitenzorg, ser. 3, 9: 56-7 (1927).

A perennial, terrestrial, lithophytic or epiphytic *herb*. *Pseudobulbs* about 7.5 cm long, 2 cm in diameter, not strongly inflated, stem-like and inconspicuous within the leaf bases, growing and flowering in an indeterminate fashion for 2-3 years before a new growth is produced, usually with about 7 fresh leaves and numerous, persistent, sheathing leaf bases which eventually become fibrous, each pseudobulb producing about 13 distichous leaves in total. *Leaves* 20–40 × 2.2–2.7 cm, ligulate, obtuse, slightly unequally bilobed at the apex, sometimes with a short mucro in the sinus, coriaceous, articulated 4–6 cm from the pseudobulb. *Scape* (13)19–30 cm long, from the axils of the leaves; peduncle erect or suberect, covered in about 9 sheaths up to 16 cm long; central sheaths mostly cylindrical in the basal half, expanded and cymbiform in the apical half; rhachis short, with (1)2–5 flowers; bracts triangular, acute, up to 1 cm long. *Flower* 5–6 cm in diameter, not opening fully; faintly scented; rhachis, pedicel and ovary green; sepals and petals pale pink, speckled white, becoming darker or faintly pink-brown with age, sometimes white; lip pale pink or white with a bright yellow patch on the base of the mid-lobe, and bright yellow callus, sometimes with spots and strong, interrupted streaks of purple on the side-lobes and callus ridges, and a few large purple spots on the mid-lobe; column purple-pink, yellowish or white, sometimes streaked purple-red below, yellow at the base; anther-cap cream. *Pedicel and ovary* 2.8–4.7 cm long. *Dorsal sepal* 4.4–4.8 × 1.3–1.6 cm, narrowly elliptic, acute, concave, porrect; lateral sepals similar, not fully spreading. *Petals* 4.0–4.7 × 1.1–1.3 cm, narrowly elliptic, slightly curved, acute, tips recurved and spreading. *Lip* 3-lobed, elongated, fused to the base of the column for about 3.5 mm; side-lobes 1.1 cm broad, weakly clasping the column, subacute to rounded and porrect, shortly hairy, margin not fringed; mid-lobe 1.4–1.9 × 1.3–1.5 cm, broadly ovate, rounded, porrect, minutely hairy with two lines of short hairs at the base in front of the callus ridges, and a dense patch of short hairs in the centre, margin entire and weakly undulating; callus ridges 2, long, tapering to near the base of the lip from inflated, converging apices, terminating well behind the base of the mid-lobe, densely hairy. *Column* 2.9–3.5 cm long, narrowly winged to the base, curved down near the apex, with some minute hairs on the margin towards the base; pollinia quadrangular. *Capsule* 3.5–6.0 × 1.5–2.5 cm, fusiform, stalked, apex with a short beak formed by the column.

ILLUSTRATIONS. Photograph 107; figure 25.1.

DISTRIBUTION. W Malaysia, Java, Sumatra (map 9A); 1500–2100 m (4920–6890 ft).

HABITAT. In high mountains on exposed rocks or steep banks, or occasionally low down on trees; flowering August-December.

C. roseum is rather a rare species with a restricted distribution. It resembles *C. eburneum* vegetatively, but has 2-5 white or pink flowers, often with a boldly marked lip, and a pair of yellow callus ridges.

It was described by J.J. Smith in 1905 from plants collected in Java. Holttum (1953) found it in the high mountains of W Malaysia, and reported that it grew on rocks or low down on trees in rather exposed places. In Java and Sumatra it is found growing as a lithophyte, or terrestrial, in full sun.

Plants with pale pink-coloured flowers and purple markings on the lip, as described by J.J. Smith, are found throughout the range of the species. A

MAP 9: The distribution of the species in section *Eburnea* (cont.):

A *C. roseum*
B *C. mastersii*
and in section *Cyperorchis*: **C** *C. elegans*
D *C. cochleare*
E *C. sigmoideum*
F *C. whiteae*

yellowish-white variant, lacking the spots on the lip, has been found in Sumatra and W Malaysia growing together with the pink form.

C. roseum is closely related to *C. parishii*, but the latter has more inflated pseudobulbs, longer leaves (about 38–53 cm (15–21 in) long), a larger flower (dorsal sepal 5.9 cm (2.3 in) in *C. parishii*, up to 4.8 cm (1.9 in) in *C. roseum*), a shorter, more sparse indumentum on the lip, acute side-lobes, a more strongly undulating mid-lobe and white rather than pinkish flowers.

C. insigne is vegetatively very distinct, producing a conspicuous annual pseudobulb, and longer, narrower leaves with acute, not bilobed, apices. The flower spike is much longer, more robust, carries many more flowers and is produced from the base of the pseudobulb. The lip is shorter but the mid-lobe is larger, and the side-lobe tips are more broadly rounded.

29. C. mastersii *Griffith ex Lindley* in Bot. Reg. 31: t.50 (1845); Anon. in Gard. Chron.: 643 (1845); Paxton, Flower Garden 3: 21-2, t.78 (1852); King & Pantling in Ann. Roy. Bot. Gard. Calcutta 8: 195-6, t.261 (1898); Seidenfaden in Opera Bot. 72: 91-3, f.52 (1983); Seth & Cribb in Arditti (ed.), Orchid Biol., Rev. Persp. 3: 303 (1984). Type: cult. *Loddiges* (holotype K!).

C. affine Griffith, Notulae 3: 336, t.291, f.3. (1851); Lindley in J. Linn. Soc. 3: 28 (1858); Reichb.f. in Gard. Chron., n.s., 10: 810-11 (1878); Warner & Williams, Orchid Album 3: t.140 (1884). Type: Assam, Khasia Hills, Churra, *Griffith* (holotype K!).

C. micromeson Lindley in J. Linn. Soc. 3: 29 (1858); Reichb.f. in Gard. Chron., n.s., 10: 810-11 (1878). Type: Assam, Khasia Hills, Churra, *Griffith* (holotype K!).

C. mastersii Griffith ex Lindley var. *album* Reichb.f. in Gard. Chron., n.s., 13: 136 (1880); Williams, Orchid Growers' Man., edn 6,: 234-5 (1885); F. Sander, Reichenbachia, ser. 1, 2: t.66 (1890). Type: not designated, **syn. nov.**

Cyperorchis mastersii Benth. in J. Linn. Soc. 18: 318 (1881); Hook.f., Fl. Brit. India 6: 15 (1890); Schltr. in Fedde, Repert. 20: 107 (1924).

A perennial, epiphytic or lithophytic *herb. Pseudobulbs* not inflated, stem-like and inconspicuous within the numerous persistent leaf bases, growing indeterminately, forming cauline roots through the lower leaf bases and occasionally producing a new growth from near the base, with about 6-17 apical leaves and numerous, distichous, alternate, sheathing, persistent leaf bases towards the base, the older growths appearing flattened and elongated. *Leaves* up to 64 × 1.8 (2.5) cm, arching, narrowly ligulate, tapering to an acute, usually unequally bilobed, forked, apex with a mucro in the sinus; articulated about 6–10 cm from the axis; leaf base with a 2–3 mm wide membranous margin. *Scape* about 25–30 cm long, from the axils of the leaves; peduncle suberect, covered in 6-8 sheaths up to 16 cm long; upper sheath cymbiform, central sheaths cylindrical in the basal half, expanded, cymbiform in the apical half; rhachis 5–10 cm long, arcuate or pendulous, not strongly exerted from the sheaths, with (2)5-10(15) closely spaced flowers; bracts 2–7 mm long, triangular, acute. *Flower* about 6 cm across, not opening fully; almond-scented; rhachis, pedicel and ovary green; petals and sepals white or faintly pink; lip white, usually with a yellow central patch at the base of the mid-lobe, and bright yellow callus ridges, sometimes with some pale to strong purple-red spots and shading on the side- and mid-lobes; column white or pale green, sometimes spotted with pale red below, with a yellow patch at the base; anther-cap cream. *Pedicel and ovary* 1.6–2.5 cm long. *Dorsal sepal* 4.3–5.2 × 0.8–1.1 cm, narrowly oblong-elliptic to narrowly obovate, acute, concave, porrect but spreading towards the tips; lateral sepals similar, slightly curved, porrect. *Petals* 3.9–4.9 × 0.5–0.7 cm, narrowly ligulate or narrowly obovate, slightly curved, porrect. *Lip* 3-lobed, elongated, fused to the base of the column for 3–5 mm; side-lobes 1.0–1.2 cm broad, lightly clasping the column, minutely hairy, apex broadly rounded to subacute and porrect, margin minutely fringed at the front; mid-lobe small, 1.0–1.3 × 1.0–1.3 cm, ovate, rounded or mucronate, porrect, minutely hairy with a dense central and basal patch of short hairs, margin undulating; callus long, with two slightly raised, well-separated ridges behind, sometimes minutely auriculate and forming a small trough-shaped depression at the base, inflated and converging towards the shortly hairy apex at the base of the mid-lobe. *Column* slender, 3.3–3.6(4.0) cm long, very narrowly winged, curved down near the apex, glabrous or sparsely shortly hairy below in a slight elliptical depression at the base; pollinia quadrangular, 2–3 mm long. *Capsule* broadly ellipsoidal, slightly pointed, stalked, about 3 × 2.5 cm with a beak almost as long as the capsule.

ILLUSTRATIONS. Plate 20, photographs 104, 105; figure 25.2.

DISTRIBUTION. N India (Sikkim, Meghalaya, Manipur), Burma, N Thailand (map 9B); 900–2200 m (2950–7220 ft).

HABITAT. On trees or rocks in evergreen forest, often in deep shade, in humus, moss or on rotting wood; flowering (September)October-December-(January).

C. mastersii was first described by Lindley (1845) from a specimen flowered by Loddiges in December 1844. He attributed the name to Griffith who had chosen it in honour of Dr Masters of the Botanical Garden in Calcutta. However, in 1851, Griffith published the name *C. affine* for the same species, but did not mention *C. mastersii*. Reichenbach (1878) distinguished *C. affine* by the purple spotting towards the front of the lip, and by the slightly more hairy lip. This latter character is variable within the species, and the name *C. affine* has since been applied to specimens with the purple pigmentation. However, the type specimen of *C. mastersii*, illustrated in the *Botanical Register*, 1845, also had red-purple spots on the lip and the two are undoubtedly conspecific.

C. mastersii was recognised to vary in its degree of pigmentation and Reichenbach (1880) named the white-lipped variant var. *album*. At the same time he rescinded his previous differentiation of *C. mastersii* and *C. affine* on the basis of colour, and instead proposed that they could be differentiated on the more upright flower spike and the lack of 'notches' (auricles) towards the base of the callus ridges in *C. affine*. However, the type specimen of *C. mastersii* does have both of these features.

Specimens of *C. mastersii* collected by one of the authors (DD) in Thailand have very faint pinkish spots on the lip, and are intermediate between the two major colour variants. Variation in the size of the auricles at the base of the lip is found in both *C. mastersii* and *C. elegans*, and seems to vary in both species from strongly expressed to absent. The angle at which the scape is held varies from erect to almost horizontal, depending on the robustness of the peduncle, and the weight of the flowers it carries.

In 1858, Lindley compounded the confusion by publishing *C. micromeson* based on a mixed collection by Griffith from the Khasia Hills. The type sheet comprises a shoot and flower spike of *C. mastersii*, a fruiting scape of *C. eburneum*, and a flattened lip of another species (probably in the genus *Coelogyne* according to J.D. Hooker on a note attached to the specimen). There is also a sketch of a lip from one of the flowers on the complete specimen which shows no obvious differences from *C. mastersii*, but is very different from the flattened lip on the sheet. The description differs from *C. mastersii* in several characters which are attributable to this alien lip.

C. mastersii can be distinguished from *C. elegans* by its inconspicuous, stem-like pseudobulb and indeterminate growth habit with the scape arising within the axils of the upper leaves, not from near the base of a swollen pseudobulb. *C. mastersii* has distinctive, unequally bilobed, forked leaf apices, with a minute mucro in the sinus, and fewer, larger and more distant flowers which open more fully. The ovate mid-lobe of *C. mastersii* is very distinct from the narrow mid-lobe expanding into two incurved apical lobes with a strongly emarginate apex typical of *C. elegans*. The pollinia of *C. elegans* are clavate rather than quadrangular. These two species often have auricles protruding at the base of the callus ridges.

C. erythraeum might also be confused with *C. mastersii*, but can be distinguished by the growth habit, its pseudobulb and leaf characters, its differently coloured and more distant flowers on a much longer rhachis which is well exerted from the sheaths, its spreading sepals and petals, its more acute, porrect, hairy side-lobe tips, and cordate to reniform mid-lobe, and its curved column not strongly deflexed near the apex.

C. eburneum is closely related and shares many of the characters of *C. mastersii*, but has a more inflated pseudobulb, which does not grow in an indeterminate manner for more than two or three years before producing a new growth. It can be distinguished by its usually single, very large flower, and its confluent, cuneate callus tip (see also the discussions under *C. eburneum* and *C. whiteae*).

C. mastersii is found at about 1500 m (4920 ft) altitude in northern Thailand. Du Puy (1984a,b) encountered it growing in the same region as *C. insigne.* Collections sent to Kew flowered in November 1984, corresponding to the end of the rainy season in Thailand. *C. insigne* flowers in February and March, in the hot, dry season, but also produces some spikes in November. These species also grow in distinct habitats. *C. insigne* grows in open, shrub-dominated maquis-like vegetation, while *C. mastersii* is usually found growing epiphytically in the taller, denser, shaded and humid stands of forest surrounding creek beds. The crowns of older plants are held well clear of the branches and plant debris in which they grow, the long stems arching upwards and outwards from their place of anchorage. The largest specimens were found on thickly moss-covered branches, although younger plants were also found on vertical tree trunks which were much less densely covered by vegetation. Occasional specimens were also found growing in humus and moss on boulders and cliffs. Kerr noted that on Doi Suthep it grew in 'shady, moist ravines in humus on dead tree trunks'.

C. mastersii produces nectar which often fills the pouch formed by the fusion of the base of the lip and the column.

There is a possible natural hybrid with *C. sanderae* or *C. insigne* collected in Annam (see discussion of *C. sanderae*) whilst *C.* × *ballianum* is a natural hybrid with *C. eburneum*, imported from Burma (*Orchid. Rev.* 12: 85, 1904).

Section **Annamaea** *(Schltr.)P. Hunt* in Kew Bull. 24: 94 (1970); Seth & Cribb in Arditti (ed.), Orchid Biol., Rev. Persp. 3: 302 (1984). Type: *Cyperorchis erythrostyla* (Rolfe)Schltr. (*C. erythrostylum* Rolfe).

Cyperorchis section *Annamaea* Schltr. in Fedde, Repert. 20: 108 (1924).

This monotypic section was established by Schlechter (1924), and was later transferred to *Cymbidium* by Hunt (1970). Its lip and column fusion, two triangular, cleft pollinia and large flower size place it in subgenus *Cyperorchis.*

FIGURE 26

1. ***C. cochleare*** (Kew spirit no. 12843)
   - a Perianth, × 0.7
   - b Lip and column, × 0.7
   - c Pollinarium, × 3
   - d Pollinium (reverse), × 3
   - e Flower, × 0.7
   - f Flowering plant, × 0.1
2. ***C. whiteae*** (Kew spirit no. 49023)
   - a Perianth, × 0.7
   - b Lip and column, × 0.7
   - c Pollinarium, × 3
   - d Pollinium (reverse), × 3
   - e Flower, × 0.7
   - f Flowering plant, × 0.1
3. ***C. elegans*** (Kew spirit no. 45555)
   - a Perianth, × 0.7
   - b Lip and column, × 0.7
   - c Pollinarium, × 3
   - d Pollinium (reverse), × 3
4. ***C. sigmoideum*** (Java, *J.J. Smith* 150)
   - a Perianth, × 0.7
   - b Lip and column, × 0.7
   - c Pollinarium, × 3
   - d Pollinium (reverse), × 3
   - e Flower, × 0.7
   - f Flowering plant, × 0.1
5. ***C. erythrostylum*** (Kew spirit no. 47871)
   - a Perianth, × 0.7
   - b Lip and column, × 0.7
   - c Pollinarium, × 3
   - d Pollinium (reverse), × 3
6. ***C. tigrinum*** (Kew spirit no. 46422)
   - a Perianth, × 0.7
   - b Lip and column, × 0.7
   - c Pollinarium, × 3
   - d Pollinium (reverse), × 3

The dorsal sepal is almost erect, the lateral sepals curved downwards and the petals are porrect, covering the column and giving the flower a narrowly triangular shape which is highly distinctive. The unusual shape of the mid-lobe of the lip, the cuneate apex of the callus, the long indumentum of the lip which is confined to the veins on the side-lobes, and the dense indumentum on the ventral surface of the column are also characteristic of the section. It is related to sections *Eburnea* and *Iridorchis.*

30. C. erythrostylum *Rolfe* in Gard. Chron., ser. 3, 38: 427 (1905), & ser. 3, 40: 265, 286, f.115 (1906), in Orchid Rev. 14: 39 (1906) & in Bot Mag. 133: t.8131 (1907); Mehlquist in Missouri Bot. Gard. Bull. 34:

1a
1b
1c
1d
1e
1f
2a
2b
2c
2d
2e
2f
3a
3b
3c
3d
4a
4b
4d
4e
4e
4f
5a
5b
5c
5d
6a
6b
6d
6e

117–18, + fig. (1946); Withner in Amer. Orchid Soc. Bull. 17: 72–3 (1948); Crosby in Amer. Orchid Soc. Bull. 20: 600–4, + fig. (1951). Type: Annam, cult. R.B.G. Glasnevin, *Micholitz* (holotype K!).

Cyperorchis erythrostyla (Rolfe)Schltr. in Fedde, Repert. 20: 427 (1924).

C. erythrostylum Rolfe var. *magnificum* Hort. ex. Gard. Chron., ser. 3, 90: 81, + fig. (1931). Type: cult. *McBeans* (type not preserved), **syn. nov.**

A perennial, epiphytic, lithophytic or terrestrial *herb. Pseudobulbs* about 6 cm long, 2 cm in diameter, produced annually, narrowly ovoid, bilaterally flattened, with 6–8 distichous leaves, and 2–3 cataphylls which become scarious with age. *Leaves* up to 45(55) × 1.5(1.7) cm, narrowly linear-obovate, arching, apex slightly unequal, apiculate, articulated 2–5 cm from the pseudobulb to a persistent, broadly sheathing, yellowish base with a 1–2 mm broad membranous margin. *Scape* 15–35 cm long, from within the axils of the cataphylls or lower leaves on immature growths; peduncle suberect, arching, slender, covered by about 6 sheaths; sheaths overlapping, cymbiform, up to 8 cm long, the middle sheaths cylindrical near the base; rhachis short, with (3)4–8(12) flowers; bracts (1.5)2.5–5 cm long, slender, narrowly lanceolate, cymbiform. *Flower* about 6 cm across, appearing narrowly triangular, not scented; sepals and petals white, the petals pale pink along the mid-vein in the basal half, sometimes spotted with pink at the base, glistening as though covered in frost and rather thin in texture; lip yellow-white, darker yellow on the mid-lobe, strongly veined with deep red, the veins becoming broken and spotted towards the margins of the side-lobes, and broader and blotched near the apex of the mid-lobe; callus cream, strongly pink-mottled; column strong purple-pink above, paler below; anther-cap cream. *Pedicel and ovary* (2.5)4–5.5 cm long. *Dorsal sepal* 4.5–5.7 × 2.2–2.6 cm, obovate-elliptic, acuminate, concave, erect; lateral sepals similar, falcate, curved downwards. *Petals* broad, 4.1–4.9 × 1.7–2.2 cm, obovate-elliptic, acuminate, slightly curved, porrect and covering the column. *Lip* 3-lobed, fused to the base of the column for about 3 mm; side-lobes 1.5–1.9 cm broad, clasping the column, with the hairs mostly confined to the veins, the hairs longest around the callus, apex broadly rounded; mid-lobe short, 1.0–1.2 × 1.3–1.5 cm, triangular, acuminate, weakly recurved, shortly hairy, margin weakly undulating; callus glabrous, broad, with (3-)5 raised ridges behind, converging into three ridges in front which become strongly inflated and confluent to form a three-lobed, cuneate apex well behind the mid-lobe of the lip, the central lobe being the largest and tapering into the mid-lobe. *Column* about 3 cm long, arcuate, very weakly winged, densely hairy beneath; pollinia triangular, 2 mm across, on a rectangular viscidium with two long, hair-like processes from the lower corners.

ILLUSTRATIONS. Plate 21; photographs 30, 108; figure 26.5.

DISTRIBUTION. Vietnam (map 8D); about 1500 m (4920 ft).

HABITAT. Not known; flowering May-July in the wild, October-December in cultivation.

C. erythrostylum was described by Rolfe (1906) from a plant cultivated at the Royal Botanic Garden, Glasnevin which had been collected by Micholitz in 1891 in Annam (Vietnam) for Messrs Sander. It is one of the most attractive of the large-flowered species with its glistening, fragile-looking white petals and sepals, and its boldly red-marked lip and red column. Its early flowering period, and its long-lasting flowers, have made it a useful species for hybridisation.

Schlechter (1924) placed this species in its own section on the basis of its having only one leaf on the pseudobulb, and a distinctive viscidium, very broad petals and sepals, and unusual lip shape. It is now known that the pseudobulb has several leaves, that the hair-like processes on the viscidium are not unusual in this subgenus, and the broad sepals and petals are also found in *C. insigne* (sect. *Iridorchis*). The shape of the lip, and especially of the short, triangular mid-lobe is indeed distinctive. Several of its other characters are, however, unique in the subgenus: the scape produced on immature growths; the long bracts; the erect dorsal sepal; the decurved lateral sepals; the porrect petals, covering the column; the side-lobes of the lip clasping over the top of the column, and with long hairs mainly confined to the veins; the very short, triangular mid-lobe of the lip; the column with a dense, ventral indumentum of long hairs; and the deep purple-pink column.

The callus structure is reminiscent of that found in *C. eburneum*: both have a broad, glabrous, ligulate basal portion, with several raised ridges which join at the apex into one inflated, wedge-shaped structure well behind the mid-lobe. However, in *C. erythrostylum* this callus apex is unique in being glabrous, and composed of three distinct lobes of which the central one is the largest, tapering well into the mid-lobe of the lip.

Section **Cyperorchis** (*Bl.*)*P. Hunt* in Kew Bull. 24: 94 (1970); Seth & Cribb in Arditti (ed.), Orchid Biol., Rev. Persp. 3: 303 (1984). Type: *C. elegans* Lindley.

Cyperorchis Bl. in Rumphia 4: 47 (1848); Hooker f., Fl. Brit. India 6: 14 (1891).

Cyperorchis section *Eucyperorchis* Schltr. in Fedde, Repert. 20: 106 (1924).

Cyperorchis in the narrow sense of Blume (1848) and P. Hunt (1970) is given sectional status within subgenus *Cyperorchis* following the treatment of Seth & Cribb (1984).

This section is characterised by its basal scape with pendulous flowers which do not open widely, its straight, narrow lip with a long hypochile, its long, slender, straight column which is abruptly deflexed near the apex, its beaked rostellum and anther-cap and its usually clavate pollinia on a rectangular viscidium (figure 14). *C. cochleare* and *C. elegans* closely fit this diagnosis, but *C. whiteae* and especially *C. sigmoidieum* show some intermediacy with section *Eburnea* to which this section is closely allied.

31. C. elegans *Lindley*, Gen. Sp. Orchid. Pl.: 163 (1833) & Sert. Orchid.: t.14 (1838); Mueller in Walp. Ann. 4: 626 (1864); Reichb.f. in Gard. Chron., n.s., 3: 429 (1875); King & Pantling in Ann. Roy. Bot. Gard. Calcutta 8: 194, t.259 (1898); Schltr. in Fedde, Repert., Beih. 4: 266 (1919); Cribb & Du Puy in Kew Bull. 38: 65-7 (1983); Seth & Cribb in Arditti (ed.), Orchid Biol., Rev. Persp. 3: 303 (1984). Type: Nepal, Gossaingsthan, *Wallich* 7354 (holotype K!).

C. longifolium D. Don, Prodr. Fl. Nepal.: 36 (1825); Hara, Stearn & Williams, Enum. Flow. Pl. Nepal 1: 37-8 (1978); Wu & Chen, in Acta Phytotax. Sin. 18: 304 (1980); Hara in Taxon 34: 690-1 (1985), *non* Lindley (1833). Type: Nepal, Gossainkunde ('Gosaingsthan'), 1819 *Wallich* s.n., (lectotype BM), **nom. rej.**

Cyperorchis elegans (Lindley) Blume, Rumphia 4: 47 (1848), in Mus. Bot. Lugd. Bat. 1: 48 (1849) & in Orcid. Archip. Ind. 1: 93, t.48c (1858); Reichb.f. in Walp. Ann. 3: 548 (1853); Hook. f., in Bot. Mag. 114: t.7007 (1888) & Fl. Brit. India 6: 14 (1890); Schltr. in Notes R.B.G. Edinburgh 5: 112 (1912) & in Fedde, Repert. 20: 107 (1924).

C. densiflorum Griffith, Notulae 3: 337 (1851); Lindley in J. Linn. Soc. 3: 28-9 (1858). Type: Assam, Khasia, Myrung, *Griffith* 229 (holotype K!).

Grammatophyllum elegans (Lindley)Reichb.f. in Walp. Ann. 3: 1028 (1853), **syn. nov.**

C. elegans Lindley var. *obcordatum* Reichb.f. in Gard. Chron, n.s., 13: 41 (1880). Type: not located, **syn. nov.**

C. elegans Lindley var. *lutescens* Hook.f., Fl. Brit. India 6: 15 (1891). Type: Icon. ex CAL (K!), **syn. nov.**

Arethusantha bletioides Finet, in Bull. Soc. Bot. Fr. 44: 179, t.5 (1897). Type: sin. loc., *Prince of Orleans* (holotype P). **syn. nov.**

Cyperorchis elegans (Lindley)Blume var. *blumei* Hort. in J. Hort., ser. 3, 54: 71 & fig. (1907). Type: cult. *Colman* (type not preserved), **syn. nov.**

A perennial, epiphyte or lithophytic *herb*. *Pseudobulbs* up to 7 cm long, 4 cm in diameter, ovoid, bilaterally flattened, with 7-13 distichous leaves, produced annually. *Leaves* up to 65(80) × 1.4(2.0) cm narrowly linear-elliptic, acute, slightly hooded and minutely unequally bilobed (but not forked) at the apex, articulated to a broadly sheathing base, 4–12 cm from the pseudobulb. *Scape* slender, 30–60 cm long; peduncle suberect to horizontal, covered in sheaths up to 10–15 cm long; rhachis slender, pendulous, usually with 20–35 closely spaced flowers; bracts triangular, 1–6 mm long. *Flowers* about 3 cm across; pendulous, bell-shaped, segments not spreading, in a large, crowded, pendulous cluster; often lightly scented; rhachis, pedicel and ovary green; sepals and petals cream or pale straw-yellow, sometimes tinged pale pink, to pale yellow-green; lip cream to pale green, occasionally sparsely red-spotted, with two bright orange-yellow callus ridges and a brown, reddish or cream depression at the base; column pale green; anther-cap cream. *Pedicel and ovary* short, 0.8–2.0(3.0) cm long. *Dorsal sepal* 3.2–4.3 × 0.7–1.1 cm, narrowly obovate, acute, concave, porrect, closely covering the column; lateral sepals similar, often mucronate, not spreading. *Petals* 3.1–4.2 × 0.4–0.8 cm, very narrow, ligulate or very narrowly obovate, obtuse, sometimes slightly curved, porrect. *Lip* deltoid, elongated, 3-lobed, fused to the base of the column for 2–3 mm; side-lobes 0.6–0.8 cm broad, triangular, erect and clasping the column, almost glabrous or minutely papillose, porrect, tapering to an acute or subacute apex, margin not fringed; mid-lobe small, 0.6–1.0 × 0.4–0.8 cm, base ligulate, expanded apically into two incurved lobes, emarginate, occasionally mucronate, porrect, with a dense patch of very short hairs in the centre, margin usually minutely undulating; callus ridges 2, converging towards their inflated, shortly hairy apices and usually with minute auricles at the base, forming a trough-shaped depression with some short marginal hairs. *Column* 2.8–3.4 cm long, slender, very sparsely shortly hairy below, with a shallow, elliptical, nectar-producing depression at the base, very narrowly winged, the wings inflated at the abruptly deflexed apex; anther-cap and viscidium elongated into a distinct protruding rostellum; pollinia clavate, deeply cleft, about 2 mm long, on a rectangular viscidium with long hair-like processes from the corners. *Capsule* 2.0–2.7 × 1.3–2.0 cm, broadly ellipsoidal, slightly pointed, stalked, with a long beak about as long as the capsule.

ILLUSTRATIONS. Plate 22; photographs 109-11; figure 26.3.

DISTRIBUTION. Nepal, N E India (Sikkim, Darjeeling, Assam, Meghalaya, Naga Hills, Lushai Hills), Bhutan, N Burma (incl. Chin Hills), S W China (Yunnan, Xizang) (map 9C); 1500–2500 m (4920–8200 ft).

HABITAT. On trees and rocks in damp, shady forest, sometimes on shaded rocks overhanging streams; flowering October-November (December).

This is one of the most striking and commonly cultivated of the *Cymbidium* species. Its large, densely crowded racemes of narrowly funnel-shaped flowers with club-shaped pollinia are very distinctive, and have led several authorities, including Blume, Hooker f. and Schlechter, to the conclusion that this should be placed in a separate genus. This is discussed under the introduction to the taxonomy of the genus. This species is well known under the name *C. elegans* Lindley, described in 1833 from a specimen collected by Wallich at Gossaingsthan, probably now known as Gossainkunde, in Nepal.

Recently, Hara *et al.* (1978) used the name *C. longifolium* D. Don for this species in their *Enumeration of the Flowering Plants of Nepal*, and this usage has subsequently been followed by Wu & Chen (1980). The name *C. longifolium* has, however, long been applied in the sense used by Lindley (1833) for another species which is also known in cultivation, to which Hara *et al.* have applied the name *C. erythraeum* Lindley.

Cribb & Du Puy (1983) attempted to conserve the traditional usage of the names *C. elegans* and *C. longifolium*, citing specimens in the Wallich Herbarium, at Kew, as types. However, Hara (1985) has shown that D. Don based his description on specimens in the Lambert Herbarium which is now at the British Museum (Nat. Hist.). The specimens there support the usage of these names as Hara *et al.* have indicated. Whilst accepting that the correct name of Lindley's *C. longifolium* is *C. erythraeum*, we consider that the use of *C. longifolium* for the well-known *C. elegans* would create considerable confusion in the orchid world. We have, therefore, applied for the name *C. longifolium* D. Don to be rejected under Article 69 of the International Code of Botanical Nomenclature. This will allow the retention of the name *C. elegans* for what is a widely grown and popular orchid (Du Puy & Cribb, *Taxon*, in press).

Several varieties of *C. elegans* have been described. Var. *obcordatum* Reichb.f. (1880) was differentiated on the obcordate, emarginate mid-lobe of the lip, and the colour of the auricles at the base of the callus. The former characters are in fact typical of this species, although the degree of their expression varies somewhat. The latter character is also highly variable within the species.

Var. *lutescens* Hook.f. (1891) was described (as a variety of *Cyperorchis elegans*) as 'a smaller plant, leaves 9 inches (23 cm), scape 7 inches (18 cm), densely clothed with imbricating sheaths 3 inches (7.5 cm) long; raceme suberect, secund, 5-flowered; flowers yellowish, 1 ¾ inches (4.5 cm) long', from a drawing in the Calcutta herbarium. This variety therefore differs from the type in its shorter leaves, and its shorter inflorescence with fewer flowers. The other characters in the description, and the drawing, fit the description of *C. elegans*. It appears that this variety is simply a smaller, rather poor specimen of *C. elegans*, and should not be recognised as a distinct variety.

Var. *blumei* Hort. (1907) was also described under *Cyperorchis elegans*, from a fine specimen shown at the Royal Horticultural Society, but neither the description nor the illustration show any characters which could be used to maintain it as distinct.

Griffith (1851) described *C. densiflorum*, based on his own collection from Meghalaya (Khasia Hills). A second specimen described in the same account is attributed to *C. mastersii*. Lindley (1858) distinguished this variant from *C. elegans* by the lack of auricles at the base of the callus in *C. densiflorum*. However, the auricles are mentioned in the type description, and dissection of the type specimen has shown them to be present. The size of the auricles varies greatly between specimens, and they may sometimes be rather obscure. Reichenbach (1875) noted that examination of the flowers on the same plant in two consecutive years showed that the size of the auricles can vary from almost absent to very conspicuous in a single specimen. He also noted that the colour of these auricles varied from orange to deep purple. The auricles may be cream, yellowish, orange, pink, reddish, purple or brown. The colour of the flower also shows some variation, from dull yellowish-green to clear cream or lightly pink tinged. The number of flowers varies, but usually there are more than 20 in a single inflorescence.

Finet (1897) described *Arethusantha bletioides* (= *C. elegans*) based on a specimen collected by the Prince of Orleans (sic) without any indication of where it was collected, or when. As both *Cyperorchis* (1848) and *Arethusantha* (1879) were based on the same species, and the former is in the earlier publication, the genus *Arethusantha* must therefore be considered synonymous with *Cyperorchis*.

C. elegans somewhat resembles *C. mastersii*, and both have auricles at the base of the callus. The key differences are outlined in the discussion of the latter. The differences with other allied species in sect. *Cyperorchis* are discussed under those species.

Hybridisation with *C. erythraeum* in Sikkim has led to hybrid swarms which display all stages of intermediacy between the two parents. This natural hybrid was first described by King & Pantling (1895, 1898) as *C.* × *gammieanum*. The natural populations are still present (C. Bailes, pers comm.).

32. C. cochleare *Lindley* in J. Linn. Soc. 3: 28 (1858); Reichb.f. in Gard. Chron., n.s., 13: 168 (1880); King & Pantling in Ann. Roy. Bot. Gard. Calcutta 8: 194–5, t.260 (1898). Type: Sikkim, *J.D. Hooker* 235 (holotype K!; isotypes K!).

Cyperorchis cochlearis (Lindley)Benth. in J. Linn. Soc. 18: 317–18 (1881); Hook.f., Fl. Brit. India 6: 15 (1891); Schltr. in Fedde, Repert. 20: 106 (1924).

Cyperorchis babae Kudo ex Masamune in J. Jap. Bot. 8: 258–60, + figs. (1932); Lin, Native Orchids of Taiwan 2: 131–4, f.254 & 60–2 (1977); Mark, Ho & Fowlie in Orchid Dig. 50: 14 (1985). Type: Taiwan, *Kudo* (holotype TI), **syn. nov.**

Cymbidium babae (Kudo ex Masamune)Masamune in Trop. Hort. 3:33 (1933); Liu & Su, Flora of Taiwan 5: 950 (1978), **syn. nov.**

C. kanran Makino var. *babae* (Kudo)Ying in Chinese Flowers 23: 7 (1976) & Coll. Ill. Ind. Orchid. Taiwan: 440 (1977), **syn. nov.**

A perennial, epiphytic *herb*, rarely terrestrial. *Pseudobulbs* small, up to 6 cm long, 2.5 cm in diameter, narrowly ovoid, slightly bilaterally flattened, produced annually, with 9–14 distichous leaves and about 5 scarious cataphylls sheathing the new growths. *Leaves* narrow, up to 50–90 × 0.6–1.0 cm, linear, tapering to an acute apex, articulated 2.5–8 cm from the pseudobulb to a broadly sheathing base. *Scape* 30–65 cm long, very slender, wiry; peduncle suberect to horizontal or arching, covered in about 6 inflated, cymbiform sheaths up to 6 cm long; rhachis slender, pendulous, with 7–20(30) flowers; bracts 1–5 mm long, triangular, scarious. *Flower* about 2.5 cm in diameter, bell-shaped, waxy, pendulous; not strongly scented; rhachis, pedicel and ovary purple;

sepals and petals greenish-brown, with a pale margin, glossy; lip yellow or orange-yellow, with numerous confluent red-brown spots on the side- and mid-lobes; callus yellow; column whitish-green, with some reddish spots ventrally; anther-cap cream. *Pedicel and ovary* 0.8–3.2 cm long. *Dorsal sepal* 3.9–4.5 × 0.6–0.9(1.1) cm, narrowly obovate, acute, porrect; lateral sepals similar, often mucronate, not spreading. *Petals* 3.8–4.4 × 0.4–0.6 cm, very narrowly obovate or ligulate, subacute, slightly curved, porrect, not spreading at least in the basal portion. *Lip* about 4–4.5 cm long, slender, elongated, deltoid, 3-lobed, fused to the base of the column for about 2 mm; side-lobes 1.0–1.2 cm broad, triangular, subacute, erect, glabrous and glossy, front margin erect; mid-lobe small, about 0.8–1.0 × 0.8–1.0 cm, cordate to elliptic, mucronate, porrect, tip deflexed, with a dense patch of short hairs in the centre, margin minutely undulating; callus of two short ridges, inflated and shortly hairy in front, tapering quickly behind, reaching about half way to the base of the lip. *Column* 3.2–3.6 cm long, slender, deflexed at the tip, with short hairs and a shallow, elliptic, nectar-producing depression at the base; anther-cap and viscidium elongated into a distinct protruding rostellum; pollinia clavate, deeply cleft, about 1.5 mm long, on a rectangular viscidium without long hair-like processes from the corners. *Capsule* about 2.5 × 1.5 cm, broadly ellipsoidal, pointed, stalked, with a long beak about as long as the capsule.

ILLUSTRATIONS. Photographs 112, 113; figure 26.1.

DISTRIBUTION. N India (Sikkim, Meghalaya), ?Burma, ?Thailand, Taiwan (map 9D); about 1500 m (4920 ft) (300–1000 m (985–3280 ft) in Taiwan).

HABITAT. In tropical valleys, in shade; flowering November-January.

This is an elegant species with very long, slender leaves, and a flower spike which resembles *C. elegans*. The rhachis of the scape is thin and wiry, purple-coloured, pendulous, and carries several bell-shaped, glossy, brown and olive-green flowers. The lip has broad side-lobes and is yellow with reddish mottling.

C. cochleare was described by Lindley (1858) from a specimen, collected by J.D. Hooker in Sikkim, which had almost finished flowering and the ovaries were starting to swell. This rather poor material led Lindley mistakenly to describe the callus as spoon-shaped at the tip, and prevented description of the mid-lobe or flower colour. In 1879 H.G. Reichenbach was given fresh flowers from the nursery of Messrs Low, and was able to give a more complete description indicating that the lip has two short but separate callus ridges. The Reichenbach plant had been collected by Boxall, reputedly from Burma, although this may not be the correct locality, there having been no further collections from that country. A cultivated specimen which was sent to Kew from Bangkok is likewise the only record for Thailand. It differs slightly in the shape of the side-lobe tips which are not as triangular and are more rounded than usual. There are many links between the floras of northern India and Thailand, and some other species show a similar disjunct distribution such as *C. mastersii* and *C. devonianum*, and *C. cochleare* may occur in the wild in Thailand.

C. cochleare is closely allied to *C. elegans*. The distributions of these two species coincide and both species flower in November. However, *C. elegans* is usually found in cooler forest at higher altitudes. The two species can be distinguished by the characters listed in table 10.8.

C. babae (Kudo ex Masamune)Masamune, from Taiwan, appears to be conspecific with *C. cochleare*. But examination of several illustrations and photographs has failed to indicate any characters by which *C. babae* might be distinguished. Lin (1977) gives detailed dissections and good colour photographs of *C. babae*, in which the only obvious distinguishing character is the absence of callus ridges. However, even in *C. cochleare* the callus ridges are very much reduced.

Wu & Chen (1980) reduced *C. babae* to synonymy in *C. elegans* (as *C. longifolium*). However, the mid-lobe of the lip of *C. babae* is elliptical and lacks the very distinctive obcordate shape of *C. elegans*, the leaves are narrower, the scape more slender, the rhachis darker coloured, the flowers less densely spaced, the segments narrower, the petals and sepals darker coloured and more lustrous, the lip darker with many fine red spots, the side-lobes less acute and the callus ridges much less strongly expressed than in *C. elegans*.

Table 10.8: A comparison of the two related species *C. cochleare* and *C. elegans*

Character	*C. elegans*	*C. cochleare*
Leaf width	up to 1.4(2.0) cm	up to 1.0 cm
Number of flowers	about 20–35	usually 7–20
Spacing of flowers	very dense	less dense
Flower colour	cream to yellow-green, with a cream to pale green lip, not or occasionally sparsely red-spotted	greenish-brown, lip yellow to orange-yellow, with numerous red spots
Flower appearance	matt, not shiny. Bell-shaped, full	waxy, lustrous. More slender, spidery
Side-lobe tip shape	porrect, tapering to acute or subacute apices	not porrect, front margins erect, with subacute apices
Mid-lobe shape	ligulate at the base, expanded into two lobes with an emarginate apex; appearing obcordate	cordate to elliptic, mucronate
Callus structure	callus long, tapering from an inflated apex to the base of the lip, usually with auricles near the base forming a trough-shaped depression	callus short, quickly tapering from an inflated apex and terminating about mid-way to the base of the lip

Lin (1977) noted that this species is widespread but not common in Taiwan, and grows ‘in the shaded and humid forests at altitudes from 300–1000 m’. It has also been reported growing as a terrestrial along streams in sand and as a semi-terrestrial in forest (Mark *et al.*, 1986). It is slightly fragrant, which explains its local name of ‘Fragrant Sand Grass.’ It should also be noted that Liu & Su (1978) do not accept this species as native to Taiwan, stating that it is probably an introduced, cultivated plant. This might explain the disjunct distribution of *C. cochleare*, but whether introduced or native, this species does appear to be present now in the wild in Taiwan.

33. C. whiteae *King & Pantling* in Ann. Roy. Bot. Gard. Calcutta 8: 193–4, t.258 (1898); Trudel in Die Orchidee 34(3): 100–2, + figs. (1983). Type: Sikkim, Gantok, *Pantling* 425 (holotype K!).

Cyperorchis whiteae (King & Pantling)Schltr. in Fedde, Repert. 20: 107 (1924), **syn. nov.**

A perennial, epiphytic *herb. Pseudobulbs* up to 10 cm long, 2.5 cm in diameter, narrowly ovoid, lightly bilaterally flattened, with about 12–14 distichous leaves, growing in an indeterminate fashion for about two years before a new growth is produced, the base covered in persistent leaf bases. *Leaves* up to 90

1.3(1.5) cm, narrowly ligulate, arching, tapering to an acute, slightly unequal apex; articulated 3–15 cm from the pseudobulb to a broad, sheathing base with a scarious margin up to 3 mm wide. *Scape* about 20–30 cm long, from near the base of the pseudobulb; peduncle suberect, arching, covered in inflated, cymbiform sheaths up to 14 cm long; rhachis pendulous, with 10-12 flowers; bracts 2–8 mm long, triangular. *Flower* about 3.5 cm in diameter, bell-shaped, suberect to pendulous; rhachis, pedicel and ovary purple; sepals and petals greenish, flushed and spotted with red-brown; lip white with numerous fine maroon or brownish spots, mid-lobe pale yellow; callus white, spotted behind with maroon; column cream, spotted ventrally with maroon; anther-cap white. *Pedicel and ovary* 1.5–2 cm long. *Dorsal sepal* 4.3–4.8 × 1.1–1.2 cm, narrowly oblong-elliptic, acute, concave, porrect, lightly recurved at the tip, covering the column; lateral sepals similar, often mucronate, somewhat spreading. *Petals* 4.1–4.5 × 0.7–0.9 cm, obovate-ligulate, acute, slightly curved, porrect, lightly spreading at the tip. *Lip* 3-lobed, elongated, fused to the base of the column for about 5 mm; side-lobes 1.1 cm broad, triangular, erect and lightly clasping the column, shortly hairy, porrect, acute; mid-lobe 1.2–1.3 × 1.2 cm, cordate, acute, mucronate, deflexed near the apex, with sparse short hairs, margin undulating; callus a long, broad, raised glabrous ridge behind, slightly swollen at the margins and forming two raised ridges which become strongly inflated and confluent at the apex (although not forming the single cuneate apex found in *C. eburneum*), terminating well behind the junction of the mid- and side-lobes, shortly hairy towards the apex. *Column* broad, 3.5 cm long, narrowly winged, sharply decurved near the apex, minutely hairy below; anther-cap elongated with a short rostellum; pollinia quadrangular-pyriform, about 2 mm long, on a rectangular viscidium with short hair-like processes from the lower corners. *Capsule* 3.5 × 1.5 cm, oblong, pointed, with a long beak which is shorter than the capsule.

ILLUSTRATIONS. Photographs 114, 115; figure 26.2.

DISTRIBUTION. Sikkim (map 9F); about 1500-2000 m (4920–6560 ft).

HABITAT. Usually on *Schima wallichii* and occasionally on *Castanopsis* sp. in evergreen forest; flowering October-November.

C. whiteae was described and illustrated by King & Pantling (1898) from a specimen collected in Sikkim, the name being given in honour of Mrs C. White who discovered this species. Pradhan (1979) suggests that it is still found in the wild, albeit rather rarely. Trudel (1983) reports new collections being made in the vicinity of Gangtok.

In a letter from King which is attached to the type specimen, it is noted that *C. whiteae* was originally thought to be a variant of *C. mastersii*. It is, however, very distinct from *C. mastersii*, lacking the strongly indeterminate growth habit and characteristic forked leaf apex of the latter, and the scape is produced from near the base of the pseudobulb. The flowers are bell-shaped and pendulous, and are held on a pendulous rhachis, while those of *C. mastersii* open more fully and are held more erect on a horizontal rhachis. The lip of *C. whiteae* is more strongly hairy, while its callus ridges are broadly separated on a glabrous, raised disc, and converge suddenly at their apices; those of *C. mastersii*, lacking the broad, glabrous disc, are closer together and converge well behind their apices.

The green-brown sepal and petal colour, the numerous small red spots on the whitish lip and the white callus ridges immediately identify *C. whiteae* in the living state. *C. cochleare* has a similar scape with pendulous bell-shaped flowers and a similar sepal and petal colour, but the lip and callus ridges are yellow, not white. It also has narrower leaves, shorter floral bracts, flower segments that are more slender, a lip with glabrous side-lobes, and callus ridges which are much shorter, closer together, and lack the broad, raised, glabrous disc.

The callus of *C. whiteae* is most strongly reminiscent of *C. eburneum*. It is possible that *C. whiteae* arose as a natural hybrid, the most likely parents being *C. cochleare*, and either *C. eburneum* or *C. mastersii*. Table 10.9 gives the characters of *C. whiteae* which could be attributed to these species if a hybrid origin were postulated.

The pollinarium is intermediate in that the pollinia are quadrangular, but are somewhat elongated and tend towards the clavate shape of *C.*

TABLE 10.9: The characters of *C. whiteae* which could be attributed to *C. cochleare* and *C. eburneum*, if a hybrid origin were postulated

C. cochleare	*C. eburneum*
pendulous rhachis	*long (to 8 mm) floral bracts
purple-colour rhachis	*oblong-elliptic sepal shape
scape from near the pseudobulb base	*broad sepals and petals
bell-shaped flowers	*white lip colour
pendulous flowers	long lip and column fusion (5 mm)
green-brown sepal and petal colour	callus ridges not well separated, strongly convergent at their apices
numerous fine red spots on the lip	callus ridges terminate well behind the mid-lobe
mid-lobe with a deflexed tip	callus with a broad, glabrous raised disc towards the base of the lip
anther-cap rostellate	*broad column
long, slender beak on the capsule	*column white
leaf tips slightly unequal	*larger capsule, longer than the beak

*Characters also found in *C. mastersii*

cochleare, and the viscidium has hair-like processes from its corners, but they are rather short.

However, several characters of *C. whiteae* could not be explained by this parentage (or by the substitution of *C. mastersii* for *C. eburneum*), including the very long leaf, the strong indumentum on the side-lobes, the lack of a dense patch of short hairs on the mid-lobe and the distinctive white callus ridges with hairy apices. These characters are significant enough to preclude the suggestion that this parentage could have produced *C. whiteae.* The proposal that *C. erythraeum* could be one of the parents would explain some of these characters, but would leave even more inexplicable characters. Therefore the suggestion of a recent hybrid origin can be discounted.

Although *C. whiteae* has some characters of section *Eburnea*, the following seem sufficient for its inclusion in section *Cyperorchis*: ovoid pseudobulbs; acute, slightly unequal leaf tips; a scape produced from near the base of the pseudobulb; slender, pendulous, many-flowered rhachis; bell-shaped, pendulous flowers; an anther-cap with a beak, somewhat clavate pollinia; and a beak almost as long as the capsule.

C. whiteae is found in evergreen forest, usually growing with other *Cymbidium* and *Bulbophyllum* species. It commonly grows on *Schima wallichii*, but is also found on *Castanopsis* sp. The forest is wet during spring, summer and autumn, but in winter it is very dry and the temperature becomes quite low (Trudel, 1983).

34. C. sigmoideum *J.J. Smith* in Bull. Agric. Indes Neerl. 13; 52-3 (1907); Comber & Sutikno in Orchid Dig. 44: 164-8, + fig. (1980). Type: Java, Loemadjang, *Connell* s.n. (holotype BO!).

Cyperorchis sigmoidea (J.J. Smith) J.J. Smith in Bull. Jard. Bot. Buitenzorg ser. 3, 9: 57, t.8 (1927); Backer, Beknopte Fl. Java 12: 346-7 (1952); Backer & Bakhuizen, Fl. Java 3: 394 (1968).

A perennial, epiphytic *herb. Pseudobulbs* about 6 cm long, 2 cm in diameter, not inflated, stem-like and inconspicuous within the leaf bases, produced annually, with about 9–12 leaves and 3 cataphylls which eventually become fibrous. *Leaves* up to 90 × 1.4–1.6(2.5) cm, linear-obovate, acute, articulated 5-8 cm from the pseudobulb. *Scape* about 25 cm long, from the base of the pseudobulb; peduncle horizontal to pendulous, covered in about 10 overlapping, cymbiform, acute sheaths up to 9 cm long; rhachis short, with 4–8(14) flowers; bracts 2–3 mm long, triangular, acute. *Flower* about 3.5 cm across, waxy, inconspicuous, pendulous to horizontal; rhachis, pedicel and ovary purple; sepals and petals deep apple-green, spotted and stained with dark or purple-brown; callus light green; column pale green, cream below; anther-cap cream. *Pedicel and ovary* 1.9–2.3 cm long. *Dorsal sepal* about 2.7–2.9 × 0.8–1.1 cm, narrowly obovate or lanceolate, acute, concave, porrect, closely covering the column; lateral sepals similar, falcate, spreading or reflexed. *Petals* about 2.4–2.5 × 0.35–0.45 cm, ligulate, acute, curved, sometimes spreading in the apical half. *Lip* short, 3-lobed, with a 5–6 mm long fusion between the lip and the base of the column; side-lobes about 6 mm broad, broadly triangular-falcate, erect and clasping the column, minutely papillose, waxy, front margin S-shaped, subacute; mid-lobe small, about 0.65–0.7 × 0.2–0.25 cm, ligulate, glabrous, waxy, strongly recurved, acute; callus short, a broad glabrous disc with 2-3 slightly raised ridges behind, with a swollen, bilobed, rounded apex. *Column* broad, 2.8–2.9 cm long, S-shaped, glabrous, the basal quarter fused to the base of the lip; anther-cap and viscidium elongated into a short, projecting rostellum; pollinia quadrangular-pyriform, about 2 mm long, cleft, on a narrow rectangular viscidium without hair-like processes from the corners.

ILLUSTRATIONS. Photographs 116, 117; figure 26.4.

DISTRIBUTION. Java, Sumatra (map 9E); 800–1600 m (2625-5250 ft).

HABITAT. On trunks and larger branches of trees in montane rainforests, often in deep shade; flowering sporadically almost throughout the year.

W. Micholitz discovered *C. sigmoideum* in central Sumatra, and sent specimens to Messrs Sander, who later, in 1905, sent one to Kew for identification, but Rolfe failed to recognise it as a new species. However, J.J. Smith described it shortly afterwards based on a Connell collection from Java.

C. sigmoideum is an epiphyte growing in the cooler montane rainforests of Java and Sumatra, often on the trunks of trees in deep shade. It is similar in growth habit to *C. roseum*, which is its closest ally on these islands. The flower is very distinctive although it does bear some resemblance to *C. whiteae* from Sikkim. It has many distinguishing characters, some of them unique in the genus. The most distinctive of these are the shiny brown-spotted, green sepals, petals and lip, the narrow, curved petals, the short lip with a relatively long fusion of the margins of the lip and column-base, broad, falcate side-lobes, a narrow, ligulate, waxy, recurved mid-lobe, the bilobed, rounded callus apex, the broad, S-shaped column and the short, backwards-projecting rostellum.

The fusion at the base of the lip and column, and the two cleft pollinia, place this species in the subgenus *Cyperorchis*. The annually produced pseudobulbs, the scape from the base of the pseudobulb, the pendulous rhachis, the narrow petals, the porrect dorsal sepal covering the column, the rostellum, the rectangular viscidium without hair-like processes from the lower corners and the quadrangular-pyriform pollinia support its inclusion in section *Cyperorchis*.

Section **Parishiella** *(Schltr.) P. Hunt* in Kew Bull. 24: 94 (1970); Seth & Cribb in Arditti (ed.), Orchid Biol., Rev. Persp. 3: 302 (1984). Type: *C. tigrinum* Parish ex Hook.

Cyperorchis section *Parishiella* Schltr. in Fedde, Repert. 20: 108 (1924).

This section contains a single, highly distinctive species. It was established by Schlechter (1924) in the genus *Cyperorchis*, and was later transferred to *Cymbidium* by P. Hunt (1970). It is very distinct vegetatively, having lens-shaped pseudobulbs which are not covered by leaf bases, and it has only 2-4 apical leaves, usually less than 17 cm long. A new pseudobulb is produced annually, and the 2–5 flowered scape is basal. The flower shape is also unusual, with a slender, spidery appearance, porrect petals, and a lip with a rectangular, cuspidate mid-lobe and highly unusual horizontal markings. It is probably most closely related to section *Iridorchis*, which the pollinarium shape closely resembles.

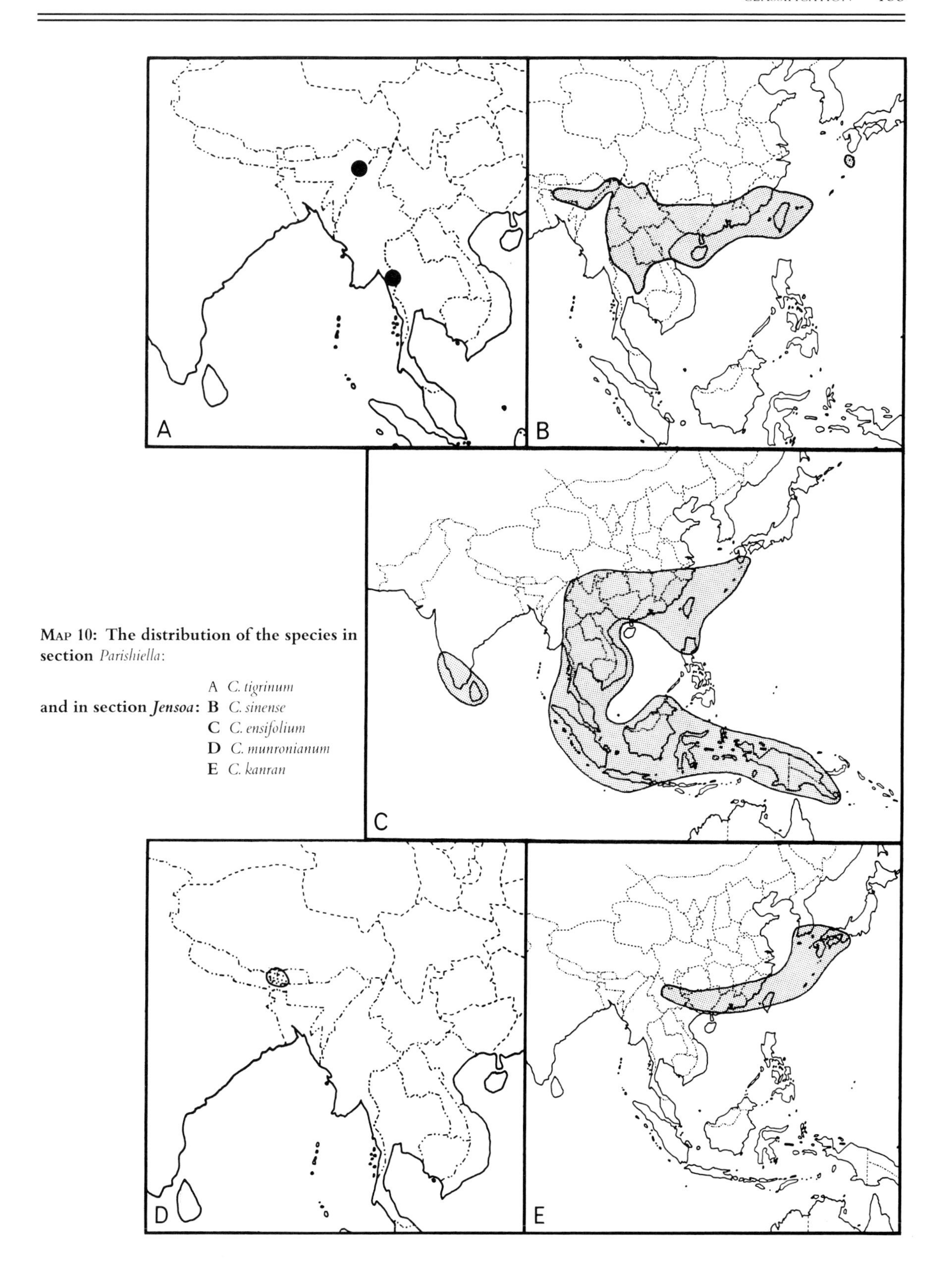

Map 10: The distribution of the species in section *Parishiella*:

A *C. tigrinum*

and in section *Jensoa*:
B *C. sinense*
C *C. ensifolium*
D *C. munronianum*
E *C. kanran*

35. C. tigrinum *Parish ex Hook.* in Bot. Mag. 90: t.5457 (1864); Parish in Mason (ed.), Burma, its people and productions 2: 171 & 179 (1883); Hook.f., Fl. Brit. India 6: 9–10 (1890); Berkeley in Orchid Rev. 3: 67 (1895); Grant, Orchids of Burma: 227 (1895); Gagnepain in Lecompte (ed.), Fl. Gen. Indo-Chine 6: 415 (1934); Ghose in Orchid Rev. 80: 187 (1972); Krishna & Shastry in Bull. Bot. Surv. India 14: 179, f.1–4 (1975); Hynniewata in Orchid Rev. 87: 219 (1979); Seidenfaden in Opera Bot. 72: 90–1, f.51 (1983); Seth & Cribb in Arditti (ed.), Orchid Biol., Rev. Persp. 3: 302 (1984). Type: Burma, Moulmein, Mulayit, *Parish* 144 (holotype K!).

Cyperorchis tigrina (Parish ex Hook.)Schltr. in Fedde, Repert. 20: 108 (1924).

A perennial lithophytic *herb. Pseudobulb* up to 3(6) × 3(3.5) cm, subrotund to broadly ovoid, strongly bilaterally compressed, lens-shaped, with (1)2–6 distichous leaves and 4–5 cataphylls; the lower leaves and cataphylls are deciduous, and the persistent leaf bases become scarious and eventually fibrous with age; mature pseudobulbs wrinkled, usually with 2–4 apical leaves, not covered by the leaf bases, with nodes towards the base and the apex only. *Leaves* up to 17(22) × 3.3 cm, narrowly elliptic, slightly twisted, articulated 1.0–1.5 cm from the pseudobulb to a broadly sheathing base with a narrow (1 mm) membranous margin. *Scape* 12–23 cm long, from the base of the pseudobulb; peduncle short, slender, suberect to horizontal, with 5–9 sheaths up to 3.7 cm long, the upper sheaths distant, spreading, cymbiform, with a short, cylindrical base; rhachis slender, usually longer than the peduncle, with 2–5 large, distant flowers; bracts triangular, acute, (2)4–9(17) mm long. *Flower* 4–5 cm across, of a spidery appearance; honey-scented; rhachis, pedicel and ovary green; sepals and petals olive-green to mustard, shaded red-brown and spotted towards the base with purple-brown; lip white, turning pink on pollination, with almost entirely purple-brown side-lobes, and spots and transverse dashes of red-purple on the mid-lobe; callus ridges white, spotted purple; column olive-green to mustard and tipped with purple-brown above, white, streaked with purple below; anther-cap cream. *Pedicel and ovary* 1.7–4.3 cm long. *Dorsal sepal* 3.7–4.2 × 0.8–1.3 cm, narrowly obovate with recurved margins, obtuse or subacute, slightly mucronate, suberect, somewhat curved over the column; lateral sepals similar, narrowly elliptic, slightly downcurved, not fully spreading. *Petals* 3.4–4.0 × 0.5–1.0 cm, narrowly elliptic, acute, often covering the column. *Lip* 3-lobed, with a relatively long mid-lobe, fused to the base of the column for about 3 mm; side-lobes 1.0–1.3 cm broad, rounded to subacute, tapering sharply towards the base, erect but not clasping the column, minutely papillose; mid-lobe 1.2–1.5 × 1.1–1.5 cm, oblong, minutely papillose, recurved, mucronate, margin undulate; callus of 2 well-defined, almost parallel, glabrous, inflated ridges extending from the base of the lip to the base of the mid-lobe. *Column* 2.5–3.0 cm long, arcuate, minutely papillose, broadly winged towards the apex; pollinia triangular, about 2 mm long; viscidium broadly triangular, tapering to 2 hair-like processes at the lower corners.

ILLUSTRATIONS. Plate 23; photographs 118,119; figure 26.6.

DISTRIBUTION. Burma, N E India (Nagaland) (map 10A); (1000)1500–2700 m ((3280)4920–8860 ft).

HABITAT. On bare rocks and in rock crevices in open situations; flowering March–July.

C. tigrinum was described by W.J. Hooker, in 1864, from specimens collected in Burma by the Rev. Charles Parish, who sent herbarium material and a watercolour sketch along with living plants to the nursery of Messrs Low. The plants were collected on the summit of Mulayit (about 2000 m, 6560 ft) near Moulmein on the Thai–Burma border. This species has since been collected several times on this and some surrounding mountains in the Dawna Range. Ghose (1972) and Hynniewata (1979) also reported it from Nagaland, on the border of N E India and N Burma. The former states that it grows on open rocks and in rock crevices, and is often subjected to frosts in winter. This is another example of a similar disjunction in the known distribution of several other *Cymbidium* species (see also *C. devonianum* and *C. mastersii*).

Parish (1883) described an interesting floral dimorphism in the flowers. He reported that the lower flowers on the spike were of a 'rich red colour', and that the column had an unusual structure. He wrote: 'The column is quite abnormal, being unusually thickened and less curved. There is no anther at all, and there are no pollen masses; but the edges of the column at the top are turned inwards so as to form a sort of hood, and underneath those edges is a small quantity of a waxy yellow substance (pollen) in an amorphous state. And, occasionally, the intermediate flowers are intermediate also in condition, having no anther, but perfect pollen masses, though without any triangular gland.' The type specimen contains a scape which is annotated by Parish as having these dimorphous flowers. An examination of these flowers, and of the observations quoted above, both suggest that the lower flowers had simply been pollinated, rather than that they were of a different structure. It is quite normal in this genus for the flower to 'blush' pink after pollination, and for the column tip to curve inwards to close the stigmatic cavity once the pollinia have been deposited. The pollinia then start to disintegrate and lose their original structure. The 'intermediate' flowers are also interesting, as it seems as though the anther-cap was dislodged and the viscidium removed by the pollinator, but the pollinia were left intact on the end of the column. It may be that these flowers were visited in an immature state, resulting in only partial removal of the pollinarium. As the upper flowers were intact, it appears that the lower flowers mature earliest (see also Seidenfaden, 1983).

C. tigrinum is very different from other *Cymbidium* species, especially vegetatively, in which respect it is more reminiscent of a species of *Coelogyne.* Indeed, Hooker (1864) was not certain that this species should be included in *Cymbidium.* However, the large flowers, the fusion of the base of the lip and the column and the two cleft pollinia place this species in the subgenus *Cyperorchis.* The relatively broad leaves, the basal flower spike, the short lip, the two distinct callus ridges and the triangular pollinia on a broad triangular viscidium suggest an affinity with section *Iridorchis,* but *C. tigrinum* has several unusual characters which suggest that it should be placed in a separate monotypic section. These characters include the small size of the plant (less than 15 cm (6 in) tall), its lithophytic habit, its lens-shaped pseudobulbs which are exposed and not covered by leaf bases, its few (2–4) short leaves less than 17(22) cm (6.7–8.7 in) long which are articulated very close to the pseudobulb, its peduncle with distant, spreading sheaths, its spreading sepals but porrect petals, its lip with almost entirely purple-brown side-lobes, its mid-lobe marked with horizontal rather than vertical markings and with a cuspidate apex, and its well-defined, parallel, glabrous callus ridges.

Subgenus Jensoa

Cymbidium *subgenus* **Jensoa** *(Raf.)Seth & Cribb* in Arditti (ed.), Orchid Biol., Rev. Persp. 3: 283–322 (1984); Seidenfaden in Opera Bot. 72: 66 (1983). Type: *Jensoa ensata* (Thunb.) Raf. (= *C ensifolium* (L.) Sw.)

Jensoa Raf., Fl. Tellur. 4: 38 (1836).

Jensoa, originally described by Rafinesque-Schmaltz, was established as a section of *Cymbidium* by Schlechter (1924), but was also given a broader definition and subgeneric status by Seth & Cribb (1984). This terrestrial subgenus was then characterised by the presence of four pollinia in two unequal pairs (figure 14F), and its distinctive callus structure where the two callus ridges converge towards the apex, forming a short tube at the base of the mid-lobe of the lip. A recently described species, *C. borneense,* also has four pollinia, but lacks the characteristic callus structure of subgenus *Jensoa,* and in its leaf surface morphology it closely resembles the species in section *Cymbidium,* alongside which it is currently placed.

In addition to the above, many other characters have been shown to be characteristic of this subgenus. The micromorphology of the abaxial leaf surface is highly characteristic; the epidermal cells are papillose, the stomatal covers project beyond the surface of the epidermis, and the apertures in these covers are almost circular (figures 5C,9). The long, thread-like seed shape is also characteristic, and the outer walls of the testa cells have transverse rather than longitudinal secondary striations (figure 3). The pods are erect rather than pendulous. The seedling is rhizomatous rather than forming a spherical protocorm. This data indicates that this is the most distinct and distantly related subgenus. However, hybridisation with other cymbidiums has shown that it is closely related, and the most satisfactory classification of this taxon is at the subgeneric level. Until leaf and seed micromorphological characters especially are more widely surveyed in the Cyrtopodiinae it is not possible to state whether or not these characters are sufficiently distinct to merit the reconsideration of the taxonomic status of subgenus *Jensoa.*

Section **Jensoa** *(Raf.)Schltr.* in Fedde, Repert. 20: 102 (1924); P. Hunt in Kew Bull. 24: 94 (1970); Seth & Cribb in Arditti (ed.), Orchid Biol., Rev. Persp. 3: 290 (1984). Type: *C. ensifolium* (L.)Sw.

Jensoa Raf., Fl. Tellur. 4: 38 (1838).

Section *Jensoa* is closely related to section *Maxillarianthe.* When Schlechter (1924) established these sections he distinguished the latter by its single-flowered scape. Further collections have shown that this character is very variable in some specimens of *C. goeringii.* Furthermore, there are strong vegetative similarities between *C. faberi* and *C. goeringii,* which suggest that they should be maintained in the same section. The leaf number can be used to divide section *Jensoa* from section *Maxillarianthe,* the former usually having up to four leaves on each shoot, while the latter has about 6–13 leaves. This division is reiterated by the presence of fibre bundles below the epidermis on both sides of the leaf in section *Jensoa,* and their absence, or sparse presence, below only one of the surfaces in section *Maxillarianthe.* Both *C. faberi* and *C. cyperifolium* are transferred to section *Maxillarianthe* on the basis of these characters.

36. C. ensifolium *(L.)Sw.* in Nov. Act. Soc. Sci. Upsal. 6: 77 (1799); Lindley, Gen. Sp. Orchid. Pl.: 162 (1833); Reichb.f. in Walp. Ann. 6: 622 (1863); Schltr. in Fedde, Repert. Beih. 4: 266 (1919); Maekawa, Wild Orchids of Japan in Colour: t.174–5 (1971); Garay & Sweet, Orchids S. Ryukyu Islands.: 141–8 (1974); Lin, Native Orchids Taiwan 2: 105–8 (1977); Wu & Chen in Acta Phytotax. Sin. 18(3): 296–7, t.1 (1980); Seth & Cribb, in Arditti (ed.), Orchid Biol., Rev. Persp. 3: 291–2 (1984). Type: China, Canton, *Osbeck* (holotype LINN!).

Figure 27

1. ***C. ensifolium*** subsp. ***haematodes*** (Kew spirit no. 47263/47265/48347)
- a Perianth, × 1
- b Lip and column, × 1
- c Pollinarium, × 3
- d Pollinia (one pair, reverse), × 3
- e Pollinium, × 3
- f Bract, × 1

2. ***C. ensifolium*** subsp. ***ensifolium*** (Kew spirit no. 47259)
- a Perianth, × 1
- b Lip and column, × 1
- c Pollinarium, × 3
- d Pollinia (one pair, reverse), × 3
- e Pollinium, × 3
- f Bract. × 1

3. ***C. cyperifolium*** subsp. ***cyperifolium*** (Kew spirit no. 6192)
- a Perianth, × 1
- b Lip and column, × 1
- c Pollinarium, × 3
- d Pollinia (one pair, reverse), × 3
- e Pollinium, × 3
- f Bract, × 1
- g Flower, × 0.8

4. ***C. sinense*** (Kew spirit no. 47253/47931)
- a Perianth, × 1
- b Lip and column, × 1
- c Pollinarium, × 3
- d Pollinia (one pair, reverse), × 3
- e Pollinium, × 3
- f Bract, × 1

5. ***C. cyperifolium*** subsp. ***indochinense*** (Thailand, *Put* 3972)
- a Perianth, × 1
- b Lip and column, × 1
- c Pollinarium, × 3
- d Pollinia (one pair, reverse), × 3
- e Pollinium, × 3
- f Bract, × 1

6. ***C. munronianum*** (Kew spirit no. 36360)
- a Perianth, × 1
- b Lip and column, × 1
- c Pollinarium, × 3
- d Pollinia (one pair, reverse), × 3
- e Pollinium, × 3
- f Bract, × 1

7. ***C. kanran*** (Kew spirit no. 40631)
- a Perianth, × 1
- b Lip and column, × 1
- c Pollinarium, × 3
- d Pollinia (one pair, reverse), × 3
- e Pollinium, × 3
- f Bract, × 1
- g Flower, × 0.8

1c
1d
1e
2b
1b
1a
2a
2f
2d
2e
1f
3e
2c
4b
3d
4c
4e
4a
3c
3f
3g
5
3a
4d
3b
4f
7b
5a
5d
5b
7d
7g
5e
5c
7e
6f
6c
6d
7f
7a
6e
7c
6a
6b

Epidendrum ensifolium L., Sp. Pl. 2: 954 (1753).

Limodorum ensatum Thunb., Fl. Japan: 29(1784) & Icon. Fl. Japan 1: 28(1974). Type: as for *C. ensifolium* (L.)Sw.

Epidendrum sinense Redouté, Liliac. 2: t.113 (1805). Type: not designated, **syn. nov.**

C. xiphiifolium Lindley in Gen. Sp. Orchid. Pl. 7: t.529 (1821). Type: China, *Hume* (holotype K!).

C. ensifolium (L.)Sw. var. *estriatum* Lindley in Bot. Reg. 13: t.1976 (1837). Type: cult. Hort. Soc. London (holotype K!).

C. ensifolium (L.)Sw. var. *striatum* Lindley in Bot. Reg. 23: t.1976 (1837). Type: China or Japan, *Fothergill* (holotype K!).

Jensoa ensata (Thunb.)Rafin., Fl. Tellur. 4: 38 (1838). Type: as for *C. ensifolium* (L.)Sw.

C. ecristatum Steud. Nomencl., edn 2, 1: 460 (1840), *nom nud.*, **syn. nov.**

C. micans Schauer in Nov. Act. Nat. Cur. 19 (suppl. 1): 433 (1843). Type: China, Macao (holotype K!).

C. albo-marginatum Mak. in Iinuma, Somoku-Dzusetsu, edn 3, 4(18): 1183 (1912). Type: not cited, but probably in MAK, **syn. nov.**

C. gyokuchin Mak. in Iinuma, Somoku-Dzusetsu, edn 3, 4(18): 1181, t.7 (1912). Type: not cited, probably in MAK, but see the above reference for an illustration of the type.

C. gyokuchin var. *soshin* Mak. in Iinuma, Somoku-Dzusetsu, edn 3, 4(18): 1182, t.8 (1912). Type: not cited, probably in MAK, but see the above reference for an illustration of the type, **syn. nov.**

C. koran Mak. in Iinuma, Somoku-Dzusetsu, edn 3, 4(18): 1179, t.3 (1912). Type: not cited, probably in MAK, but see the above reference for an illustration of the type, **syn. nov.**

C. niveo-marginatum Mak. in Iinuma, Somoku-Dzusetsu, edn 3, 4(18): 1183 (1912). Type: not cited, probably in MAK, **syn. nov.**

C. shimaran Mak. in Iinuma, Somoku-Dzusetsu, edn 3, 4(18): 1183, t.10 (1912). Type: not cited, probably in MAK, but see the above reference for an illustration of the type.

C. yakibaran Mak. in Iinuma, Somoku-Dzusetsu, edn 3, 4(18): 1182, t.9 (1912). Type: not cited, probably in MAK, but see the above reference for an illustration of the type.

C. arrogans Hay. in Icon. Pl. Formos. 4: 76 (1914) & in *op. cit.* 6: 79, t.12 (1916); Mark, Ho & Fowlie in Orchid Dig. 50: 31 (1986). Type: Taiwan, Kusukusu, *Hayata & Sasaki* (holotype TI).

C. misericors Hay. in Icon. Pl. Formos. 4: 79, t.386 (1914); Mark, Ho & Fowlie in Orchid Dig. 50: 32 (1986). Type: Taiwan, Mt. Kwannonzan, near Tamsui; cult. in seminario Taihoku, *Hayata & Soma* (holotype TI).

C. rubrigemmum Hay. in Icon. Pl. Formos. 6: 81, t.15 (1916); Mark, Ho & Fowlie in Orchard Dig. 50: 31 (1986). Type: Taiwan, cult. in seminario Taihoku, *Soma* (holotype TI).

C. gonzalesii Quisumbing in Philippine J. Sci. 72: 485 (1940). Type: Philippines, Luzon, *Quisumbing* 5783E (holotype PNH), **syn. nov.**

C. ensifolium (L.)Sw. var. *misericors* (Hay.) T.P. Lin, Nat. Orchids Taiwan 2: 105, t.36–39 (1977); Liu & Su in Fl. Taiwan 5: 942 (1978).

C. gyokuchin Mak. var. *arrogans* (Hay.)Ying in Col. Ill. Indig. Orchids Taiwan 1: 126, t.43 (1977), **syn. nov.**

C. kanran Mak. var. *misericors* (Hay.)Ying, Col. Ill. Indig. Orchids Taiwan 1: 440 (1977).

C. ensifolium(L.)Sw. var. *rubrigemmum*(Hay.)Liu & Su in Fl. Taiwan 5: 940–1 (1978).

C. ensifolium (L.)Sw. var. *yakibaran* (Mak.)Wu & Chen in Acta Phytotax. Sin. 18(3): 296 (1980).

C. ensifolium (L.)Sw. var. *susin* Yen, Icon. Cymbid. Amoyens. D.b. I (1964), Wu & Chen in Acta Phytotax. Sin. 18: 297 (1980). Type: China, Fukien, An-shee Hsien, *Yen* 2039 (holotype not located).

A short to medium-sized, perennial, terrestrial *herb*. *Pseudobulbs* small, to 3 × 1.5 cm, ovoid, often inconspicuous, occasionally subterranean, small, covered in leaf bases and scarious cataphylls which become fibrous with age, both with very narrow membranous margins about 1 mm broad. *Leaves* 2–4(5), up to 29–94 × 0.8–2.5 cm, distichous, not merging with the cataphylls, erect, weakly to strongly arching, linear-elliptic, acute, the margin sometimes serrulate towards the apex, articulated 1.5–5.0 cm from the pseudobulb. *Scape* 15–67 cm long, slender to medium, produced from the base of the pseudobulb inside the cataphylls, with 3–9 flowers in the apical third; peduncle covered in cymbiform sheaths up to

5–6.5 cm long, the uppermost sheath often distant from the others; bracts 0.4–2.2(2.9) cm long, triangular or linear-ovate. *Flowers* 3–5 cm across; often strongly scented; rhachis, bracts, pedicel and ovary greenish, often stained red-brown; petals and sepals straw-yellow to green with 5–7 more or less obvious longitudinal red or red-brown veins, the petals often with a stronger central stripe and red-brown spots and blotches towards the base; lip pale yellow or green, occasionally white, side-lobes streaked red, with a red margin, mid-lobe with red blotches or transverse spots; column pale yellow, with red dashes beneath; anther-cap cream; occasionally the red-brown pigment is absent from the flowers, which are then pale green and white in colour. *Pedicel and ovary* 1–3.7 cm long. *Dorsal sepal* (16)19–31 × 4.6–8.8(9.8) mm, narrowly elliptic, acute to oblong, obtuse or mucronate, erect; lateral sepals similar, spreading, horizontal or slightly drooping. *Petals* 14–26 × 5.5–9.0 mm, ovate-elliptic, subacute, shorter and either almost equal in breadth, or broader than the sepals, sightly porrect, but not closely covering the column. *Lip* 1.4–2.2 cm long when flattened; side-lobes rounded, minutely papillose; mid-lobe 6–12 × (5)6–10 mm, ovate to triangular, mucronate or acute, minutely papillose, margin kinked to undulate, entire; callus of 2 ridges, converging in the apical half to form a short tube at the base of the mid-lobe. *Column* 1–1.5(1.8) cm long, arching, narrowly winged; pollinia 4, broadly ovate, in two unequal pairs, on a semi-circular viscidium. *Capsule* about 6 × 2 cm, fusiform, strongly ridged, tapering at each end, with a 1–1.5 cm long beak, held erect and parallel to the rhachis.

DISTRIBUTION. Sri Lanka, S India, China, Hong Kong, Taiwan, Ryukyus (S Japan), Indo-China, Thailand, W Malaysia, Sumatra, Java, Borneo, New Guinea, Philippines (Luzon) (map 10C); 300–1800 m (985–5905 ft).

HABITAT. Terrestrial in lightly shaded, broad-leaved forest, often in damp situations; flowering January–April, but in equatorial regions flowering is sporadic throughout the year.

C. ensifolium was originally described by Linnaeus (1753) as *Epidendrum ensifolium*, based on a specimen bought in Canton by Osbeck, who noted that it was cultivated in houses in China for its scent, as it is to the present day. Swartz (1799) transferred it to the genus *Cymbidium*. Thunberg (1784), in *Flora Japonica*, named this species *Limodorum ensatum*, but later used *Epidendrum* in the accompanying book of illustrations (Thunberg, 1794). This species is probably not native to Japan, but it is widely cultivated there. Rafinesque transferred Thunberg's species to the genus *Jensoa* in 1838, and this publication is the source of the sectional and subgeneric name *Jensoa*.

C. ensifolium has been in cultivation in Japan and China for over two thousand years, and is prized for its shape and the perfume of its flowers. The variants which lack red pigment in the flowers are highly valued, and have been described several times under different names including *susin*, *soshin*, *gyokuchin* and *misericors*, and popular names such as Goddess of Mercy and Iron bone/Iron stick and the Japanese name 'Kwannon'. Of these only var. *susin* (Yen, 1964; Wu & Chen, 1980) has been published as a variety of *C. ensifolium*, whilst the others have been published as distinct species or varieties of species now synonymised under subsp. *ensifolium*. *C. xiphiifolium* is also a white- and green-flowered variant, which Lindley (1821) described from a specimen collected in Hong Kong. The name refers to the leaves which appear rather stiff and are reminiscent of *Iris xiphium*. Several specimens from Hong Kong have been examined, and they are similar to those from mainland China, all being referable to subsp. *ensifolium*. These variants occur rarely in the wild, without forming distinct populations, and have been selectively maintained in cultivation. This type of mutation also occurs sporadically in many other *Cymbidium* species.

Apart from this type of colour variation, there is also some variation in the amount of red-brown striation of the petals and sepals, in the ground colour of the petals and sepals, and in the amount of red pigment on the lip. Lindley recognized this variation, and published var. *estriatum* in 1837 to include the specimens with weak striations on the sepals and petals, the striations being distinct only at the base of the segments instead of extending to near the apex. Variants with all degrees of striation occur, and it seems undesirable to recognise these as distinct taxa.

Several Makino (1912) names are used in Japan to distinguish cultivated variants, but are not applied to wild plants. The form of the plant is important horticulturally in Japan, and although many of the names are accompanied by a line drawing, the flowers and scape are often omitted and the form of the plant is highly stylised. Reference to other works which use these names illustrates how some of them have been applied. Following the treatment of other authors (Garay & Sweet, 1974; Wu & Chen, 1980; Seidenfaden, 1983; Seth & Cribb, 1984), the names published by Makino (1912) have been placed under the synonymy of *C. ensifolium*. Wu & Chen (1980) recognised var. *yakibaran* as distinct, but so little information is available about this taxon that it is at present included in the synonymy.

Hayata (1914, 1916) distinguished three species from Taiwan. *C. arrogans* was reported to differ from the specimens of *C. ensifolium* from mainland China in its falcate, semi-oblong lateral sepals. The sepal shape is variable, and the lateral sepals tend to droop in *C. ensifolium*. Consequently, the lateral sepals are usually oblique and tend to curve, so neither of these characters excludes this variant from *C. ensifolium*. *C. misericors* Hayata (1914) was described from a green- and white-flowered specimen collected in Taiwan. This is another of the horticulturally desirable variants, lacking the red and brown pigmentation normally present in the flower, which occur sporadically in wild populations. Finally, Hayata described *C. rubrigemmum* in 1916. T.P. Lin (1977) considered it to be conspecific with *C. misericors*, but with red pigmentation in the flowers, and distinguished these Taiwanese plants from *C. ensifolium* in mainland China by their slender, arcuate leaves and their reddish-coloured cataphylls of the former. These all fall within the range of variation of *C. ensifolium* in China and the Taiwanese taxa are therefore included in *C. ensifolium*, although further study and comparison of specimens from China and Taiwan may indicate a more suitable treatment at the infraspecific level. The photographs of *C. ensifolium* and these three taxa by Mark *et al.* (1986) illustrate the variation in the strength of red and brown colouring in the flowers of *C. ensifolium* in Taiwan, but otherwise the flowers and plants are remarkably similar.

C. ensifolium also occurs on Luzon in the Philippines, where it was named *C. gonzalesii* by Quisumbing (1940). He suggested that this was allied to *C. faberi* but did not compare it with *C. ensifolium*, which the description and plate given very closely resemble. The petals are narrower than the sepals, and the margin of the lip flat, allowing its inclusion in the type subspecies of *C. ensifolium*.

Two further names in this complex of synonyms are *C. micans* Schauer (1843) from Macao and *C. ecristatum* Steud. Both agree well with subsp. *ensifolium*.

Cultivated specimens from Japan and China include variants with variegated leaves. The names 'setsugetsu', 'hohrai' and 'gyokuryu' do not appear to have been published, but they are commonly used with reference to these variants in Japan (Nagano & Nagano, 1955).

A specimen from Hong Kong, known there as 'Golden Line', and collected from the wild (*Fowlie* s.n.) which has very erect leaves and pure white flowers with a single maroon stripe up the centre of each tepal, and a white lip spotted with red, is a striking departure from the rather drab colours normally found in *C. ensifolium*. This population is possibly the result of hybridisation, perhaps with *C. lancifolium* (or *C. dayanum*) as the other parent, donating the clear colours to the flower.

C. ensifolium has only three to four, occasionally five, leaves on each pseudobulb, and the leaves are very distinct from the cataphylls which surround the base of the plant, a habit shared by *C. sinense*, *C. munronianum* and *C. kanran*. *C. faberi* and *C. cyperifolium* are easily distinguished by their more numerous leaves. *C. kanran* differs from *C. ensifolium* in having longer leaves and longer, more slender, acuminate sepals, and petals which are about twice as broad as the sepals. *C. sinense* has broader leaves, dark green in colour and glossy on the upper surface, more numerous flowers on the scape, strongly porrect petals forming a hood over the column and a larger mid-lobe to the lip. *C. munronianum*, which replaces *C. ensifolium* in the Himalaya of northern India, has a robust habit and broad leaves, similar to *C. sinense*, but has a smaller flower with a smaller mid-lobe to the lip than *C. ensifolium*.

In this study, two subspecies of *C. ensifolium* are recognised but further research, especially of living material, will be necessary to clarify the taxonomy of this variable species. The characters which distinguish these subspecies are listed in table 10.10.

Key to the subspecies of *C. ensifolium*

Leaves arching, usually less than 50 cm long and 0.8–1.5 cm broad, usually with an entire margin; petals and sepals almost equal in breadth; mid-lobe of the lip ovate, with a lightly kinked margin, blotched with red .... *subsp.* **ensifolium**

Leaves almost erect, usually longer than 50 cm, and 1.5–1.9(2.5) cm broad, usually with a serrated margin: petals broader than the sepals; mid-lobe of the lip triangular to narrowly elliptic, with an undulating margin, finely spotted with red .... *subsp.* **haematodes**

subsp. **ensifolium**

Subsp. *ensifolium* is a small, clump-forming plant with short (to 55 cm), narrow (0.8–1.6 cm broad, but usually up to 1.1 cm), and arching leaves, often without serrations on the margin. The flower spike is usually held clear of the foliage, and has up to 9 flowers. The flowers are 3.5–4 cm across, with drooping lateral sepals. The sepals and petals are straw-yellow to light brown in colour, with a strong central stripe of red-brown, and several weaker stripes often only distinct towards the base. The petals are as broad as the sepals, or slightly narrower. The lip has an ovate or slightly elongate-ovate mid-lobe which is obtuse or mucronate at the apex, with a few kinks in the margin, but not strong undulations.

ILLUSTRATIONS. Photographs 4, 120, 121; figure 27.2.

DISTRIBUTION. China, Taiwan, Ryukyu Islands, Philippines (Luzon).

Subsp. *ensifolium* is widely distributed throughout southern China. Wu & Chen (1980) indicate that it is found as far west as Yunnan and Sichuan, and as far north as Henan (see map 10C). It is also found in Hong Kong and Taiwan, but has not been collected on Hainan. Despite the large number of names and descriptions published from Japanese material, it seems unlikely that this species is native to Japan. Garay & Sweet (1974) give the most northern locality as Okinawa, where it is uncommon. Nagano & Nagano (1955) note that specimens are found in the wild in Kyushu, the southernmost of the Japanese islands, but these may be naturalised seedlings which have escaped from cultivation.

In China, this subspecies is commonly found between 500 and 1000 m (1640 and 3280 ft) altitude but it probably also occurs at higher altitudes. It flowers in autumn and early winter, and is most commonly found in flower during August. Little is known concerning its habitat on mainland China. In Hong Kong, *C. ensifolium* is found from 250 m (820 ft) to the highest peaks in the territory at about 1000 m (3280 ft), growing in partial shade near streams or water seepage on sloping land, in humus-rich soil. Consequently it is most common in ravines, in broad-leaved or bamboo forest. Occasionally, specimens are found in grassland or steep, exposed mountain slopes where the forest cover has retreated or has been burnt, although these are usually rather weak plants (G. Barretto, pers. comm.). *C. ensifolium* is uncommon in Taiwan, where it grows in hardwood forest in dry conditions, at altitudes of 300–3000 m (985–9850 ft), flowering mainly between July and October, but also sporadically throughout the rest of the year (Lin 1977, Liu & Su 1978).

subsp. **haematodes** *(Lindley) Du Puy & Cribb*, **comb. et stat. nov.** Type: Sri Lanka, *Macrae* 12 (holotype K!).

C. ensifolium (L.)Sw. var. *haematodes* (Lindley)Trimen, Cat. 89 (1885) & Handb. Fl. Ceylon 4: 180, t.90 (1898); Jayaweera in Dassanayaka, Fl. Ceylon 2: 186 (1981).

C. ensifolium sensu J.J. Smith, Orchid. Java: 478 (1905).

C. sundaicum Schltr. in Fedde, Repert., Beih. 4: 266 (1919). Type: Java, *J.J. Smith* (cited in J.J. Smith, Orchid. Java: 478, 1905).

C. sundaicum Schltr. var. *estriata* Schltr. in Fedde, Repert., Beih. 4: 266 (1919). Type: Java ?. cult. Bot. Gard. Buitenzorg.

C. munronianum sensu Ridley, Fl. Malay Peninsula 4: 46 (1924), *non* King & Pantling.

C. siamense Rolfe ex Downie in Kew Bull.: 382 (1925). Type: Thailand, Doi Suthep, *Kerr* 242 (holotype K!), **syn. nov.**

C. munronianum sensu Holttum, Fl. Malaya 1: 515 (1953) *non* King & Pantling.
C. ensifolium sensu J.J. Wood in Orchid Rev. 85: 94–6 (1977).
C. ensifolium sensu Comber in Orchid Dig. 164–8 (1980).
C. ensifolium sensu Seidenfaden in Opera Bot. 72: 71–3, f.38, t.4A (1983).
C. ensifolium sensu Du Puy & Lamb in Orchid Rev. 92: 352, f.291 (1984).

Subsp. *haematodes* is commonly a strong plant, usually larger than subsp. *ensifolium*. Its leaves are long, often about 2 cm broad, stiff and erect, arching near the tip, and they usually have a finely serrulate margin. The flower spike is weak in comparison, with up to 8 but usually fewer flowers which are small (3–3.5 cm across) with almost horizontal, spreading lateral sepals. The sepals and petals are pale straw-yellow or light brown to green in colour, usually with about five red lines over the veins. The petals are broader than the sepals. The lip is small, often with a triangular mid-lobe which is strongly recurved, terminating in an acute apex, and has a finely but strongly undulating margin.

ILLUSTRATIONS. Plate 24; photographs 122, 123; figure 27.1.

DISTRIBUTION. Sri Lanka, S India, Thailand, W Malaysia, Sumatra, Java, Borneo, New Guinea.

Subsp. *haematodes* is widely distributed, and variable over this range, especially the presence and strength of the leaf serrations, the length and breadth of the leaf, the number and size of the flowers, the breadth of the petals and the shape of the mid-lobe of the lip.

The area of distribution is divided into several more or less geographically isolated islands and peninsulas. Some of these have populations which contain a dominant variant which appears to be distinct in one or more characters. The plants with the broadest leaves with very light serrations on the margins are found on Java and Sumatra, and have been named as *C. sundaicum* Schltr. Specimens from New Guinea typically have very stiff, erect leaves with a strongly serrated margin, and the mid-lobe of the lip is triangular with a strongly undulating margin. Plants from Thailand are comparatively small, with narrow leaves, but the petals are sometimes very broad in comparison to the sepals, and the leaves are long and erect. This variant has been named *C. siamense* Rolfe (1925).

The type of *C. haematodes* was collected on Sri Lanka by Macrae. This is one of the more isolated islands in the range. Here, *C. ensifolium* is represented by a variant which appears similar in some respects to subsp. *ensifolium*. It has 4–8 rather large flowers (up to 4 cm across), and decurved lateral sepals. The sepals are larger than is normally found in subsp. *haematodes*, and they are narrowly elliptic rather than oblong in shape. However, in all other respects it agrees well with S E Asian material of subsp. *haematodes*.

Subsp. *haematodes* has broader leaves than the typical subspecies, and the petals are broader than the sepals (figure 28). The scattered cluster of subsp. *haematodes* reflects the large variation found, but there is no break in variation between the specimens from Thailand (1), and Sri Lanka (2) which are most similar to subsp. *ensifolium*, and the more extreme variants from Sumatra (3). It can be seen in figure 29 that subsp. *ensifolium* from China is distinctly smaller in all three parameters used. Table 10.10 lists the differences between these two subspecies.

The two subspecies cannot be reliably distinguished by any single character. The characters vary in their degree of expression in both subspecies, and there is some overlap in some characters. Further study is necessary to clarify the patterns of variation in both subspecies.

Specimens from southern China have rather broad, erect leaves, and the flowers appear to be typical of subsp. *haematodes*, with an undulating lip margin, but the petals are similar in breadth to the sepals indicating subsp. *ensifolium*. Examination of further specimens from southern China and Taiwan is necessary to determine the variation present in the regions where the distributions of the two subspecies meet.

Colour variation is encountered in subsp. *haematodes*, equivalent to that found in subsp. *ensifolium*. In Sabah, two colour variants have been found. One, from lower altitudes (600 m, 1970 ft), has the striped petals and sepals typical of the species. The other, from higher altitudes (1800 m, 5905 ft), has green sepals, and petals with a single central stripe. This

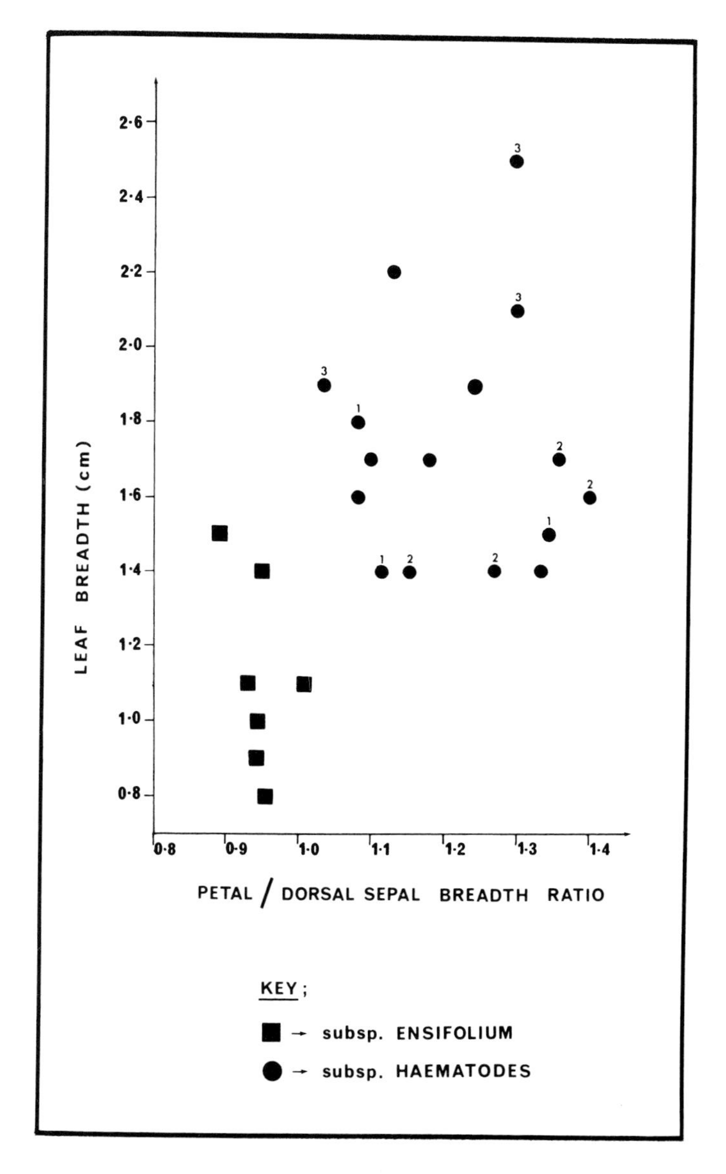

FIGURE 28: Scatter diagram displaying the range of variation in leaf breadth and comparative breadth of the petals in *C. ensifolium*

Two groups are apparent, those indicated as subsp. *ensifolium* originating in China and Taiwan, while those indicated as subsp. *haematodes* are from S India, Indo-China, Malesia and New Guinea; in particular
1 Sri Lanka
2 Thailand
3 Sumatra

Subsp. *haematodes* usually has broader leaves than subsp. *ensifolium*, and its petals are 1–1.4 times as broad as its sepals.

latter does not appear to differ from the other specimens of subsp. *haematodes*, except in colour, and cannot be treated as a separate taxon (Du Puy & Lamb, 1984).

The trans-equatorial distribution of subsp. *haematodes* reduces the effect of seasonality upon flowering time, and specimens may flower at any time of year in various parts of the range. However, Comber (1980) reports that in Java the most common flowering time is at the end of the wet season, in February–May. In the more seasonal climate of Sri Lanka, Jayaweera (1981) records December–April as the flowering time and Trimen (1885) gives January–March, although a specimen from southern India was in flower during July. The flowering period in Thailand is also during February and March, at the start of the warm, dry season.

Jayaweera (1981) and Trimen (1885) give the habitat in Sri Lanka as the submontane or mid-country tropical, wet, evergreen forests, from 300 to 1800 m (985 to 5905 ft) altitude. Comber (1980) reports that it is more common in western Java, where the dry season is less severe, than in the east. It is found in semi-open submontane forest in good

TABLE 10.10: A comparison of the characters which can be used to distinguish the two subspecies of *C. ensifolium*

	subsp. *ensifolium*	**subsp. *haematodes***
Leaves	short (29–56 cm) up to 0.8–1.6 cm broad (normally about 1.1 cm) arching leaf margin usually entire leaf base with membranous margin about 1 mm broad	long ((44)52–94 cm) broader up to 1.4–2.5 cm broad (normally 1.5–1.9 cm) erect, arching at the tip leaf margin usually serrated leaf base with membranous margin about 2 mm broad
Sepals	dorsal sepal 27–31 × (5.8)7.0–8.8(9.8) mm sepals narrowly-elliptic lateral sepals curved downwards	dorsal sepal 16–26(31) × 4.6–8.7 mm sepals oblong lateral sepals horizontal
Petals	21–26 × 7.1–7.7(8.8) mm petals and sepals almost equal in breadth	14–26 × 5.5–9 mm petals broader than sepals
Lip	mid-lobe 9–12 × (5)6.2–10 mm mid-lobe usually ovate mid-lobe margin kinked mid-lobe with few large red blotches	mid-lobe 6–11 × (5.1)6.2–8.2(9.1) mm mid-lobe triangular to ligulate-elliptic mid-lobe margin undulating mid-lobe with more numerous, finer, transverse red spots
Distribution	China, Taiwan, Hong Kong, Ryuku Isl. (Japan), Philippines (Luzon)	S India, Sri Lanka, Thailand, Malaya, Sumatra, Java, Borneo, New Guinea

FIGURE 29: Map illustrating the average leaf length, leaf breadth and comparative petal breadth of specimens from various parts of the distribution of *C. ensifolium*

Subsp. *ensifolium* (from China) is distinctly smaller in all three parameters than subsp. *haematodes* (from the remainder of the distribution).

light, from 500–1300 m (1640–4265 ft). In Sabah it is found from 600–1800 m (1970–5905 ft). The plants found at the lower altitudes grow in light to heavy shade in deep leaf litter, often under *Casuarina sumatrana*, while those from the higher altitudes prefer peaty or mossy ground in cool, damp moss forest, often near a river, and have occasionally been noted growing in thick moss on the bases of tree trunks (Du Puy & Lamb, 1984).

In Thailand, subsp. *haematodes* grows in tall grass in semi-shaded positions in open oak and dipterocarp forest, at altitudes of 300–1500 m (985–4920 ft). The plants were only visible amongst the dry and brown grass because the leaves of the orchid remained green. The pseudobulbs grew about 3 cm (1.2 in) below the surface of the heavy clad soil, and were covered in fibrous leaf sheaths and cataphylls, giving protection from forest fires which were frequently started to clear the undergrowth or for hunting.

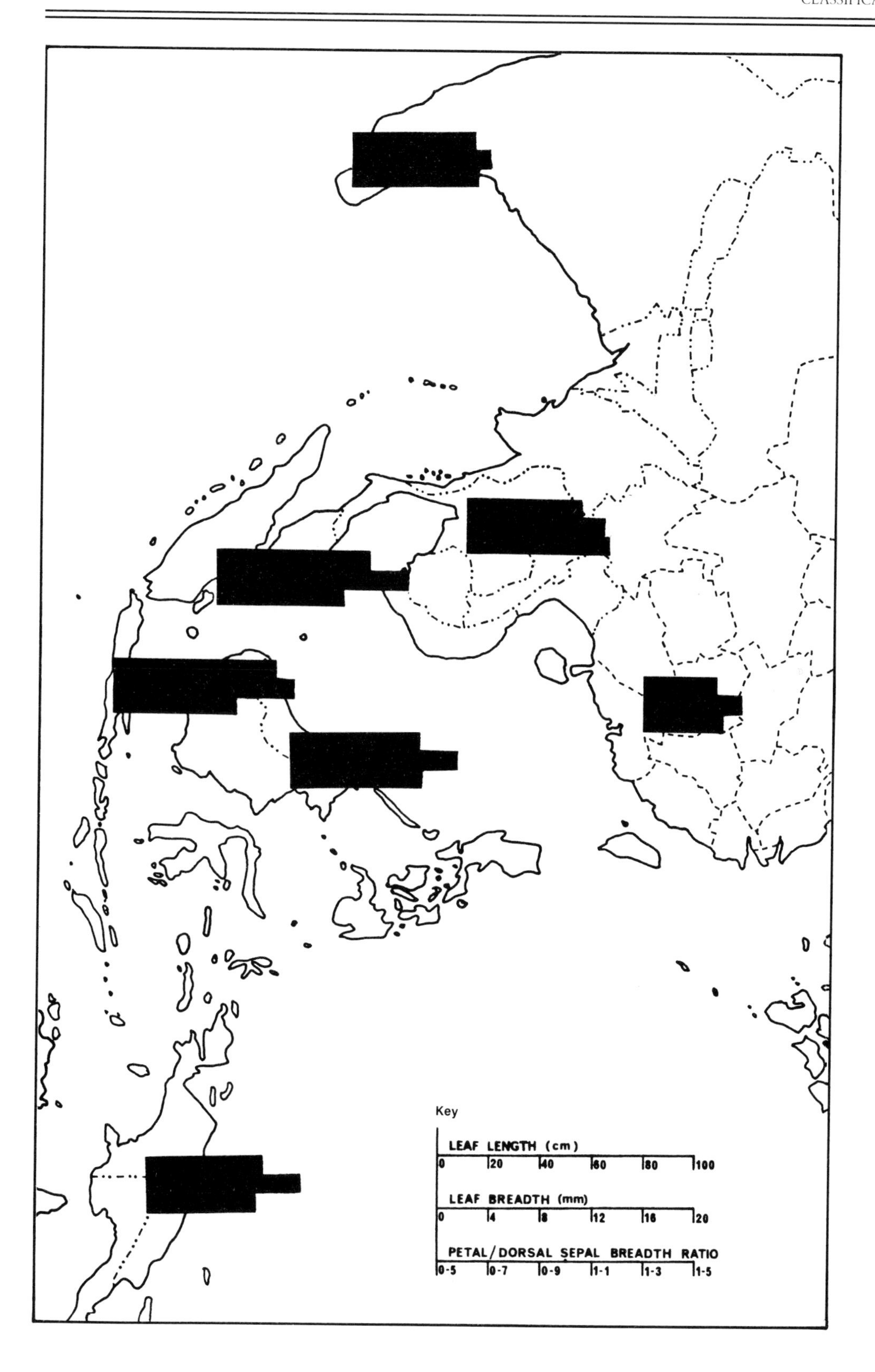
Key
LEAF LENGTH (cm)
0 20 40 60 80 100
LEAF BREADTH (mm)
0 4 8 12 16 20
PETAL/DORSAL SEPAL BREADTH RATIO
0·5 0·7 0·9 1·1 1·3 1·5

37. C. munronianum *King & Pantling* in J. Asiatic Soc. Bengal 64: 338 (1895) & in Ann. Roy. Bot. Gard. Calcutta 8: 187, t.249 (1898). Type: Sikkim, Teesta Valley, *Pantling* s.n. (holotype CAL).

C. ensifolium var. *munronianum* (King & Pantling)Tang & Wang in Acta Phytotax. Sin. 1: 91 (1951).

A medium-sized, perennial, terrestrial *herb. Pseudobulbs* small, up to 3 × 2 cm, ovoid, prominent, covered in scarious cataphylls up to 11 cm long, which become fibrous with age, and broad sheathing leaf bases, both with a 2 mm broad membranous margin, with 3–4(5) distichous leaves. *Leaves* broad, 60–80 × 1.8–2.7 cm, linear-elliptic, erect, articulated 3–5 cm from the pseudobulb, the lowest leaves long, and distinct from the cataphylls. *Scape* up to 60 cm tall, erect, arising basally, with 8–13 flowers produced in the apical third of the spike; peduncle with about 5 short, amplexicaul sheaths which are closely adpressed to the peduncle, up to 5 cm long, distant, only the lowest sheaths overlapping each other; bracts small, 0.2–0.8 cm long, triangular. *Flowers* comparatively small, 2.5–3.5 cm across; scented; rhachis, pedicel and ovary green; sepals and petals pale green or yellow to cream, with about five pale, purple-brown, broken longitudinal lines; lip pale yellow, side-lobes pink, streaked with red, with a solid red margin, mid-lobe blotched with red; column pale green, streaked maroon below; anther-cap cream. *Pedicel and ovary* 1.2–2.5 cm long. *Dorsal sepal* 16–26 × 5–6.5 cm, oblong-elliptic, rounded and shortly mucronate at the apex, erect; lateral sepals similar, spreading, almost horizontal. *Petals* 16–22 × 6–9.8 mm, ovate, subacute, slightly porrect but not closely shading the column, broader than the sepals. *Lip* 1.2–2.0 cm long when flattened; side-lobes erect, prominent, minutely pubescent; mid-lobe small, 7.1–8(10) × 4.2–7.1 mm, ligulate or oblong, much narrower than the side-lobes when the lip is flattened, not strongly recurved, minutely papillose, margin entire and weakly undulating or kinked; callus of 2 ridges, converging in the apical half to form a small tube at the base of the mid-lobe. *Column* 7–11 mm long, arching, narrowly winged; pollinia 4, broadly ovate, in two unequal pairs. *Capsule* about 5 cm long, fusiform, ridged, held erect and parallel to the rhachis.

ILLUSTRATION. Figure 27.6.

DISTRIBUTION. N E India (Sikkim), Bhutan (map 10D); about 500 m (1640 ft).

HABITAT. No habitat data available; flowering December–May.

C. munronianum was collected by Pantling in the Teesta Valley in Sikkim, and was described and illustrated by King & Pantling in 1895, being dedicated to Mr Munro, resident in Sikkim at that time. J.D. Hooker (1891) did not differentiate between the Himalayan specimens and *C. ensifolium*, but it now appears that *C. munronianum* replaces *C. ensifolium* in northern India. Further collections from Nepal, Bhutan, Meghalaya (Khasia Hills) and eastern Assam (Nagaland) are needed to determine the complete distribution of this species.

C. munronianum is closely related to *C. ensifolium* subsp. *haematodes*. Their flowers are very similar, being small (3–3.5 cm across) and usually pale yellow in colour with about five red stripes on the sepals and petals, and similar red markings on the lip. The mid-lobe of the lip of *C. munronianum* is narrower than that of *C. ensifolium* subsp. *haematodes*, but both have an undulating margin. *C. munronianum* has more numerous flowers in its inflorescence, and is similar to *C. sinense* in this respect. Vegetatively it is closer to *C. sinense* which also grows in N E India (Meghalaya), but is not recorded from the Himalaya. All three taxa have comparatively broad leaves, and although *C. sinense* has the broadest leaves they can be difficult to distinguish on this character alone. The flower of *C. munronianum* is very distinct from that of *C. sinense* in size, colour and the manner in which the petals are held (table 10.11).

C. munronianum can be further distinguished from both of these related taxa by the characters of the inflorescence. Its peduncular sheaths are short and distant from each other, they clasp the peduncle very closely, and do not spread out even at their tips. Its floral bracts are very short, and its numerous flowers are small. Table 10.11 compares these three taxa.

C. cyperifolium is also found in the same region, but it is readily distinguished by its more numerous, narrower leaves, its much longer floral bracts, its larger flowers, its green petal and sepal colour lacking red stripes and its mid-lobe with inflated papillae.

Very few habitat notes are available for this species. It flowers between December and May. The type description notes its habitat as 'in the Teesta Valley on dry knolls; at an elevation of 1500 ft (c. 500 m)'.

38. C. sinense *(Jackson in Andr.)Willd.*, Sp. Pl., ed. 4, 111 (1805); Lindley, Gen. Sp. Orchid. Pl.: 162 (1833); Makino in Iinuma, Somoku-Dzusetsu 4: 1063 & 1184 (1912); Hayata, Icon. Pl. Formos. 4: 83–4 (1914); Garay & Sweet, Orchids South. Ryukyu Islands: 146 (1974); Su, Nat. Orchids Taiwan, ed. 2,: 126–8 (1975); Hu, Gen. Orchid. Hong Kong: 96 (1977); Lin, Nat. Orchids Taiwan 2: 126–8, t.53–55, + fig. (1977); Liu & Su, Fl. Taiwan 5: 948 (1978); Wu & Chen in Acta Phytotax. Sin. 18: 292, t.1 & 2 (1980); Seidenfaden in Opera Bot. 72: 73–4, f.39 (1983); Seth & Cribb in Arditti (ed.), Orchid Biol., Rev. Persp. 3: 294 (1984); Mark, Ho & Fowlie in Orchid Dig. 50: 22 (1986). Type: Icon. in Andr. Bot. Rep. 3: t.216 (1802)!

Epidendrum sinense Jackson in Andr. Bot. Rep. 3: t.216 (1802); Sims in Bot. Mag. 23: t.888 (1805).
C. fragrans Salis. in Trans. Hort. Soc. 1: 298 (1812). Type: as for *C. sinense.*
C. chinense Heynh., Nomencl. 2: 179 (1846). *Sphalm.* for *C. sinense.*
C. ensifolium auct. non (L.)Sw., Hook.f., Fl. Brit. India 6: 13 (1891), in part.
C. hoosai Mak. in Bot. Mag. Tokyo 16: 27 (1902). Type: Japan, Musashi Prov., cult. Tokyo Bot. Garden, *Makino* s.n. (holotype TI or MAK).

Table 10.11: A comparison of the characters which separate the three closely related taxa *C. munronianum*, *C. ensifolium* subsp. *haematodes* and *C. sinense*

Character	*C. munronianum*	*C. ensifolium* subsp. *haematodes*	*C. sinense*
Leaf breadth	1.8–2.7 cm	1.4–2.5 cm	1.5–3.5 cm
Spike length	up to 60 cm	17–67 cm	41–72 cm
Number of flowers	8–13	3–8	(6)8–26
Sheath length	up to 5 cm	up to 6.5 cm	up to 10 cm
Sheath arrangement	sheaths distant, amplexicaul	sheaths overlapping, cymbiform at the apex	sheaths overlapping, cymbiform at the apex
Floral bract length	0.2–0.8 cm	0.2–2.0 cm	0.4–2.5 cm
Flower size	2.5–3.5 cm across	3–3.5 cm across	about 5 cm across
Tepal colour	pale straw-yellow, with longitudinal red-brown stripes	pale straw-yellow to pale green, with longitudinal red-brown stripes	dark, chocolate-brown (in northern India)
Dorsal sepal length	16–26 mm	16–26(31) mm	26–39 mm
Lip length	12–20 mm	14–22 mm	21–29 mm
Mid-lobe size	7.1–8(10) × 4.2–7.1 mm	6–11 × (5.1)6.2–8.2(9.1) mm	12–16 × 9–13 mm
Column length	7–11 mm	10–14(18) mm	12–16 mm
Petal arrangement	petals spreading, not covering the column	petals spreading, not covering the column	petals usually closely covering the column
Flowering period	December–May	January–March, but also sporadic throughout the year (in northern India)	October–November (in northern India)

C. albo-jucundissimum Hay., Icon. Pl. Formos. 4: 74 (1914) & op. cit. 6: 80, t.13 (1916). Type: Taiwan (Formosa), cult. Taihoku, *Hayata* s.n. (holotype TI).

C. sinense (Jackson in Andr.)Willd. var. *margicoloratum* Hay. in Icon. Pl. Formos. 6: 82, t.16B & 17 (1916). Type: Taiwan, cult. Taihoku, *Soma* (holotype TI).

C. sinense (Jackson in Andr.)Willd. var. *albo-jucundissimum* (Hay.) Masam, in Trop. Hort. 3: 31 (1933).

C. ensifolium var. *munronianum auct. non* King & Pantling, Tang & Wang in Acta Phytotax. Sin. 1: 91 (1951).

C. sinense (Jackson in Andr.)Willd. var. *album* Yen, Icon. Cymbid. Amoy. (1964). Type: China, Kwangtung, *Yen & Hong* 1006 (holotype not located).

C. sinense (Jackson in Andr.)Willd. var. *bellum* Yen in Icon. Cymbid. Amoy. (1964). Type: China, Fukien, *Cheng & Cheng* 1019 (holotype not located).

A medium-sized, perennial, terrestrial *herb. Pseudobulbs* small, up to 3 × 2 cm, ovoid, prominent, covered in scarious cataphylls up to 13 cm long, which become fibrous with age, and broad, sheathing leaf bases, both with a 2 mm broad membranous margin, with 2–4(5) distichous leaves. *Leaves* broad, (25)40–103 × (1.5)2–3.2 cm, linear-elliptic, arching, dark green, glossy, articulated 4–9 cm from the pseudobulb, the lowest leaves long and distinct from the cataphylls. *Scape* 40–80 cm tall, erect, robust, arising basally from inside the cataphylls, usually with (6)8–26 flowers produced in the apical half to third of the scape and held above the leaves; pendunc1e covered in 3–5 or more sheaths up to 10 cm long, usually overlapping each other; bracts 0.4–2.5(4.1) cm long, narrowly triangular, acuminate. *Flowers* about 5 cm across; strongly scented; peduncle, rhachis, floral bracts, pedicel and ovary greenish, often stained purple-brown; sepals usually purple-brown, sometimes very dark, petals slightly paler, often showing darker veins, occasionally tepals straw-yellow with dark red veins; lip cream or pale yellow, side-lobes streaked red with a solid red margin, mid-lobe heavily spotted and blotched with dark red; column reddish above, cream with red spots below; anther-cap cream. *Pedicel and ovary* 1.9–4.0 cm long. *Dorsal sepal* 26–39 × 5.7–8.7 mm, narrowly elliptic, acute, erect to porrect, margins lightly revolute; lateral sepals similar, spreading. *Petals* 23–34 × 7.1–11.0 mm, narrowly ovate, broader than the sepals, porrect, often forming a hood over the column. *Lip* 2.1–2.9 cm long when flattened; side-lobes erect, rounded, often reduced, minutely papillose or minutely pubescent; mid-lobe broad, 12–16 × 9–13 mm, oblong with an obtuse mucronate apex, to ovate, with an acute apex, recurved, often as broad as the side-lobes when the lip is flattened, minutely papillose or with minute papillose hairs, margin entire and weakly undulating or kinked; callus in two ridges, converging in the apical half to form a small tube at the base of the mid-lobe. *Column* 1.2–1.6 cm long, arching, narrowly winged; pollinia 4, broadly ovate, in two unequal pairs. *Capsule* about 6 cm long, fusiform, ridged, beaked, held erect and parallel to the rhachis.

ILLUSTRATIONS. Plate 25, photographs 124, 125; figure 27.4.

DISTRIBUTION. Meghalaya (Khasia Hills), Burma, N Thailand, China, Hong Kong, Taiwan, Ryukyus (map 10B); 300–2300 m (985–7545 ft).

HABITAT. Open to dense mixed or evergreen forest, in shade or semi-shade; flowering October–March.

C. sinense is a robust plant, usually with 3–4 arching, dark green shiny leaves which are comparatively broad (usually 2.5–3.5 cm, 1–1.4 in), and large cataphylls (to 13 cm (5 in) long). The flower spike is tall and robust, commonly with 8–15 flowers which are held well clear of the foliage. It was one of the earliest *Cymbidium* species to be described, by Jackson (1802) as *Epidendrum sinense*, based on a specimen grown by Hibbert which flowered in September 1801. He recorded that the species had been in cultivation, in Britain, since 1793, but had not flowered previously. No type specimen has been preserved, and the illustration in the *Botanical Repository* must therefore serve as the type.

Salisbury (1812) had concluded that the *Epidendrum sinense* of Jackson, and also the specimen figured in the *Botanical Magazine* (1805), should be transferred to the genus *Cymbidium*. He named it *C.*

fragrans, not realising that the transfer had already been made by Willdenow, earlier the same year.

C. hoosai Makino (1902), based on a Chinese specimen cultivated in Japan, has been sunk into *C. sinense* by several authors, including Wu & Chen (1980). The original description agrees well with *C. sinense*.

C. sinense is a widespread orchid, variable over this range particularly in flower colour, which ranges from deep purple-brown through pale and striped to green and lacking any red pigmentation. Several of these variants have been described as distinct taxa. The most common variant in China and Taiwan has dark, purple-brown petals and sepals, with a yellowish, red-spotted lip but lighter brown variants are known which have dark veins visible, especially in the petals. This latter colour is typical of the specimens from northern India. Specimens from Thailand have light straw-coloured petals and sepals, all with red veins.

Hayata (1914) described *C. albo-jucundissimum* from a specimen cultivated in Taiwan which had green and white flowers, and a more hirsute mid-lobe of the lip. This is an albino variant, uncommon in the wild, but of horticultural value and therefore quite common in cultivation. This type of colour variation is found in many species in section *Jensoa*. Although Wu & Chen (1980) recognise this as a distinct variety, it is treated here as a synonym of *C. sinense*.

Var. *marginatum* Hayata (1916) differs in the colour of the mid-lobe of the lip. It is not spotted, but has a purple-red margin instead. This is a minor colour variant which was again described from a cultivated specimen.

C. sinense has been cultivated for several centuries in China and Japan, and is particularly popular for its attractive foliage, its tall, elegant flower spikes which carry many strongly scented flowers and for its ease of cultivation. Colour variants are highly prized, and are consequently maintained in cultivation, but may be rare or absent in the wild. Several of these variants have been named as species or varieties, but although they may merit horticultural names, they are not taxonomically significant. Lin (1977) and Su (1975) record that there is also a variant with variegated leaves which is of great horticultural value in Taiwan, and is known as the 'Golden thread orchid'.

A specimen from Hong Kong, ascribed to *C. ensifolium* var. *munronianum* by Tang & Wang (1951) has a spike somewhat similar to *C. munronianum*, with short clasping peduncular sheaths and short, triangular floral bracts. However, the leaves are broader than in *C. munronianum*, and the large size of the flower, and especially of the lip, indicates that the specimen should be included in *C. sinense*. In fact, the length of the floral bracts is a somewhat variable character in *C. sinense*. Specimens collected in Meghalaya of northern India and in Yunnan have long, acuminate bracts up to 3.5 cm (1.4 in) long, while those from eastern China and Taiwan are usually shorter, up to 2.0 cm (0.8 in), and those already noted from Hong Kong are only 1 cm (0.4 in) long or less.

C. sinense is most readily distinguished from the other species in section *Jensoa* by its vegetative habit and by characters of the scape. *C. munronianum* also has broad leaves, but they are usually narrower than in *C. sinense*, and the scape is weaker, and the individual flowers are much smaller. *C. sinense* flowers are usually larger than those of *C. ensifolium* or *C. munronianum*, especially in the size of the lip. The sepals of *C. sinense* are long, narrow and elliptic; the petals are broader than the sepals and are strongly porrect, closely shading the column. The lip is large, with a broad mid-lobe, and has an entire margin which may be kinked, but is never strongly undulating.

C. kanran has similar narrow sepals and broader petals, but the sepals are much longer, more slender and more acuminate than those of *C. sinense*. Its leaves are longer and narrower, and the scape has fewer flowers.

C. cyperifolium and *C. faberi* can both be distinguished by their more numerous, much narrower leaves, and by the presence of shiny, inflated papillae on the mid-lobe of the lip. *C. cyperifolium* can usually be further distinguished by its clear green sepals and petals; and its very long floral bracts.

It can be difficult to distinguish the species in section *Jensoa* from the flowers alone. Those of *C. ensifolium* and *C. sinense* are both variable, and are often rather similar to each other, so that the broad, dark green, shiny leaf of *C. sinense* is usually the easiest character by which these species can be separated. The large, numerous flowers on a long and robust scape, the long, narrow sepals, the petals which are broader than the sepals (see also *C. ensifolium* subsp. *haematodes*) and are porrect, covering the column, and the large, broad mid-lobe of the lip of *C. sinense* usually serve to separate the flowers of the two species.

C. sinense flowers during January–March in Taiwan and China, and in October–November in northern India. Therefore, the flowering time in northern India does not appear to coincide with the flowering of *C. munronianum* to which it is clearly related. These two species are not recorded as being sympatric, but further study may show that their distributions coincide (see also the discussion under *C. munronianum*, and table 10.11 for a comparison between these two species and *C. ensifolium* subsp. *haematodes*).

In Hong Kong, Taiwan, and eastern China, *C. sinense* occurs at altitudes of 250–1000 m (820–3280 ft). In northern India, Burma, Yunnan (and probably Thailand) it occurs at higher altitudes, usually between 1400–2300 m (4595–7545 ft). In China and Taiwan it grows in humus-rich soil in dense or partial shade, often near streams or water seepages (Lin 1977; Ying 1977; Barretto pers. comm.). In Burma, it has been found in mixed deciduous and evergreen forest containing species of *Cephalotaxus*, *Podocarpus*, *Alnus* and *Shorea*. In Thailand, it occurs in semi-shaded spots in lower montane forest.

C. sinense occurs in the southernmost provinces of China, including Fujian, Guangdong, Guangxi, Hainan, Sichuan and Yunnan, and also in Hong Kong and Taiwan (Wu & Chen, 1980).

39. C. kanran *Makino* in Bot. Mag. Tokyo 16: 10 (1902) & in Iinuma, Somoku-Dzusetsu, ed. 3, 4(18): t.4, 5 & 6 (1912); Makino & Nemoto, Fl. Japan: 1628–32 (1931); Makino, Fl. Japan: 682 (1948); Ohwi, Fl. Japan: 355 (1965); Maekawa, Wild Orchids Japan in Colour: 406–9, t.166 & 167 (1971); Garay & Sweet, Orchs. Ryukyu Islands: 144-5 (1974); Lee, Ill. Fl. & Faun. Korea 18: 771, 826, t.183 (1976); Lin, Native Orchids Taiwan: 116–17, t.118, 45 & 46 (1977); Liu & Su, Fl. Taiwan 5: 946–7 (1978); Wu & Chen in Acta Phytotax. Sin. 18(3): 298, t.1 (1980); Chow Cheng, Formosan Orchids, ed. 3: 46 (1981); Mark, Ho & Fowlie in Orchid Dig. 50: 34–5 (1986). Type: Japan, Musashi prov., cult. Bot. Gard. Koishikawa, *Makino* s.n. (holotype MAK).

C. kanran var. *latifolium* Mak., in Iinuma, Somoku-Dzusetsu 4: 5, t.5 (1912). Type: not indicated, but probably in MAK.

C. oreophyllum Hayata, Icon. Pl. Formos. 4:80, t.38c (1914). Type: Formosa; cult. Tiahoku, *Hayata* (holotype TI).

C. misericors Hayata var. *oreophyllum* (Hayata)Hayata, Icon. Pl. Formos. 4: 81 (1914).

C. purpureo-hiemale, Hayata, Icon. Pl. Formos. 4: 81 (1914). Type: Formosa, cult. Taihoku, *Hayata* (holotype TI).

C. linearisepalum Yamam. in Trans. Nat. Hist. Soc. Formosa 20: 40 (1930). Type: not located.

C. tosyaense Masamune in Trans. Nat. Hist. Soc. Formosa 25: 14 (1935). Type: Japan, Takao-Syu Heitogum Tosga, cult. *Watanabe* (holotype MAK), **syn. nov.**

C. sinokanran Yen, Icon. Cymid. Amoy., G–1 (1964). Type: Fukien, Nan-gien Hsien, *Cheng* 4001 (holotype not located).

C. sinokanran Yen var. *atropurpureum* Yen, Icon. Cymbid. Amoyens., G–2 (1964). Type: China, Fukien, Nan-gien Hsien, *Cheng* 4002 (holotype not located).

C. omiense Wu & Chen in Acta Phytotax. Sin. 11: 32, pl. 5, t.4–8 (1966). Type: China, Szechuan, Omei-shan, *Fee* 2009 (holotype PE), **syn. nov.**

C. tentyozanense T.P. Lin, Nat. Orch. Taiwan 2: 112 (1977): *sphalm.* for *C. tosyaense* Masamune.

C. faberi Rolfe var. *omiense* (Wu & Chen)Wu & Chen in Acta Phytotax. Sin. 18: 299 (1980), **syn. nov.**

C. quibiense Feng & Li in Acta Bot. Yunnan. 2(3): 334–6 (1980). Type: Yunnan, Quibei Xian, *Qui Pin-yun* 59140 (holotype KUN), **syn. nov.**

A medium-sized, perennial, terrestrial *herb*. *Pseudobulbs* 4–6 × 1–1.8 cm, narrowly ovoid, often conspicuous, covered in scarious cataphylls and broad, sheathing leaf bases, with 3–4(5) almost distichous leaves. *Leaves* narrow, up to (20)30–90(120) × 0.5–1.5(2) cm, linear to linear-elliptic, acuminate, arching, dark green, shining, margins often minutely serrulate towards the apex, articulated about 5 cm from the pseudobulb, the lowest leaves long and distinct from the cataphylls. *Scape* (15)25–60(80) cm tall, erect, arising basally from inside the cataphylls, with (3)5–12(20) rather distant flowers produced in the apical half to one-third of the scape and held above the leaves; peduncle covered in about 3–6 distant sheaths up to 4.5 cm long; bracts (0.8)1.5–3.5(5) cm long, narrowly ovate, acuminate. *Flowers* 5–7 cm across, spidery in appearance; strongly scented; rhachis, pedicel and ovary green to purple; flower colour variable, but usually sepals and petals are olive to clear green, with a short maroon stripe over the mid-vein at the base; lip pale yellow or pale green, side-lobes streaked with red with a solid red margin, mid-lobe spotted and blotched with red, with a narrow cream margin; column usually pale green with purple spots towards the apex, and cream streaked with red below; anther-cap cream. *Pedicel and ovary* 2–3.5 cm, usually longer than the bracts. *Dorsal sepal* (25)30–45(50) × (3)3.5–5(7) mm, narrowly linear-elliptic, strongly acuminate, erect; lateral sepals similar, spreading. *Petals* 22–30(35) × (3)5–8(10) mm, narrowly ovate, acuminate, porrect, almost forming a hood over the column. *Lip* (1.6)2–2.7 cm long when flattened; side-lobes erect, rounded, minutely papillose or minutely pubescent; mid-lobe (10)12–16 × 7–12 mm, oblong-ovate to triangular-ovate, obtuse to subacute, recurved, minutely papillose, minutely hairy or with some papillae, margin entire or slightly erose, kinked but not strongly undulating; callus of 2 ridges, converging in the apical half to form a small tube at the base of the mid-lobe. *Column* 10–15(17) mm long, arching, narrowly winged; pollinia 4, broadly ovate in two unequal pairs, on a small crescent-shaped viscidium. *Capsule* about 5 cm long, fusiform, ridged, beaked, held erect and parallel to the rhachis.

ILLUSTRATIONS. Photographs 126, 127; figure 27.7.

DISTRIBUTION. S China, Hong Kong, Taiwan, Ryukyus, S Japan, S Korea (map 10E); 800–1800 m (2625–5905 ft).

HABITAT. In open hardwood forest, in shade; (October)November–February.

C. kanran was described by Makino in 1902, from a wild-collected Japanese specimen which had been introduced into cultivation. The specific epithet means 'orchid which flowers in winter', referring to the flowering time of this species. The leaves are slender, glossy and deep green in colour, rather like those of *C. sinense*, but much narrower and often longer. The flowers are elegant and scented and are known in a wide range of colours, and combined with the graceful form of the plant make it a desirable plant for pot cultivation. It has been cultivated for many centuries in Japan and Taiwan, and several unusual variants are prized.

C. kanran is closely allied to *C. sinense*, both having 3–4 glossy, dark green leaves and scapes with several comparatively large flowers, and their petals are strongly porrect, forming a hood over the column. *C. kanran* can be distinguished from *C. sinense* and from all other related species by its long, narrow, acuminate leaves, its relatively long floral bracts which are almost equal in length to the pedicel and ovary in the lowest flower in the spike, but are shorter in the upper flowers, and its very long, narrow, finely tapering sepals which give the flower a spidery appearance. The sepals are usually seven or more times as long as they are broad, whereas in *C. sinense* or *C. ensifolium* they are only about four times as long as broad.

Four taxa, *C. oreophilum* Hayata (1914), *C. purpureo-hiemale* Hayata (1914), *C. linearisepalum* Yamamoto (1930) and *C. sinokanran* Yen (1964), are considered by most authors to be synonyms of *C. kanran* (Su, 1975; Lin, 1977; Ying, 1977; Liu & Su, 1978; Wu & Chen, 1980).

The flower colour of *C. kanran* varies enormously, although the most common and widespread colour is olive-green with some red-brown on the central vein of the petals and sepals, and a pale green or yellow lip lightly spotted with red. Nagano & Nagano (1955) list six colour variants recognised in cultivation in Japan, and these are well illustrated by Maekawa (1971). These variants can be grouped into the following categories on the basis of the colour of the petals and sepals: 1. green with weak to strong red-brown veins, and red-brown markings on the lip — the most widespread colour; 2. purple-brown; 3. pink; 4. pale yellow; 5. green, without any red-brown markings on the tepals or the lip; and 6. multi-coloured, known as the 'sasara' colour variant. This wide colour range has led to the naming of several taxa. Makino named four forms in Japan to cover this colour variation; f. *purpurascens* (1902), f. *viridescens* (1912), f. *rubescens* (1912) and f. *purpureo-viridescens*

(1912). When Yamamoto (1930) described *C. lineari-sepalum* he also described two colour variants, naming them f. *atropurpureum* and f. *atrovirens*. Masamune (1933) later changed their taxonomic status from forms to varieties. Yen (1964) also noted two varieties of his *C. sinokanran* from Taiwan; the typical green variety and var. *atropurpureum*.

Hayata (1914) described the purple variant from Taiwan as a separate species, *C. purpureo-heimale*, meaning the winter-flowering purple *Cymbidium*. Ying (1976) reduced this to *C. kanran* f. *purpureo-heimale*, and later (1977) raised it again to varietal status. This variation in flower colour in Taiwan is illustrated by Mark *et al.* (1986). These are all simply colour variants, and although it may be horticulturally useful to distinguish them with cultivar names, they are not taxonomically distinct. *C. purpureo-heimale* was further distinguished by the presence of lines of short hairs on the mid-lobe of the lip.

C. oreophilum Hayata (1914) was described as having a 'botryoideo-tuberculate' mid-lobe. Specimens of *C. kanran* usually have some thickening of the veins in front of the callus ridges, and the indumentum varies from minutely papillose to very shortly pubescent, or with some swollen papillae. This variation in the indumentum seems to vary independently of flower colour, and it is impossible to recognise distinct taxa on the basis of this character.

C. kanran normally has slender leaves less than 1.5 cm (0.6 in) broad, but there are variants with leaves up to 2 cm (0.8 in) broad. In 1912, Makino published var. *latifolium*, a variant with broader leaves. However, this variation is continuous, and does not appear to have any geographic pattern, and is probably not taxonomically significant.

Further variation, in cultivated plants, includes those with variegated leaves with narrow white margins which are known as 'Takachiko' or 'Nangoku' in Japan (Nagano & Nagano, 1955).

C. tosyaense Masamune (1935) was described (as 'tosyaenus') from a specimen cultivated in Japan. This name has possible been corrupted to *C. tentyozanense* by Lin (1977). The description of this taxon does not differ from that of *C. kanran*, and it is now placed in the synonymy of that species.

There are two variants known in Hong Kong. One has the normal, very long sepals, but the other has shorter, less acuminate petals and sepals, but is unmistakably close to *C. kanran* in the other characters of the vegetative plant and the flower spike (G. Barretto, pers. comm.). *C. omiense* Wu & Chen (1966), from Sichuan (China), appears to be a similar variant. The description includes many characters which are characteristic of *C. kanran*. Wu & Chen describe this variant as more 'feathery' than *C. faberi*, a description which fits the spidery flowers of *C. kanran*. Although this variant is a smaller plant with shorter leaves, smaller flowers and less acuminate sepals, nevertheless it has the very narrow sepals, long floral bracts and porrect petals characteristic of *C. kanran*, and the description is otherwise very similar. *C. omiense* is therefore treated here as a new synonym of *C. kanran*.

Similarly, *C. quibiense* Feng & Li (1980) from Yunnan appears to refer to this variant, although the sepals are even broader. This taxon was originally stated to have some affinity with *C. faberi*. However, the leaf number, its short, ovate mid-lobe with a margin which is not undulating, and its indumentum of minute papillae (not strongly inflated), all prevent its inclusion in *C. faberi*. The description closely fits a small specimen of *C. kanran*, and although it has shorter sepals than is normal in *C. kanran* they are still slender and acuminate. The variation is continuous between the more extreme *C. quibiense* and more normal specimens of *C. kanran* and consequently this variant is not recognised as distinct in this study. Further investigation may clarify the status and distribution of this variant.

The possibilities of hybridisation between *C. kanran* and either *C. ensifolium* or *C. goeringii* cannot be discounted as the possible origin of *C. quibiense*. Maekawa (1971, p. 411, 479, t.168) illustrates two specimens which are hybrids between *C. kanran* and *C. goeringii*. The name *C.* × *nishiuchianum* was reputedly given to these by Makino, but was never published, and the hybrid does occur in the central part of the province of Tosa, in Japan. Cheng (1981) also illustrates several natural hybrids of *C. kanran*, with *C. ensifolium*, *C. goeringii* and *C. sinense* found in Taiwan. These may even account for the variation in *C. kanran* in Hong Kong, as *C. kanran* and *C. sinense*

25. *Cymbidium sinense.* China, *Klehm* s.n., cult. Kew

26. *Cymbidium faberi* var. *szechuanicum*. Nepal, *Bailes* 1040, cult. Kew

are usually found growing together there (G. Barretto, pers. comm.).

C. kanran is found in southern Japan, in the south-eastern tip of Honshu, and in Shikoku and Kyushu. It is also found in the Ryukyu Islands, including Yakusima, Amami-Osima and Okinawa (Garay & Sweet, 1974; Ohwi, 1965). It has been reported from one locality in Korea (Lee, 1976). This northerly distribution limit reflects its relatively high tolerance of cold. Wu & Chen (1980) give its distribution in China as the southern provinces from Sichuan and Yunnan to Taiwan and Hong Kong.

Makino (1902) reports that in the warmer parts of southern Japan, *C. kanran* grows in shady forests. In Taiwan it grows at 800–1400 m (2625–4595 ft) in ravines or on mountain tops, often near the ridge of south-east facing slopes, usually in hardwood forests. In Hong Kong it is also found at high altitudes, usually in association with *C. sinense*. It is probably a high altitude plant in southern China also.

Section **Maxillarianthe** Schltr. in Fedde, Repert. 20: 101 (1924); P. Hunt in Kew Bull. 24: 94 (1970); Seth & Cribb in Arditti (ed.), Orchid Biol., Rev. Persp. 3: 289 (1984). Lectotype: *C. goeringii* (Reichb.f.)Reichb.f.), chosen by P. Hunt (1970).

The sectional limits as defined by Schlechter (1924) and Seth & Cribb (1984) have been extended here to include *C. faberi* and *C. cyperifolium* (see discussion of section *Jensoa*). This section is now characterised by having 5–13 leaves on each shoot, and by the complete absence of subepidermal fibre bundles in *C. faberi* and *C. goeringii*, and their restriction to below the adaxial (upper) epidermis only in *C. cyperifolium*. There are 4 pollinia.

40. C. cyperifolium *Wall. ex Lindley*, Gen. Sp. Orchid. Pl.: 163 (1833); Hook.f. in J. Linn. Soc. 3: 28 (1858) & Fl. Brit. India 6: 13 (1891); King & Pantling in Ann. Roy. Bot. Gard. Calcutta 8: 186–7, t.248 (1898); Seth & Cribb in Arditti (ed.), Orchid Biol., Rev. Persp. 3: 291 (1984). Type: India, Khasia Hills, Sylhet, *Wallich* 7353 (holotype K! – herb. Lindley).

C. viridiflorum Griffith, Itin. Not. Bhotan: 53 (1835). Type: Bhutan, *Griffith* s.n. (holotype K!).

C. carnosum Griffith, Notul. 3: 339–40 (1851). Type: India, Khasia Hills, *Griffith* 185 (holotype ?CAL).

Cyperorchis wallichii Bl., Orchid. Archip. Ind.: 92 (1858). Type: As for *C. cyperifolium*.

C. aliciae Quisumbing in Philippine J. Sci. 72: 486–7, t.2 & 7 (1940) & in Orchid J. 3: 64, f.28 (1954). Type: Philippines, Luzon, Nueva Vizcaya Province, cult. *Mrs K.B. Day*, *Quisumbing* (holotype PNH), **syn. nov.**

A medium to short, perennial, terrestrial *herb. Pseudobulbs* small, inconspicuous, covered by several cataphylls and sheathing leaf bases. Cataphylls leaf-like, with a 2 mm broad membranous margin, becoming scarious and fibrous with age. *Leaves* 5–10, distichous, up to (30)50–90 × 0.9–1.5 cm, linear-elliptic, acute, erect, margin entire, conduplicate at the base but not petiolate, obscurely articulated 4–6 cm from the base, the lowest leaves short, making a gradual transition between the cataphylls and the true leaves. *Scape* 23–43 cm tall, medium or slender, erect, arising basally from within the cataphylls, with (2)4–7 flowers produced in the apical third; peduncle covered by slender sheaths up to 8 cm long, which are distant except towards the base; bracts subulate or narrowly ovate, (0.3)1.5–4.0 cm long, often exceeding the pedicel and ovary in length, becoming scarious. *Flower* about 4–5 cm across; lemon-scented; rhachis, pedicel and ovary greenish, often stained dull purple; sepals and petals apple-green, fading to yellow-green, occasionally pale yellow or straw-coloured with 5–7 longitudinal red-brown lines (in Indo-China); lip pale green or whitish, sometimes pale yellow, with red-purple streaks on the side-lobes which become confluent at the margin, and red-purple spots and blotches on the mid-lobe; column green or yellow, spotted purple below; anther-cap cream. *Pedicel and ovary* 1.3–3.5(4.3) cm long. *Dorsal sepal* (20)25–35 × (4)5–8(11) mm, narrowly oblong-elliptic, acute, erect; lateral sepals similar, spreading. *Petals* 19–29 × (5)6–9(10) mm, ovate to elliptic, acute, usually broader than the sepals, porrect, usually closely shading the column. *Lip* (14)17–22(24) mm long when flattened; side-lobes erect, rounded, usually slightly angled at the apex, minutely pubescent or minutely papillose; mid-lobe (7)9–13 × (6)8–11 mm,

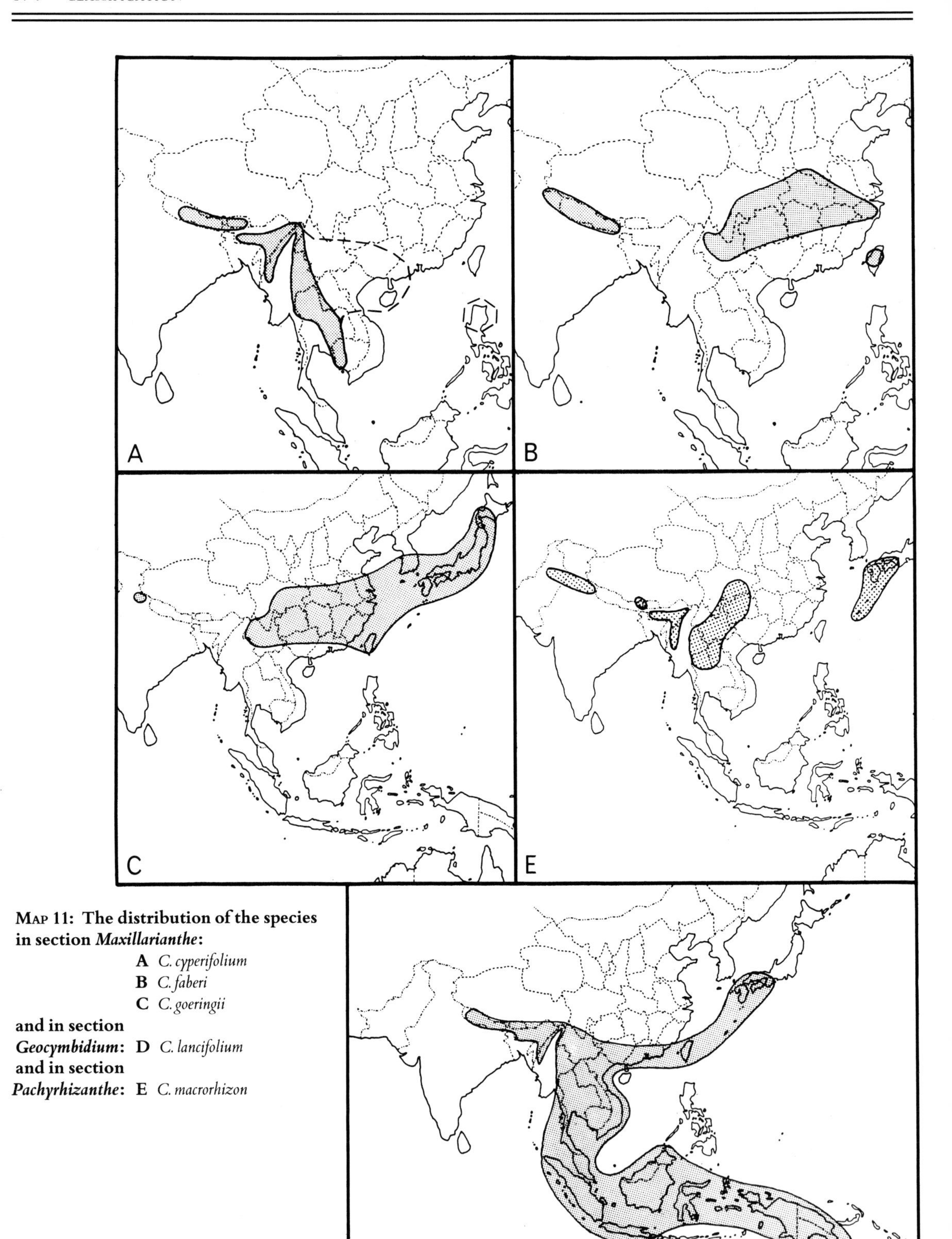

Map 11: The distribution of the species in section *Maxillarianthe*:
A *C. cyperifolium*
B *C. faberi*
C *C. goeringii*
and in section *Geocymbidium*: **D** *C. lancifolium*
and in section *Pachyrhizanthe*: **E** *C. macrorhizon*

oblong to broadly ovate, obtuse or subacute, often as broad as the side-lobes when the lip is flattened, strongly recurved, with some small papillae which are occasionally sparse and almost confined to the apical region, margin entire; callus ridges converging in the apical half to form a short tube at the base of the mid-lobe. *Column* 1–1.5 cm long, arching, narrowly winged; pollinia 4, in two pairs, broadly ovate, on a broadly crescent-shaped viscidium. *Capsule* 5–7 cm long, fusiform, held erect and parallel to the rhachis, retaining the column as a short, apical beak.

DISTRIBUTION. Nepal, N E India (Nagaland, Manipur, Mizoram, Meghalaya, Sikkim), Bhutan, S China, Burma, Thailand, Cambodia, Philippines (Luzon) (map 11A); 1500–2750 m (4920–9020 ft) (300–900 m (985–2950 ft) in Indo-China).

HABITAT. Temperate rainforest or bamboo forest, on steep banks of boulders and loam, in shade (Himalaya, Meghalaya); flowering November–January (Himalaya, Meghalaya), May–July (Thailand, Cambodia).

C. cyperifolium was first collected in the Khasia Hills and named by Wallich, but it was formally described and published by Lindley in 1833. King & Pantling (1898) described and illustrated a specimen collected in Sikkim. Griffith (1835) also collected this species, in Bhutan, and described it under the name *C. viridiflorum*. The type specimen agrees well with that of *C. cyperifolium* and the flower colour is described as green with a reddish-spotted lip. A Griffith specimen (5264) in the Edinburgh herbarium is annotated with the name '*C. tesserte*', but this name does not appear to have been validly published, and the specimen at Kew with the same number is annotated *C. viridiflorum*.

Griffith (1851) described *C. carnosum* from a specimen which he had collected in the Khasia Hills. This is also attributable to *C. cyperifolium*. The description includes several characters which are diagnostic of *C. cyperifolium* subsp. *cyperifolium*, notably its long bracts which exceed the ovary in length, its connivent petals, its green sepals and petals, its connivent callus ridges and its four sessile pollinia.

C. aliciae, described from Luzon in the Philippines by Quisumbing (1940), agrees closely with *C. cyperifolium*. It has 8–10 leaves placing it in section *Maxillarianthe* with *C. faberi* and *C. cyperifolium*. The flowers differ slightly from *C. cyperifolium* in their very narrow, acuminate sepals (3–4 mm wide) and petals which are yellowish in colour with some purplish staining, in their yellow callus ridges and in their much shorter floral bracts, but the lip is similar to that of *C. cyperifolium*, with somewhat angled apices to the side-lobes, although the mid-lobe is more triangular in shape. Despite having seen little material of this species for study we are convinced that it is conspecific with *C. cyperifolium*.

C. cyperifolium has been recorded by various authors from the Himalaya, Meghalaya and western China. Wu & Chen (1980) state that the distribution of *C. cyperifolium* in China includes the provinces of Yunnan, Hainan, Guangdong, Guangxi and Guizhou, but some of these collections may be attributable to *C. faberi*.

C. cyperifolium is best known from specimens collected in the Himalaya and Meghalaya of northern India. However, the distribution of this species continues south into Indo-China, where a distinctive variant occurs which is recognised here as a separate subspecies.

Key to the subspecies of *C. cyperifolium*

Leaves 50–90 cm long; floral bracts 18–35(43) mm long; sepal and petals green with red-brown at the base over the mid-vein; petals broader than the sepals . . . . *subsp.* **cyperifolium**

Leaves 32–51 cm long; floral bracts 3–22(25) mm long; sepals and petals pale yellow with 5–7 red-brown longitudinal lines; petals equal to or narrower than the sepals . . . . *subsp.* **indochinense**

subsp. **cyperifolium**.

This subspecies is a medium-sized, terrestrial, clump-forming herb. Its leaves are numerous (usually 7–10 per pseudobulb) and their strongly distichous arrangement gives a fan-like appearance. They are narrow, grass-like, stiffly arching and have an entire margin. The lowest leaves are usually short, making a gradual transition between the leaves and the cataphylls which surround the pseudobulb. The slender scape has up to six flowers with apple-green sepals and petals, with a short red-brown streak over the mid-vein at the base. The petals usually point forward, covering the column. The ovate mid-lobe of the lip has an entire margin, is pale green or whitish, with a few red-brown blotches, and is somewhat papillose. The flowers are subtended by very long, slender floral bracts which usually equal or exceed the pedicel and ovary in length.

ILLUSTRATIONS. Photographs 128, 129; figure 27.3.

DISTRIBUTION. Nepal, N E India, Bhutan, ?S China.

The numerous leaves, which show a gradual transition to the cataphylls, the green flowers, and the papillae on the lip, distinguish this subspecies from *C. ensifolium*, *C. sinense*, *C. munronianum* and *C. kanran*, which have only 3–4 leaves. The long, slender floral bracts are usually characteristic of subsp. *cyperifolium*. *C. kanran* also has rather long floral bracts, and may have green flowers and a papillose mid-lobe, but its sepals are much more slender and acuminate and it has fewer leaves per pseudobulb.

C. faberi, which has a similar number of leaves to *C. cyperifolium*, is distinguished by its often finely serrated leaf margins, its usually much shorter floral bracts, its usually more robust scape with more numerous (up to 20) flowers, and its longer, narrowly ligulate or tapering mid-lobe of the lip, which has a strongly undulating and minutely fimbriate margin. The mid-lobe is covered by inflated papillae in both *C. faberi* and *C. cyperifolium* subsp. *cyperifolium*, but those in *C. faberi* are more numerous, larger and more conspicuous.

Specimens of *C. faberi* from Nepal, and the western Himalaya of N India often have bracts which are similar in length to those of *C. cyperifolium* subsp. *cyperifolium*, and they closely resemble each other in habit and flower colour. This has led to the confusion of these two species by several authors including Duthie (1906), Banerjee & Thapa (1978) and Raizada, Naithari & Saxena (1981). The distinguishing characters are discussed more fully under *C. faberi* var. *szechuanicum*.

C. cyperifolium *subsp.* **indochinense** *DuPuy & Cribb*, **subsp. nov.**, e subspeciei typica foliis brevioribus plerumque minus 50 cm longis, bracteis floralibus brevioribus 3–22 mm longiis, petalis sepalisque aureis rubro-brunneis striatis ornatis, et sepalis quam petalis latiis distinguendo. Typus: Thailand, Chiengrai, *Put* 3972 (holotypus K!).

A medium to small, terrestrial herb which resembles *C. ensifolium* in its habit and flower colour (pale yellow with 5–7 longitudinal red-brown lines). However, it has 5–8 leaves, and the transition between the full-size leaves and the cataphylls is gradual.

Its habit is similar to var. *cyperifolium*, although it is usually a smaller plant (leaf length 32–51 cm in this variety, 50–90 cm in var. *cyperifolium*). The mid-lobe of the lip of subsp. *indo chinense* has a similar shape to that of subsp. *cyperifolium*, with angled side-lobe apices, and it has similar papillae, although they are more sparsely scattered and are often only evident towards the apex. Otherwise, var. *indo chinense* differs from var. *cyperifolium* as indicated in the key.

ILLUSTRATION. Figure 27.5.

DISTRIBUTION. Burma, Thailand, Cambodia, Philippines

Subsp. *indochinense* differs from *C.faberi*, which has a similar leaf number, in its few-flowered scape, and several characters of the mid-lobe of the lip, which is ovate, rather than ligulate or tapering. The indumentum of inflated papillae is much less pronounced than in *C. faberi*. The margin of the lip of *C. faberi* is characteristically minutely fimbriate and strongly undulate, whereas the lip of subsp. *indo chinense* is entire.

The long floral bracts, which are characteristic of subsp. *cyperifolium*, are much shorter in subsp. *indo chinense*. Flower colour may also cause confusion with related species. In Thailand, two related species occur; *C. sinense* and *C. ensifolium*. These three species are very similar in habit, size and flower colour. The differences between them are summarised in table 10.12. The key characters of each species are indicated by an asterisk.

Seidenfaden (1983) highlights the confusion surrounding this group of related species, in Thailand. *Cymbidium sinense* (his figure 39) appears to be correctly identified. *C. ensifolium* (his figure 38) shows many of the features of *C. ensifolium* subsp. *haematodes*, except that the flowers drawn on the spike differ from the flower drawn in close-up in having broader petals. The plant referred to as *C. siamense* (his figure 40) is probably referable to *C. cyperifolium* var. *indochinense*, with six leaves, and the petals the same breadth as the sepals. *C. siamense*, usually distinguished by its broad petals, is actually a synonym of *C. ensifolium* subsp. *haematodes* in the present work.

Subsp. *indochinense* is found in Burma, Thailand and Cambodia, at altitudes of 300–900 m (985–2950 ft). It grows in open, probably deciduous forest, in grassy undergrowth, and flowers from May to July. In common with some specimens of *C. ensifolium* subsp. *haematodes* in Thailand, the pseudobulbs are sometimes produced below soil level, and are further protected, by the fibrous cataphylls and sheathing leaf bases, from the numerous fires in their habitat. All of the specimens of this taxon which have been examined have lost their old leaves and show signs of scorching on the older pseudobulbs, with only the new growths bearing leaves.

41. C. faberi *Rolfe* in Kew Bull.: 198 (1896); Schltr. in Fedde, Repert., Beih. 4: 266 (1919); Su, Native Orchids Taiwan: 114–15, t.14.3 (1975); Lin, Native Orchids Taiwan 2: 109–10, t.40 + fig. (1977); Liu & Su in Fl. Taiwan 5: 942–3 (1978); Wu & Chen in Acta Phytotax. Sin. 18(3): 299 (1980); Seth & Cribb in Arditti (ed.), Orchid Biol., Rev. Persp. 3: 292 (1984). Type: China, Chekiang (Zhejiang), Mt. Tientai, *Faber* 94, in part (lectotype K!, chosen here).

C. scabroserrulatum Makino in Jap. Bot. Mag. 16: 154(1902). Type: China, Musashi, cult. Tokyo, *Makino* s.n. (holotype MAK).

C. oiwakensis Hayata, Icon. Pl. Formos. 6: 80, t.14 (1916), Mark, Ho & Fowlie in Orchid Dig. 50: 22 (1986). Type: Taiwan, Gokwanzan, Oiwaka, *Hayata* s.n. (holotype TI).

C. cerinum Schltr. in Fedde, Repert., Beih. 12: 350–1 (1922). Type: E Tibet, Xizang Province, *Limpricht* 1392 (holotype B).

C. fukiense Yen, Icon. Cymid. Amoyens. AI (1964). Type: China, Fukien, Changchow, *Yen* 3001 (holotype not located).

Table 10.12: A comparison of three similar species which occur sympatrically in N Thailand

Character	***C. cyperifolium* subsp. *indochinense***	***C. ensifolium* subsp. *haematodes***	***C. sinense***
Leaf number	*(5)6–8	2–4(5)	2–4(5)
Leaf breadth (mm)	*9–14	14–17	*(15)20–35
Leaf length (cm)	32–51	52–94	40–100
Flower number	5–7	3–6	*usually 8–15
Dorsal sepal length (mm)	(20)26–33	*16–26(31)	26–39
Petal/dorsal sepal breadth comparison	*petals almost equal to or narrower than the sepals	petals broader than the sepals	petals broader than the sepals
Lateral sepal orientation	slightly drooping	*almost horizontal	slightly drooping
Petal orientation	porrect, forming a hood over the column	*weakly porrect, not forming a hood over the column	porrect, forming a hood over the column
Mid-lobe breadth (mm)	(6)7–9	8–10	9–13
Mid-lobe margin	entire or slightly kinked	*tightly undulating	entire or slightly kinked
Mid-lobe indumentum	*some inflated papillae especially near the apex	minute papillae only	minute papillae only
Flowering period	May–July	(January)February–March	October–March

*indicates key characters for each taxon.

A short to medium-sized, perennial, terrestrial *herb. Pseudobulbs* small, inconspicuous, covered in leaf-like cataphylls which become fibrous with age, and sheathing leaf bases, both with a narrow (less than 1 mm) membranous margin, with 5–9(13) distichous leaves. *Leaves* up to (30)40–100 × 0.4–1.1 cm, linear-elliptic, acute, arching, often grey-green in colour, often with a serrulate margin, conduplicate at the base but not strongly tapering into a petiole, obscurely articulated 2–6 cm from the pseudobulb, the shortest leaves merging with the cataphylls. *Scape* 26–62 cm tall, erect, slender or robust, arising basally from within the outermost cataphylls, with 4–20 flowers produced in the apical half to third of the scape; peduncle covered in sheathing sterile bracts up to 6.5 cm long, usually overlapping towards the base of the spike; bracts narrowly triangular, (0.8)1.0–3.0(4.0) cm long. *Flowers* about 6 cm across, often drooping and not opening fully; lightly scented; rhachis dull green, pedicel and ovary greenish, stained red-brown; petals and sepals green to yellowish, sometimes stained reddish, especially over the mid-vein of the petals; lip yellowish or green, often with a narrow white margin, side-lobes lined red, usually with a red margin, mid-lobe with many reddish spots and blotches; column yellowish; anther-cap pale yellow. *Pedicel and ovary* 1.4–2.9(3.9) cm long. *Dorsal sepal* (22)26–36(44) × 5.8–10.4 mm, narrowly obovate to elliptic or oblong-elliptic, acute to acuminate, suberect; lateral sepals similar, spreading. *Petals* 20–33 × 6.4–10.9 mm, slightly shorter than the sepals, but almost equal in breadth, similar in shape or more ovate, usually somewhat porrect and covering the column. *Lip* 1.9–3.3 cm long when flattened, with a long mid-lobe; side-lobes erect, rounded, often reduced, minutely papillose or minutely pubescent; mid-lobe long, (9)11–17 × 5.1–12.2 mm, ligulate, often tapering to a mucronate apex, occasionally oblong with a broad, mucronate apex, strongly recurved, often as broad as the side-lobes when the lip is flattened, covered in glossy, inflated papillae, margins minutely fimbriate or erose and strongly undulating and crisped; callus in two ridges, converging in the apical half to form a small tube at the base of the mid-lobe. *Column* 1.2–1.9 cm long, arching, narrowly winged; pollinia 4, broadly ovoid. *Capsule* 5 cm long, fusiform or oblong-fusiform, erect with a short, apical beak.

Figure 30

1. ***C. faberi* var. *szechuanicum*** (Sichuan, *Henry* 5515)
   - a Perianth, × 1
   - b Lip and column, × 1
   - c Pollinarium, × 4.5
   - d Pollinia (one pair, reverse), × 4.5
   - e Pollinium, × 4.5
2. ***C. faberi* var. *faberi*** (Yunnan, *Cavalerie* 2233)
   - a Perianth, × 1
   - b Lip and column, × 1
   - c Pollinarium, × 4.5
   - d Pollinia (one pair, reverse), × 4.5
   - e Pollinium, × 4.5
3. ***C. faberi* var. *szechuanicum*** (Nepal, *Bailes* 1040; Kew spirit no. 49390)
   - a Perianth, × 1
   - b Lip and column, × 1
   - c Pollinarium, × 4.5
   - d Pollinia (one pair, reverse), × 4.5
   - e Pollinium, × 4.5
   - f Bract, × 1
4. ***C. lancifolium*** (Kew spirit no. 48293)
   - a Perianth, × 1
   - b Lip and column, × 1
   - c Pollinarium, × 4.5
   - d Pollinia (one pair, reverse), × 4.5
   - e Pollinium, × 4.5
5. ***C. goeringii*** (Kew spirit no. 48256)
   - a Perianth, × 1
   - b Lip and column, × 1
   - c Pollinarium, × 4.5
   - d Pollinia (one pair, reverse), × 4.5
   - e Pollinium, × 4.5
6. ***C. goeringii*** (Kew spirit no. 26171)
   - a Perianth, × 1
   - b Lip and column, × 1
   - c Pollinarium, × 4.5
   - d Pollinia (one pair, reverse), × 4.5
   - e Pollinium, × 4.5
7. ***C. macrorhizon*** (Kew spirit no. 40631)
   - a Perianth, × 1
   - b Lip and column, × 1
   - c Pollinarium, × 4.5
   - d Pollinia (one pair, reverse), × 4.5
   - e Pollinium, × 4.5
   - f Flower, × 1

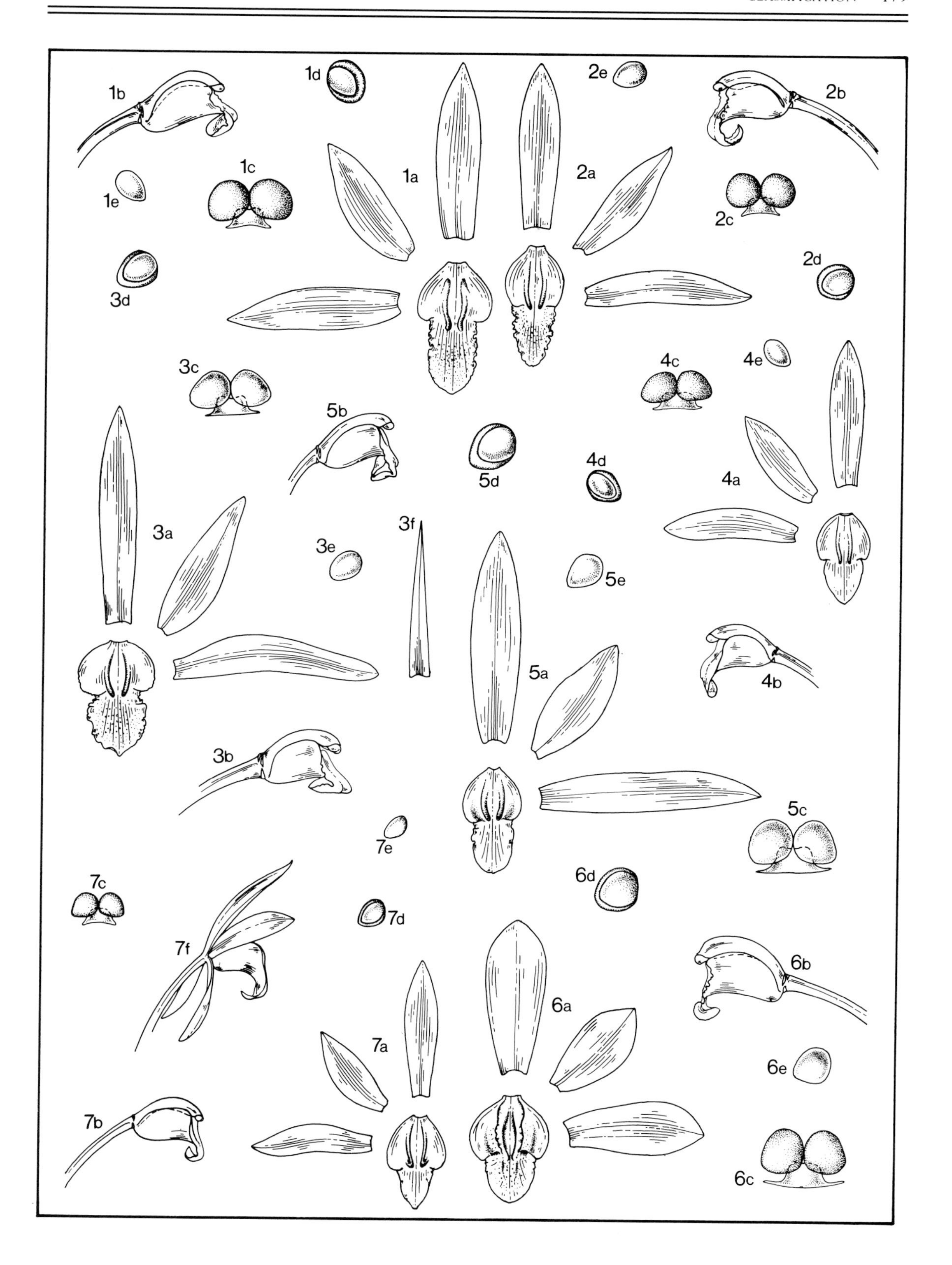
1b
1d
2e
2b
1e
1c
1a
2a
2c
2d
3d
3c
5b
4c
4e
5d
4d
4a
3a
3f
3e
5e
5a
4b
3b
5c
7e
7c
6d
7d
7f
6b
6a
7a
6e
7b
6c

DISTRIBUTION. Nepal, N India (N Uttar Pradesh), China (Henan, Anhui, Zhejiang, Jiangxi, Hunan, Guizhou, Yunnan, Sichuan), Taiwan (map 11B); (700)1000–2900 m ((2295) 3280–9515 ft).

HABITAT. On steep land or cliffs, often amongst *Miscanthus*, in open situations or in shade in low, scrubby forest; flowering (January)March–June.

The habit of *C. faberi*, the 6–9 slender leaves without subepidermal strands of lignified sclerenchyma, the slightly drooping flowers and the porrect petals (and often the dorsal sepal also) forming a close covering over the column, place *C. faberi* along with *C. goeringii* in section *Maxillarianthe*.

C. faberi was described from two specimens, *Faber* 94 and *Henry* 5515, both from S China. The former is a mixed collection from Zhejiang, composed of *C. goeringii* (right-hand specimen) and *C. faberi* (left-hand specimen), which has been selected here as the lectotype. Both it and *Henry* 5515 agree with the type description but the latter specimen has broader leaves, longer floral bracts, and flowers with more acuminate segments and a broader mid-lobe of the lip (see under var. *szechuanicum*).

C. scabroserrulatum is considered here to be conspecific with *C. faberi*. It was described by Makino (1902) from a plant cultivated in Japan, but originally imported from China. In the leaf number (up to nine), the minutely serrulate leaf margins, the flower number (5–10) and colour, and lip features it agrees well with var. *faberi*. The habit of the plant is likened, by Makino, to *C. virescens* (= *C. goeringii*) which is also in section *Maxillarianthe*, and the description again agrees well with *C. faberi*. Flower colour, shorter bract length, larger flower number and the undulating, fimbriate lip margin exclude this from *C. cyperifolium*.

Hayata (1916) described *C. oiwakense* from a plant collected in Taiwan, differentiating it from all other Taiwanese species by its 'manifestly denticulate lips'. This is described more fully in the type description as a crisped, undulating and minutely erose lip margin, a set of characters diagnostic of *C. faberi*. The rest of the description also agrees well with *C. faberi*, and Su (1975), Lin (1977), Liu & Su (1978) and Wu & Chen (1980) have all included *C. oiwakense* as a synonym.

Schlechter (1922) described *C. cerinum* from a specimen collected by Limpricht in east Tibet (Xizang Province), where it was in cultivation. Schlechter noted that it was without doubt closely related to *C. faberi*, but that it differed in the waxy yellow flower colour, the narrower sepals, the remarkably wide, parallel (not S-shaped) keels and in the mid-lobe of the lip. The colour is not unusual in *C. faberi* and the other features all fall within the range of variation of *C. faberi*. This variant is therefore placed in the synonymy of *C. faberi*, following the treatment of several previous authors, notably Wu & Chen (1980).

Its wide distribution, and variation over this range, have undoubtedly led to this species being described several times under different names. Wu & Chen (1980) recognise three varieties of this species; var. *faberi*, var. *szechuanicum* and var. *omiense*. Var. *szechuanicum* appears to be distinct. Var. *omiense* has fewer leaves (4–5) than *C. faberi*, a shorter scape not exceeding the leaves in length, smaller and fewer flowers with a short, ovate not ligulate, acute mid-lobe with a margin that is not minutely fimbriate, nor is it undulating. These characters prevent its inclusion in *C. faberi*, but are similar to some specimens of *C. kanran*, under the synonymy of which this taxon is now tentatively placed. The description of *C. omiense* as 'feathery' corresponds well with the 'spidery' appearance of the flowers of *C. kanran*.

Key to the varieties of *C. faberi*

Plant small to medium; leaves 5–8, usually up to 8 mm broad, grey-green; scape robust, usually with 9–20 flowers; floral bracts mostly much shorter than the pedicel and ovary; sepals somewhat obovate, obtuse to acute, mucronate; mid-lobe of the lip narrowly ovate, tapering to an acute apex (see figure 31B) .... *var.* **faberi**

Plant medium to robust; leaves 8–9(15), 8–11 mm broad, green; scape slender to robust, usually with 2–8 flowers; floral bracts exceeding the pedicel and ovary in the lower flowers; sepals elliptic, long and acuminate; mid-lobe of the lip oblong, with a broad, mucronate apex (see figure 31A) .... *var.* **szechuanicum**

var. **faberi**

This is typically a small plant resembling *C. goeringii*, with 6–9 short, narrow, grey-green, arching leaves, often with serrulate margins. It has a surprisingly robust scape with numerous (up to 20) flowers which are olive-green to yellow-green in colour, and are held above the foliage. The sepals are usually narrowly obovate, and the petals are porrect and cover the column. The mid-lobe of the lip is yellow or green, usually with red markings, elongated, covered in many small, glossy, inflated papillae and has a tightly undulating and erose margin.

ILLUSTRATIONS. Photograph 130; figures 30.2, 31.

DISTRIBUTION. China, Taiwan.

This variety is somewhat similar to *C. cyperifolium* in that they both usually have 5–9 or so leaves, the lowest leaves short and merging with the cataphylls, and the mid-lobe has some inflated papillae. It is easily distinguished by its long mid-lobe in comparison with the total length of the lip, its strongly undulating and minutely fimbriate mid-lobe margin, its much more numerous and dense, inflated, shiny papillae on the mid-lobe, its larger number of flowers on a more robust spike and its much shorter floral bracts.

C. ensifolium is found in China in the same regions as var. *faberi*, but usually at lower altitudes, and is easily distinguished by its fewer (usually broader) leaves (about four), which are all long and distinct from the cataphylls, its weaker, fewer flowered spike and the lip without either inflated papillae or an undulating and fimbriate mid-lobe margin.

Var. *faberi* is a high altitude plant, growing up to 3000 m (9840 ft) in mountainous regions in the southern provinces of China and in Taiwan. It is often found growing in open, sunny situations amongst stands of *Miscanthus* which has leaves similar in appearance. It often grows on steep land or on cliffs, often near streams in soil-filled crevices and ledges, and it has also been collected in more shaded, moist forests. The distribution, and the high altitude preference, suggests that it is cold-tolerant, probably as hardy as *C. goeringii*, which will survive outdoors in a sheltered position in warmer parts of the British Isles.

var. **szechuanicum** *(Y.S. Wu & S.C. Chen)Y.S. Wu & S.C. Chen* in Acta Phytotax. Sinica 18: 299 (1980). Types: Sichuan, Chion-lai-shan, *Wu* 2040, *Wu & Fee* 2055, *Fee* 2061 (syntypes PE).

C. cyperifolium sensu Duthie in Ann. Roy. Bot. Gard. Calcutta 9: 135 (1906), *sensu* Banerjee & Thapa, Orchids of Nepal: 89 (1978) & *sensu* Raizada, Naithani & Saxena, Orchids of Mussoorie: 39–40 (1981), *non* Lindley.

C. szechuanicum Y.S. Wu & S.C. Chen in Acta Phytotax. Sinica 11: 33 (1966).

Var. *szechuanicum* differs from var. *faberi* in that it often has fewer flowers, on a weaker scape, its floral bracts are longer and often exceed the ovary in the lower flowers, it has longer, elliptic sepals tapering to an acuminate apex and the lip is broader and more oblong in shape, with a broad, mucronate apex (figure 31A). The plant is larger and more luxuriant in its growth habit, closely resembling *C. cyperifolium*.

ILLUSTRATIONS. Plate 26; photograph 131; figures 30.1, 30.3, 31.

DISTRIBUTION. China (Sichuan, Yunnan, Guizhou), Nepal, India (N Uttar Pradesh).

Wu & Chen (1966) originally described this as a distinct species, but later (1980) reduced it to varietal status within *C. faberi*. The type specimen was collected in Sichuan, and the second specimen cited in the description of *C. faberi*, *Henry* 5515, was also collected there, and it is included in var. *szechuanicum*.

Several other collections, from W China and the Himalaya of N India and Nepal, which had previously been identified as *C. cyperifolium*, are included here in *C. faberi* var. *szechuanicum* (Duthie, 1906; Banerjee & Thapa, 1978; Raithada, Naithani & Saxena, 1981). Vegetatively they closely resemble the sympatric *C. cyperifolium*, but their leaves are slightly narrower and have a minutely serrulate margin. The scapes of specimens from N India and Nepal are slender and usually carry 2–5 flowers, while those from China are usually robust, and more similar to those of var. *faberi*. The few-flowered scapes, the long floral bracts and the similar-shaped green flowers of var. *szechuanicum* are very suggestive of *C. cyperifolium*,

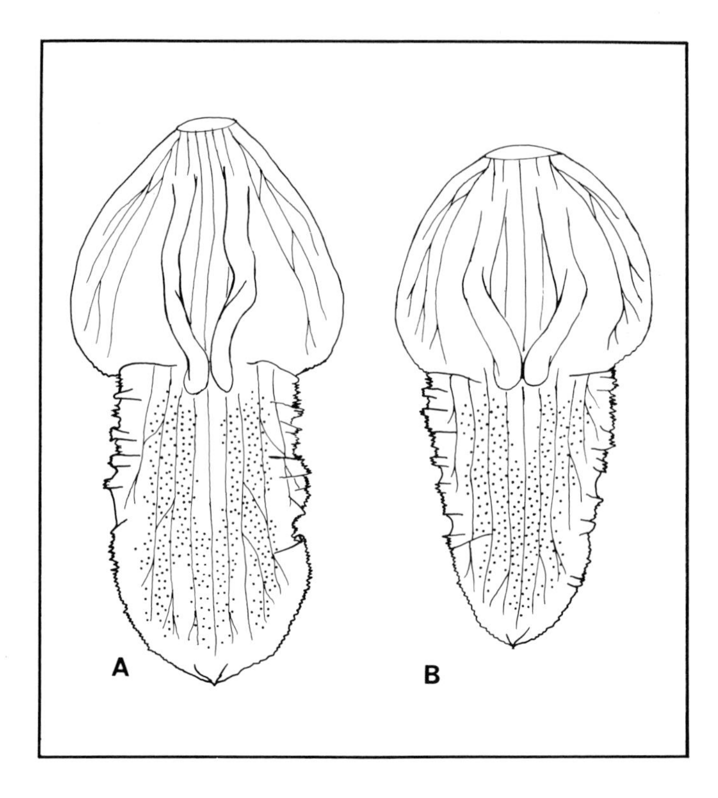

FIGURE 31: The lip shapes of the two varieties of *C. faberi*, partially flattened, × 2.5
A – *C. faberi* var. *szechuanicum*
B – *C. faberi* var. *faberi*

but the mid-lobe of the lip has a tightly undulating and minutely erose margin, and is ornamented with numerous, inflated, shiny papillae in *C. faberi*. These characters of the mid-lobe are the most constant characters which distinguish these two species. They further differ in their flowering period; from January until April for *C. faberi* and during November and December for *C. cyperifolium* in this region.

Recent collections made from the Kathmandu Valley in E Nepal (*Bailes* 1040), at about 2200 m (7220 ft), in temperate oak/rhododendron forest, and also in the N W Himalaya of N India, at about 2000 m (6560 ft) near Mussoorie (*Du Puy* 558), in similar vegetation which have proved to be identical with those collected by Duthie (1906). The Mussoorie collection was made in early January at which time there was frost at night, and patchy snow lay on the ground, suggesting that they may be hardy in sheltered areas of Britain. In this region *C. faberi* occurs on steep north or north-west facing slopes in deep shade under dense scrubby woodland cover. The major constituent species in the woodland were the evergreen oak *Quercus leucotricophora*, *Rhododendron arboreum*, *Pieris ovalifolia*, *Viburnum coriaceum* and *Berberis* species, with a wild *Rosa* species and a white-leaved bramble, *Rubus paniculatus* scrambling through these low trees. *C. faberi* was often found on rocky outcrops, or at the base of *Mahonia nepaulensis* shrubs which provided protection from grazing animals. Ground cover was sparse, including ferns and sedges amongst the leaf litter. Plants flowered at Kew during February, at the same time as the specimens collected in E Nepal.

42. C. goeringii *(Reichb.f.)Reichb.f.* in Walp. Ann. 3: 547 (1852); Summerhayes in Bot. Mag. 174: t.413 (1963); Ohwi, Fl. Japan: 355 (1965); Garay & Sweet, Orchids Southern Ryukyu Islands: 143 (1974); Y.N. Lee, Ill. Fl. Fauna Korea 18: 771, 826, t.183 (1976); Wu & Chen in Acta Phytotax. Sin. 18: 299 (1980); Seth & Cribb in Arditti (ed.), Orchid Biol., Rev. Persp. 3: 290 (1984). Type: Japan, *Goering* 592 (holotype W).

Maxillaria goeringii Reichb.f. in Bot. Zeit. 3: 334 (1845).
C. virescens sensu Lindley in Bot. Reg. 24: misc. 37 (1838); Makino in Iinuma, Somoku Dzusetsu 4(18): 1186–7, t.15 (1912); Schltr. in Fedde, Repert.

Beih. 4: 272 (1919); Y.N. Lee, Ill. Flor. Fauna Korea 18: 771, 826, t.183 (1976); *non* Willd. (1805). Type: Japan, cult. *Rollinsons, Siebold* (holotype K!).

C. virens Reichb.f. in Walp. Ann. 6: 626 (1861). Type: as *C. virescens* Lindl. [*sphalm.* for *C. virescens* Lindl.].

C. mackinnoni Duthie in J. Asiat. Soc. Bengal 71(2): 41 (1902). Type: N W India, nr. Mussoorie, *Mackinnon* s.n. (isotype K!), **syn. nov.**

C. formosanum Hayata in J. Coll. Sci. Tokyo 30: 335 (1911) & in Mater. Fl. Formos.: 335 (1911); Schltr. in Fedde, Repert. Beih. 4: 267 (1919); Lin, Nat. Orchids Taiwan 2: 112–16, + figs. (1977); Liu & Su, Fl. Taiwan 5: 943, t.1572 (1978); Mark, Ho & Fowlie in Orchid Dig. 50: 27, + figs. (1986). Type: Taiwan, *Nakahara* s.n. (holotype TI).

C. forrestii Rolfe in Notes Roy. Bot. Gard. Edinburgh 8: 23, t.11 (1913); Schltr. in Fedde, Repert. Beih. 4: 267 (1919). Type: S W China, Yunnan, *Forrest* 415 (holotype E!).

C. yunnanense Schltr. in Fedde, Repert. Beih. 4: 74 (1919). Type: China, Yunnan, E of Tong-Tchouan, *Maire* 6425 (holotype B†).

C. pseudovirens Schltr. in Fedde, Repert. Beih. 12: 351 (1922). Type: China, Zhejiang, Ningpo, *Limpricht* 304 (holotype B†).

C. tentyozanense Masam. in Trans. Nat. Hist. Soc. Formosa 25: 14 (1935), **syn. nov.**

C. uniflorum Yen, Icon. Cymbid. Amoy., A2 (1964). Type: China, Fukien, *Lin* 7001 (holotype not located).

C. chuen-lan C. Chow, Formosan Orchids ed. 1: 21 (1968); *nom. inval.*, holotype not indicated.

A terrestrial *herb* with thick, fleshy roots. *Pseudobulbs* small, ovoid, enclosed in about 6 scarious cataphylls. *Leaves* 5–7(8), up to 80 × 0.2–1 cm, but usually less than 40 cm long, the shortest merging with the cataphylls, linear-elliptic, acute, usually V-shaped in section, arching, usually with a serrated margin, narrowed towards the base. *Scape* erect, 1-(occasionally 2–4) flowered; peduncle up to 15(21) cm, covered in 4–8 sheaths; sheaths up to 6(9) cm long, becoming scarious, cylindrical in the basal half, expanded and cymbose in the upper half, acute; bract 2.5–6 cm, usually exceeding the ovary, scarious, cymbose, acute, often red-tinted. *Flower* about 4–5 cm across, porrect or slightly nodding; sometimes scented; pedicel and ovary green to purplish; sepals and petals apple-green to red-brown, stained red towards the base, especially over the mid-vein of the petals; lip cream with crimson spots and margin on the side-lobes and sparse red blotches on the mid-lobe; callus cream to pale yellow; column usually pale green, cream towards the base, spotted lightly above and densely below with maroon; anther-cap cream, sometimes purple or yellow below. *Pedicel and ovary* 2.5–6 cm, curved behind the flower. *Sepals* usually obovate to elliptic, obtuse or apiculate, occasionally narrowly elliptic, acute, margins often incurved; dorsal sepal often porrect, 2.5–3.9 × 0.85–1.25 cm; lateral sepals similar, spreading. *Petals* 1.85–3 × 0.85–1.1 cm, oblong-ovate to oblong-elliptic, obtuse, oblique, usually closely covering the column. *Lip* 1.7–2.6 × 1–1.6 cm when flattened, subentire to 3-lobed; side-lobes sometimes much reduced, erect, rounded, sometimes slightly angled at the front, papillose; mid-lobe 0.7–1 × 0.75–1.1 cm, broadly ovate to oblong, obtuse or rounded, strongly recurved, papillose, the margin sometimes minutely undulate; callus 2-ridged, convergent towards the apex and forming a short tube which extends into the base of the mid-lobe, with two large pads of callus on the side-lobes adjacent to the callus ridges. *Column* 1.3–1.9 cm long, broadening into two wings towards the apex; pollinia 4, in two unequal pairs. *Capsule* up to 8 cm long, fusiform, erect, and parallel to the rhachis, pedicellate, beaked.

DISTRIBUTION. Japan, Korea, Ryukyu Islands, widespread in southern China and Taiwan, rare in N W India (map 11C); 500–3000 m (1640–9840 ft).

HABITAT. Terrestrial in open forest, usually on lightly shaded cliffs or slopes, often in coniferous forests near the sea in Japan; flowering during January–March.

This species was first described by Lindley as *C. virescens*, based on a specimen collected in Japan in 1838. However, this name had previously been used by Willdenow (1805) for a South American orchid. The next available name for this species is *C. goeringii* originally described as *Maxillaria goeringii* by H.G. Reichenbach (1845), based on a specimen collected in Japan by Goering. He later transferred it to *Cymbidium* (1852).

The name *C. virens* Reichb.f., which is based on the same type specimen as *C. virescens* Lindl., appears to be a mis-spelling of the latter name introduced by Reichenbach into circulation, and unfortunately taken up by later authors.

C. goeringii varies considerably over its entire range. Some of the most distinct variants are found in Taiwan, including that described by Fukuyama as *C. formosanum*. This name was based on a variant with slender, elliptic, acute sepals, and is often used for large-flowered variants from Taiwan. Wu & Chen (1980) regard this as synonymous with *C. goeringii*, with no discontinuities in variation to suggest that it is a separate taxon. Similar variants are found in S China and Japan. The variability of var. *goeringii* in Taiwan is well demonstrated by Mark *et al.* (1986), who illustrate it with four photographs (as *C. formosanum*).

C. yunnanense Schltr. from S W China is a variant with small flowers, a glabrous lip and a supposedly distinct callus structure. The first two characters are highly variable in *C. goeringii*, and the description of the callus agrees well with that of *C. goeringii*. Wu & Chen (1980) also consider this to be synonymous with *C. goeringii*.

Similarly, Schlechter (1922) distinguished *C. pseudovirens* from *C. goeringii* by the presence of two outgrowths on the outer side of the keels, and a distinct mid-lobe shape. The pads of callus on the side-lobes of the lip which are typical of *C. goeringii* but vary in prominence. The mid-lobe, from its description, falls within the range of variation normally found in *C. goeringii*.

There has been some discussion of the differences between Chinese and Japanese specimens, and the possibility of recognising two distinct taxa. Nagano (1955) suggests that they differ in that the sepals and petals of the Chinese plants are more slender and red-brown in colour, and the flowers have a stronger fragrance. Summerhayes (1963) reiterates these differences, stating that the Chinese specimens 'may be distinguished by the sepals which are broadest in the middle instead of the upper part ...'. In general, such patterns of variation are encountered, but specimens with obovate or elliptic petals have been collected in both countries. Variation in colour and scent also occurs, but these characters are not constant and cannot be used to separate two taxa.

An apparently widely disjunct portion of the distribution is in the Himalaya of northern India, west of Nepal. The plants collected from around Mussoorie, and described by Duthie (1902) as *C. mackinnoni*, are similar to the type of *C. goeringii*, except that the leaf margins are not serrated. Various degrees of serration can be found in *C. goeringii* elsewhere in its range and therefore this character cannot be used reliably to differentiate these taxa.

Rolfe described *C. forrestii* from a specimen collected in Yunnan, and differentiated it from the Japanese *C. goeringii* by the prominent side-lobes of its lip, and the prominent mounds of tissue on the side-lobes adjacent to the usual two callus ridges. In the type specimen of *C. virescens* the lip is also three-lobed and the mounds on the side-lobes are present but not as prominent. Therefore, these characters are considered insufficient to distinguish *C. forrestii* as a distinct species.

C. goeringii has been cultivated in Japan and China for several centuries, and many variants have been selected there and maintained in cultivation. Some of these, such as those with variegated leaves, or albino or peloric flowers, are very distinct and highly prized.

Wu & Chen (1980), in their revision of *Cymbidium* in China, recognised four distinct varieties of *C. goeringii*. Three of these are accepted here, var. *longibracteatum* being considered insufficiently different from var. *tortisepalum* to be given taxonomic recognition. This is a very variable species, in flower shape, flower colour, number of flowers and in the breadth of the leaves. Many particularly interesting variants occur in Taiwan.

Key to the varieties of *C. goeringii*

1. Flowers (1)2–4(5); bract usually shorter than the ovary .... *var.* **tortisepalum**
   Flowers solitary, occasionally two; bract usually longer than the ovary 2
2. Leaves 2–4 mm broad .... *var.* **serratum**
   Leaves 5–10 mm broad .... *var.* **goeringii**

var. **goeringii**

Var. *goeringii* is typically a small, tufted species with 5–7 short, arching leaves. Its flowers are usually borne singly, on a short scape which is covered by overlapping, scarious bracts which may be tinged with red. The flowers are often slightly nodding, and do not open widely, the petals remaining porrect, often closely covering the column. The sepals and petals are often green, with a red line over the base of the mid-vein of the petals, but variants with various degrees of red-brown pigmentation are known. The sepals are often widest near the apex, although variants with oblong or elliptic sepals are not unusual, and the size of the flower can also vary considerably.

ILLUSTRATIONS. Plate 27; photographs 5, 6, 132–136; figures 30.5, 30.6.

DISTRIBUTION. Japan, Korea, Ryukyu Islands, Taiwan, S China, N W India (N Uttar Pradesh).

Wu & Chen give the distribution of var. *goeringii* in China as including all of the southern provinces as far north as Sichuan, Henan and Jiangsu, and including Taiwan but not Hainan Island or Hong Kong (Anhui, Fujian, Guangdong, Guangxi, Guizhou, Hubei, Henan, Hunan, Jiangsu, Sichuan, Taiwan, Yunnan and Zhejiang). Y.N. Lee (1976) notes that it is abundant in forests in the southern part of Korea, and it has also been collected on Tsushima in the Straits of Korea.

Maekawa (1971) notes that *C. goeringii* occurs in southern Japan in the warm, temperate vegetation zone, in *Castanopsis cuspidata* and *Machilus thunbergii* dominated broad-leaved, evergreen forest, often with *Podocarpus* and *Cephalotaxus* (both Coniferae) and *Camellia japonica*. It is usually found in open forest, on lightly shaded cliffs or rocky slopes, often amongst grasses or bamboos.

var. **serratum** *(Schltr.)Y.S. Wu & S.C. Chen* in Acta Phytotax. Sinica 18: 300 (1980). Type: China, Guizhou, *Esquirol* (holotype B).

C. serratum Schltr. in Fedde, Repert. Beih. 4:73–4 (1919).

C. gracillimum Fukuyama in Trans. Nat. Hist. Soc. Formosa 22: 413–15, t.1 & 2 (1932); Mark, Ho & Fowlie in Orchid Dig. 50: 24, + figs. (1986). Type: Taiwan, prov. Sintiku, Mt. Tyotui-zan *Fukuyama* 3220 (holotype KANA).

C. goeringii Reichb.f. var. *angustatum* F. Maekawa, The Wild Orchids of Japan in Colour: 416, t.171 (1971), *nom. nud.*

C. formosanum Hayata var. *gracillimum* (Fukuyama)Lui & Su in Flora of Taiwan 5: 943 (1978).

This variety is distinguished from var. *goeringii* by its very narrow, strongly serrated leaves. The flowers are usually solitary, and often have a somewhat spidery shape, with rather slender sepals and petals, although they may be short and obovate. The greenish flowers have some red veining on the sepals and especially strongly on the petals, and the lip is variously blotched with red. A variant without any red pigmentation in the flower is also known. It is usually sweetly scented.

ILLUSTRATION. Photograph 137.

DISTRIBUTION. China (Guizhou), Taiwan, Japan.

Var. *serratum* was described by Schlechter (1919) as *C. serratum*, from a specimen collected in southern China. Wu & Chen (1980) recognised it as a distinct variety of *C. goeringii*, and their treatment is followed here. It also occurs in Taiwan, where it has variously been named *C. gracillimum*, *C. formosanum* var. *gracillimum*, and *C. goeringii* var. *angustatum*. The variability in flower shape and colour in Taiwan is illustrated by Mark *et al.* (1986). They also note that it has an earlier flowering season than var. *goeringii*. It has also been reported from Japan (F. Maekawa, 1971). It occurs mainly at high altitudes (up to 3000 m, 9840 ft) in Taiwan, in steep rocky localities, often in broad-leaved forest (Mark *et al.*, 1986).

var. **tortisepalum** *(Fukuyama)Y.S. Wu & S.C. Chen* in Acta Phytotax. Sin. 18: 300 (1980). Type: Taiwan, *Fukuyama* 3983 (holotype KANA).

C. tortisepalum Fukuyama in Jap. Bot. Mag. 48: 304–6, t.1 (1934); Ying in Quart. J. Chinese Forestry 11(2): 100–1 (1976); Lin, Nat. Orchids Taiwan 2: 129, t.56–59, + fig. (1977); Mark, Ho & Fowlie in Orchid Dig. 50: 24, + figs. (1986).

C. longibracteatum Y.S. Wu & S.C. Chen in Acta Phytotax. Sin. 11(1): 31, t5 (1966). Type: China, Sichuan, Chiou-lai-shan, *Fee* 2064 (holotype PE).

C. tsukengensis C. Chow in Taiwan Orchid Bull. 8: no. 2 (1970) & Formosan Orchids, 2nd ed: 41 (1974); Mark *et al.* in Orchid Dig. 50: 27, 29, + figs. (1986); *nom. inval.*, holotype not indicated.

C. tortisepalum Fukuyama var. *viridiflorum* Ying, Coloured Ill. Ind. Orch. Taiwan 1: 415 (1977). Type: Taiwan, prov. Taichung, Lishan, *Ying* 5282 (holotype not located).

C. goeringii var. *longibracteatum* (Y.S. Wu & S.C. Chen)Y.S. Wu & S.C. Chen in Acta Phytotax. Sin. 18(3): 300 (1980).

This variety was originally described as *C. tortisepalum*, from Taiwan, and is distinct in that the spike commonly bears two to four flowers instead of the single flower of other varieties of *C. goeringii*. Furthermore, its bracts are usually slightly shorter than the pedicel and ovary, and the sepals are slender and elliptic, usually with a slight twist in them and often cream in colour. It has minutely but sharply serrulate leaf margins, and vegetatively the plant resembles *C. goeringii*.

DISTRIBUTION. China (Sichuan), Taiwan.

It seems possible that var. *tortisepalum* is a product of introgression following hybridisation with one of the species from sect. *Jensoa* such as *C. ensifolium* or *C. kanran*, especially when its wide variation in bract length and flower colour are taken into consideration (see also table 10.13). Field studies are required to investigate this possibility. There is evidence in Taiwan of hybridisation between *C. goeringii* and several other species. Chow (1979) illustrates *C. tortisepalum* and several other plants which appear to be intermediate between *C. goeringii* and *C. ensifolium*.

Wu & Chen described *C. longibracteatum* from Sichuan, but later reduced it to varietal rank. They distinguished var. *longibracteatum* by its longer floral bracts which exceed the ovary in length, by the absence of the slight twist in the sepals and by its different mid-lobe shape. However, its similarity to var. *tortisepalum* is striking, and var. *longibracteatum* is therefore reduced to synonymy here. A hybrid origin can again be postulated.

The name *C. tsukengensis* was first used by Chow Chen (1970), but was not validly published. It has recently been resurrected by Mark *et al.* (1986), but no attempt was made to publish it formally. It is applied to specimens from high altitudes in Taiwan, near the highest elevations which var. *tortisepalum* occurs. It differs from this variety in its 1- or 2-flowered scape, and is therefore intermediate between var. *goeringii* and var. *tortisepalum*. Its flowers do not appear to be outside of the range of variation expected in either variety, and this variant is certainly not distinct enough to be regarded as a distinct species.

The variability of the flowers of this variety in Taiwan is well illustrated by Mark *et al.* (1980, as *C. tortisepalum* and *C. tsukengensis*), who have photographed several dissimilar specimens. They also note that it has an attractive scent.

In Taiwan it occurs mainly on steep slopes up to about 1500 m (4920 ft). It prefers an open habitat, often growing in full sun, in association with *Miscanthus* and other grasses in rocky meadows.

Section **Geocymbidium** *Schltr.* in Fedde, Repert. 20: 101 (1924); P. Hunt in Kew Bull. 24: 94 (1970); Seth & Cribb in Arditti (ed.), Orchid Biol., Rev. Persp. 3: 288 (1984). Type: *C. lancifolium* Hook., lectotype chosen by P. Hunt (1970).

The single species in this section is highly distinctive vegetatively, and is characterised by having elongated, narrowly fusiform pseudobulbs which have about 2–3 apical leaves, and a lateral scape. The leaves have an oblanceolate lamina, which narrows towards the base to a slender petiole.

43. C. lancifolium *Hook.*, Exot. Bot. 1: t.51 (1823); Loddiges, Bot. Cab. 10: t.927 (1824); Lindley, Gen. Spec. Orchid. Pl.: 164 (1833) & in J. Linn. Soc. 3: 30 (1858); Hook. f., Fl. Brit. Ind. 6: 9(1891); King & Pantling in Ann. Bot. Gard. Calcutta 8: 185, t.247 (1898); J.J. Smith, Orchid. Java: 476–7 (1905); Maekawa in J. Jap. Bot. 33: 320 (1958); T.P. Lin, Nat. Orch. Taiwan: 119–21, t.47–50, + figs. (1977); Seidenfaden in Opera Bot. 72: 68–71, f.36–37, t.8D (1983); Du Puy & Lamb in Orchid Rev. 92: 352–3, fig. 294 (1984); Seth & Cribb in Arditti (ed.), Orchid Biol., Rev. Persp. 3: 288–9 (1984). Type: Nepal, *Wallich* s.n., cult. *Shepherd* (holotype K!).

C. cuspidatum Blume, Bijdr. Fl. Nederl. Ind. 8: 379 (1825); Lindley, Gen. Sp. Orchid. Pl.: 170(1833);

Table 10.13: Illustrating the intermediacy of *C. goeringii* var. *tortisepalum* between *C. goeringii* var. *goeringii* and *C. ensifolium*

Character	*C. goeringii* var. *goeringii*	*C. goeringii* var. *tortisepalum*	*C. ensifolium*
Leaf number and arrangement	5–6(8), the lowest leaves short, and merging with the cataphylls	5–8, the lowest leaves short and merging with the cataphylls	3–4, all leaves long, cataphylls and leaves well separated
Peduncle and sheaths	peduncle covered in overlapping sheaths	sheaths overlapping in the basal half of the peduncle	sheaths overlapping towards the base of the peduncle
Floral bract length	bract longer than the ovary	bract almost equalling or shorter than the ovary	bract shorter than the ovary
Ovary	ovary erect, bent towards the tip	ovary almost erect, almost parallel to the rhachis, bent towards the tip	ovary angled away from the rhachis, not bent at the tip
Flower number	1(2)	2–4(5)	3–9
Petals	petals porrect, closely covering the column	petals porrect, closely covering the column	petals forward pointing, slightly spreading, not closely covering the column

Griffith, Icon. Pl. Asiat.: t.300 (1851); Type: Java, Salak and Cereme Mts, *Blume* s.n. (holotype L!).

C. javanicum Blume, Bijdr. Fl. Nederl. Ind. 8: 380, t.19 (1825); Maekawa in J. Jap. Bot. 33: 320 (1958); Wu & Chen in Acta Phytotax. Sin. 18: 304 (1980). Type: Java, Mt. Seribu, *Blume* s.n. (holotype L!).

C. gibsonii Lindley Paxt. Fl. Gard. 3: 144 (1852–3); Reichb.f., Walp. Ann. 6: 623 (1861). Type: Assam, Khasia Hills, cult. *Chatsworth* (holotype K!).

C. papuanum Schltr. in Fedde, Repert., Beih. 1: 952 (1913) & in *op. cit.* 21: t.336, no. 1296 (1928); Reeve, T.M. in Orchadian (1984), Type: New Guinea, Bismark Mts, *Schlechter* 18680 (holotype B†), **syn. nov.**

C. caulescens Ridley in J. Fed. Mal. St. Mus. 5: 167 (1915); Seidenfaden in Opera Botanica 72: 67, f.36 (1983). Type: Thailand, Ko Samui Isl., *Robinson* s.n. (holotype SING, isotype K!), **syn. nov.**

C. kerrii Rolfe in Kew Bull.: 381–2 (1925); Seidenfaden in Opera Bot. 72: 68–71, figs. 36 & 37 (1983). Type: Thailand, Doi Suthep, *Kerr* 227 (holotype K!).

C. nagifolium Masamune in Jap. Bot. Mag. 44: 220 (1930); Nackejima, Enum. Orch. Ryukyus, pt. 1: 23, 52, t.29 (1971); Mark, Ho & Fowlie in Orchid Dig. 50: 19 + figs. (1986). Type: Japan, Yakushima, *Masamune* s.n. (holotype TI).

C. aspidistrifolium Fukuyama in Bot. Mag. Tokyo 48: 438, t.213 (1934); Mark, Ho & Fowlie in Orchid Dig. 50: 17, + figs. (1986). Type: Taiwan, Mt. Syoagyoku-san, *Fukuyama* 4137 (holotype KANA).

C. syunitianum Fukuyama in Bot. Mag. Tokyo 49: 757–8 (1935); Mark, Ho & Fowlie in Orchid Dig. 50: 17, + figs. (1986). Type: Taiwan, Kwarenko, Mt. Taroko-taizan, *S. Sasaki* 4688 (holotype KANA), **syn. nov.**

C. javanicum Blume var. *aspidistrifolium* (Fukuyama)F. Maekawa in J. Jap. Bot. 33: 320 (1958) & in Wild Orchids of Japan in Colour: 400, 479, t.163 (1971); Ohwi, Fl. Japan, ed. 2: 454 (1965); Lui & Su, Fl. Taiwan 5: 943 (1978), **syn. nov.**

C. javanicum Blume var. *pantlingii* F. Maekawa in J. Jap. Bot. 33: 320 (1958). Type: Sikkim, near Sureil, *Pantling* 75 (holotype K!), **syn. nov.**

C. maclehoseae S.Y. Hu in Chung Chi J. 11: 15, f.2 (1972) & Gen. Orchid. Hong Kong: 96 (1977). Type: Hong Kong, New Territories, *S.Y. Hu* 9369 (holotype A), **syn. nov.**

C. lancifolium var. *aspidistrifolium* (Fukuyama)Ying, Col. Ill. Indig. Orchids Taiwan 1: 439 (1977), **syn. nov.**

C. lancifolium Hook. var. *syunitianum* (Fukuyama)Ying, Col. Ill. Indig. Orchids Taiwan 1: 439 (1977), **syn. nov.**
C. robustum Gilli in Ann. Naturhist. Mus. in Wein 84: 22–3 (1983). Type: Papua New Guinea, Kompiam, *Gilli* G546 (holotype W), **syn. nov.**
C. bambusifolium Fowlie, Mark & Ho in Orchid Dig. 50: 19, + figs. (1986). Type: Taiwan, nr. Taichung City, *Ho Fu-Shun* s.n., cult. *Fowlie* et al. FMH 83 T8 (holotype UCLA), **syn. nov.**

A medium- to small-sized, perennial, terrestrial *herb. Roots* thick, fleshy, whitish, often visible as stilt-like supports keeping the rest of the plant well above the substrate. *Pseudobulbs* 3–15 × 0.4–1.5 cm, narrowly fusiform, slightly bilaterally flattened, erect, closely spaced and often crowded towards the apex of a rhizomatous stem; the new pseudobulb is formed annually from a shoot produced slightly above the base of the mature pseudobulb, causing the plant to grow at an angle of about 45° to the horizontal. The strongly bilaterally flattened new growths are initially formed of 6–9 folded, sharply keeled, acute, distichous, overlapping cataphylls which are largest and most leaf-like towards the apex of the growth, where 2–4(5) true leaves are eventually produced. The cataphylls are green at first, but became scarious and then fibrous as the pseudobulb inflates, eventually disintegrating, leaving the old pseudobulbs exposed. *Leaves* up to 9–50(60) × (1.3)1.9–5.5 cm, narrowly obovate to elliptic, the margin sometimes finely serrated towards the acute to acuminate apex, suberect to horizontal, narrowing to a slender conduplicate petiole, articulated to an expanded, sheathing base 0.5–8 cm from the pseudobulb. *Scape* 7–35 cm long, with (2)4–8 flowers, erect, produced laterally on the pseudobulb from the axils of the cataphylls; peduncle covered in 5–7 overlapping sheaths; sheaths up to 1.2–3.4 cm long, cymbiform, acute, with a cylindrical, sheathing base and an inflated, spreading apex; bracts 0.4–1.6(3.5) cm long, lanceolate, cymbiform, acute, becoming scarious. *Flowers* 2.5–5 cm across; not usually scented; rhachis, pedicel and ovary usually pale green; sepals and petals usually white to pale green, occasionally apple-green, with a central maroon stripe, and spotting over the mid-vein which does not reach the apex, and may be weak or absent in the sepals; lip white, pale green or pale yellow, with red spots and blotches on the mid-lobe and purple-red stripes on the side-lobes which become confluent at the margin; callus white, sometimes finely speckled with red; column pale green, streaked purple-red below; anther-cap cream. *Pedicel and ovary* 1-9–4.0 cm long. *Dorsal sepal* 1.7–3 × 0.3–0.8 cm, narrowly oblong to obovate, acute or apiculate, erect; lateral sepals similar, oblique, spreading. *Petals* 1.5–2.3(3) × 0.4–0.8(1) cm, oblong to narrowly elliptic, oblique, acute or apiculate, tending to cover the column. *Lip* 1.4–2 × 0.8–1.6 cm, subentire to strongly 3-lobed, minutely papillose; side-lobes erect, rounded to obtuse at the apex, usually not well differentiated from the mid-lobe; mid-lobe 0.6–1.2 × 0.6–1.4 cm, broadly rounded or ovate to ligulate, usually acute to mucronate, recurved, occasionally hooded, margin entire; callus ridges 2, convergent towards the apex, forming a short tube which extends into the base of the mid-lobe, occasionally crenulate at the apex. *Column* 1.0–1.5 cm long, slender, arching, winged towards the apex; pollinia 4, on a broadly crescent-shaped viscidium. *Capsule* 4.5–5 × 1–1.4 cm, fusiform to club-shaped, pedicellate, held erect and parallel to the rhachis, retaining the column which forms a short apical beak.

ILLUSTRATIONS. Plate 28; photographs 138–140; figure 30.4.

DISTRIBUTION. N India (Himalaya, Meghalaya, Sikkim), Nepal, Bhutan, China, Hong Kong, Taiwan, Ryukyu Islands, Japan, Burma, Indo-China, W Malaysia, Java, Sumatra, Borneo, Moluccas, New Guinea (map 11D); 300–2300 m (985–7445 ft).

HABITAT. In deep shade in broad-leaved forest, usually in rich soil and deep humus and leaf litter, often in conjunction with tree roots or rotting wood; the flowering period is very variable, but is generally April–October in the more northern, seasonal localities, and is sporadic, throughout the year in the southern, tropical localities.

C. lancifolium is the most widespread species in the genus, and is distributed from Nepal and N India to Japan and Taiwan, and through Indo-China, Malesia and the Philippines, to New Guinea. It is character-

27. *Cymbidium goeringii* var. *goeringii.* Japan, *Kuyama* s.n., cult. Kew

28. *Cymbidium lancifolium.* Andrew s.n., cult. Kew

ised by its distinctive, petiolate leaves with a relatively broad, elliptic lamina, its superposed, cigar-shaped pseudobulbs, its ascending habit, and its delicate, white or greenish flowers in a short, erect spike produced from the central nodes of the pseudobulb. Its habit and flowers indicate a relationship with the saprophytic *C. macrorhizon* which, however, lacks leaves and chlorophyll and is apparent only when in flower. *C. lancifolium* is rather morphologically variable, as might be expected from such a widespread species, and these two factors are undoubtedly responsible for its extensive synonymy. Variation is particularly apparent in the Taiwanese and New Guinea representatives of this species.

Hooker (1823) based his original description on a cultivated plant which had been collected by Nathaniel Wallich in Nepal. Shortly afterwards, Blume (1825) described two species from Java, *C. cuspidatum* and *C. javanicum*, differing slightly in their leaf shape, but he compared neither with *C. lancifolium*. Comparison of the types of these Javanese species shows them to differ little from each other or from *C. lancifolium*, and these are considered to be conspecific with *C. lancifolium*, following J.J. Smith (1905).

C. gibsonii, described by Lindley from a cultivated specimen imported from Meghalaya in N E India, was distinguished by its 'fusiform, jointed, naked stem'. However, although the pseudobulbs in *C. lancifolium* are covered by cataphylls when they are young, as the pseudobulbs swell the cataphylls split and become scarious, soon disintegrating and often leaving the pseudobulbs exposed.

Rolfe (1925) based his *C. kerrii* on a broad-lipped specimen collected by Kerr in northern Thailand. Seidenfaden (1983) compared the Thai material of *C. kerrii* and *C. lancifolium*, and concluded that the variation in lip shape was continuous and did not justify the recognition of two species. A plant with a similarly broad lip which was consequently obscurely 3-lobed led Masamune (1930) to describe *C. nagifolium*. It is also considered here to be conspecific with *C. lancifolium*. Broad-lipped specimens have also been collected elsewhere, for example at high altitudes in Sabah (*Lamb* SAN 91582).

A robust variant from Taiwan was described by Fukuyama in 1935 as *C. syunitianum*. Similarly *C. maclehoseae* S.Y. Hu (1972) from Hong Kong and *C. robustum* Gilli (1983) from New Guinea are based on unusually large specimens. It is evident that these variants fall within the range of variation encountered in *C. lancifolium* in several distant parts of its range, and they cannot be considered as specifically distinct.

Of more interest is the recent rediscovery by Reeve (1984) of plants agreeing with Schlechter's (1913) description of *C. papuanum* from New Guinea. Unfortunately the type specimen, in the Berlin herbarium, was destroyed during the war, leaving some doubt as to the true identity of this variant. Schlechter stated that his species differed from *C. lancifolium* by the much smaller habit of the plant, its pale yellow flowers and by the lip. Although his description of the lip is not unusual for this species, his later (1928) drawing of the flower shows the lip as having a concave, hooded apex. These three characters are found in the specimens collected by Reeve, and he further differentiated this variant by its rather large inflorescence in relation to the size of the plant, its slightly drooping flowers which do not open fully, and by the production of distinctive, long, creeping rhizomes which extend through the leaf in which the plants grow. However, most of these characters are not outside of the range of variation found in *C. lancifolium*. Plants of a similar small size have been collected from several countries, and indeed specimens from the northern extremes of the distribution, in Japan, are usually very small. The inflorescence is large relative to the small size of the plant, but is in fact similar to that of ordinary specimens of *C. lancifolium* collected both on New Guinea and in other parts of the range. The cream-coloured flowers are somewhat unusual, but are also known from other specimens of normal size from New Guinea, which are otherwise indistinguishable from *C. lancifolium*. Drooping, not fully open flowers with a hooked apex have been noted in specimens from other countries, and in specimens which have flowered under poor conditions in cultivation. The creeping habit which at first seems to be so distinctive (as in *Reeve* 437 and *Stevens* LAE 58147) is also found in specimens of *C. lancifolium* from high altitudes in Burma (*Kingdon-Ward* s.n.), in Yunnan, China (*Handel-Mazzetti* 9414) and Meghalaya (*Hooker & Thompson* s.n.). Furthermore, it is also found in larger, more typical specimens of *C. lancifolium* from lower altitudes in New

Guinea (*Reeve* 702) and Burma (*Baldwell* 13546). It seems likely that many of the characters used to distinguish Schlechter's *C. papuanum* are high altitude adaptations of *C. lancifolium*, recognisable as ecotypes rather than as distinct species.

The name *C. aspidistrifolium* is currently widely used for plants in cultivation originally from Taiwan. It was discovered by Fukuyama in 1934, and was distinguished from *C. lancifolium* by its green rather than white flowers with oblong, thick-textured sepals and petals, entire leaf margins, and autumn rather than summer flowering period. Leaf serration is, however, very variable. An example occurs in the four plants which constitute Blume's type of *C. javanicum*, where one has entire leaf margins, while the others are serrated towards the apex. Maekawa's *C. lancifolium* var. *pantlingii* from Sikkim also has entire leaf margins, while Seidenfaden (1983) reports both types of leaf margin in Thailand. Maekawa (1958), Ohwi (1965) and Su (1975) and Liu & Su (1978) all treat the Taiwanese taxon as *C. lancifolium* var. *aspidistrifolium*, distinguished mainly by its green flower colour and its late flowering season. However, the intermediate nature of *C. syunitianum* and other specimens from Taiwan (Mark *et al.*, 1986), and *C. maclehoseae* (from Hong Kong) in these features suggest that *C. aspidistrifolium* is perhaps best treated as a synonym of *C. lancifolium*.

The variation of *C. lancifolium* in Taiwan has been examined and documented by Mark *et al.* (1986). They recognised four species, all of which are considered here to be conspecific with *C. lancifolium*. The largest variants with the strongest scapes were named as *C. syunitianum* (see previously), the medium-sized plants with greenish flowers as *C. aspidistrifolium* and those with entire leaf margins and whitish flowers were named as *C. nagifolium*. These were further differentiated by the season of the year at which they usually flowered. Furthermore, they described a new species, *C. bambusifolium*, to include the smallest variants with very weak scapes and pale green flowers. There is undoubtedly great variation in plant size, flower colour and shape, and flowering time in Taiwan, but the distinctions used by Mark *et al.* to recognise distinct taxa appear to vary continuously from one extreme to the other, and become particularly confused when compared with specimens from outside Taiwan. Critical study of this species in Taiwan, particularly of the green-flowered specimens with somewhat spreading petals, may produce a more useful infraspecific classification.

C. lancifolium is found growing at between 300 m and 1800 m (985–5905 ft) elevation in its more northerly localities in Japan, Taiwan and China. In N India, S W China, Burma and Thailand it occurs between 1000 and 2300 m (3280–7445 ft). In Malaya, the Malay Archipelago and New Guinea it has been recorded from a wide altitudinal range, from as low as 300 m (985 ft) in Sabah, up to 2000 m (6560 ft) in New Guinea.

It usually grows in deep shade, in leaf litter on the floor of broad-leaved forest, for example in *Castanopsis* forest in New Guinea (Reeve, 1984). In Java, it grows in montane forest in similarly shaded positions in leaf litter, often on steep slopes and ridge tops. In Thailand it has been found growing near creek beds, in very deep shade where there is little competition from other herbs. It has often been reported that its roots are anchored on mossy tree roots or on rotting wood buried in the substrate. The plant is often supported above the substrate on long, stilt-like roots, with each new pseudobulb being produced slightly above the previous one, allowing the plant to survive in the accumulating leaf litter. The broad leaves also seem to be well adapted to the low light intensities of the forest floor.

Section **Pachyrhizanthe** *Schltr.* in Fedde, Repert. Beih. 4: 73 (1919); Seth & Cribb in Arditti (ed.), Orchid Biol., Rev & Persp. 3: 288 (1984). Lectotype: *C. aberrans* (Finet)Schltr. (=*C. macrorhizon* Lindley) chosen by Seth & Cribb, 1984.

Pachyrizanthe (Schltr.)Nakai in Bot. Mag. Tokyo 45: 109 (1931).

Cymbidium section *Macrorhizon* Schltr. in Fedde, Repert. 20: 99–101 (1924); P. Hunt in Kew Bull. 24: 93 (1970). Type: *C. macrorhizon* Lindley.

Section *Pachyrhizanthe* was established by Schlechter (1919), although he later substituted the name *Macrorhizon* (Schlechter, 1924). Hunt (1970) followed this change, but Seth & Cribb (1984) reinstated the previous name under the current rules in the International Code of Botanical Nomenclature.

The single species of this section is a saprophyte, with an underground rhizome and leaves reduced to scarious scales. The scape is held above ground. Although this species lacks true leaves, and micromorphological and anatomical data are therefore unavailable, the flowers have four pollinia, and the callus ridges converge to form a short tube at the base of the mid-lobe of the lip, both characteristic of subgenus *Jensoa*. It is closely related to section *Geocymbidium* which has a similar scape.

In justification of his removal of these variants to a new genus, *Pachyrhizanthe*, Nakai (1931) emphasised the underground rhizome destitute of leaves, pseudobulbs and usually roots, the terminal scape, and the two lateral furrows on the column. We can find no evidence of the last of these characters. Garay & Sweet (1974) stated in their discussion of these characters, that 'the saprophytic habit alone upon which he based his decision is not acceptable. There are many genera with both autotrophic and saprophytic species.' Furthermore, the similarity of the scape and the flowers to other species in subgenus *Jensoa*, and especially to *C. lancifolium*, strongly suggests that they are closely related and should be maintained in the same genus.

44. C. macrorhizon *Lindley*, Gen. Sp. Orchid. Pl.: 162 (1833); Hooker f., Fl. Brit. India: 9 (1891); Duthie in Ann. Roy. Bot. Gard. Calcutta 9 (2): 134–5, t.114 (1906); Pradhan, Indian Orchids: Guide to Identification & Culture 2: 470 (1979); Wu & Chen in Acta Phytotax. Sin. 18: 305 (1980); Hashimoto & Kanda, Jap. Indig. Orch. in colour: 185 (1981); Seidenfaden in Opera Bot. 72: 66–7, f.35 (1983). Type: India, Kashmir, *Royle* (holotype K!).

Bletia nipponica Franchet & Savatier in Enum. Pl. Jap. 2: 511 (1879). Type: Japan, *Savatier* s.n. (holotype P).

C. nipponicum (Franchet & Savatier)Rolfe in Orchid. Rev. 3: 39 (1895); Ohwi, Fl of Japan: 155 (1965); Maekawa, The Wild Orchids of Japan in Colour: t.164 (1971); Garay & Sweet, Orch. S. Ryukyu Islands: 142 (1974); Mark, Ho & Fowlie in Orchid Dig. 50: 14, + fig. (1986).

C. pedicellatum Finet in Bull. Soc. Bot. Fr. 47: 268, t.9A (1900); Rolfe in Orchid Rev. 12: 303 (1904). Type: as for *Bletia nipponica.*

Yoania aberrans Finet in Bull. Soc. Bot. Fr. 47: 274, t.9B (1900). Type: Japan, *sine coll.* (holotype P).

Aphyllorchis aberrans (Finet)Schltr. in Engl. Bot. Jahrb. 45: 387 (1911).

C. aberrans (Finet)Schltr. in Fedde, Repert. Beih. 4: 264 (1919); Maekawa, The Wild Orchids of Japan in Colour: t.165 (1971).

C. aphyllum Ames & Schltr. in Fedde, Repert. Beih. 4: 73, 265 (1919), *non* (Roxb.)Sw. (1799) = *Dendrobium pierardii.* Type: China, Sichuan, *Wilson* 4712 (holotype AMES!).

Pachyrhizanthe aberrans (Finet)Nakai in Bot. Mag. Tokyo 45: 109 (1931).

Pachyrhizanthe aphyllum (Ames & Schltr.)Nakai, l.c. (1931).

Pachyrhizanthe macrorhizon (Lindl.)Nakai, l.c. (1931).

Pachyrhizanthe nipponicum (Franchet & Savatier)Nakai, l.c. (1931).

Pachyrhizanthe sagamiense Nakai in Bot. Mag. Tokyo 45: 110 (1931). Types: Japan, Hondo, *Musashi* (florum), *Hisauchi* s.n.; Hondo, *Sagami* (fructum), *Hisauchi* s.n. (syntypes not located), **syn. nov.**

C. szechuanensis Hu in Quart. J. Taiwan Mus. 26: 140 (1973). Type: as for *C. aphyllum.*

A small, perennial, terrestrial *saprophyte*, without leaves or pseudobulbs. *Roots* short, often absent, occasionally produced towards the apex of the rhizome. *Rhizome* 3–8 mm in diameter, soft, fleshy, tuberculate, with nodes 2–15(19) mm apart, usually branching, whitish, subterranean. Tubercles closely spaced, each with a tuft of short hairs. Scales up to 5 mm long, scarious, often disintegrating leaving an indistinct scar; branches originating in the axils of the scales. *Scape* 7–32 cm long, erect, with (1)3–6(8) flowers, usually produced terminally on the rhizome or on its side branches; peduncle covered with 4–8, mostly overlapping sheaths, of which 1–4 are above ground; sheaths 13–27 mm long, ovate, cymbose, with a cylindrical sheathing base and a loosely sheathing, somewhat spreading, cymbiform apex; bracts 2–17 mm long, oblong to lanceolate, acute, somewhat keeled, becoming scarious. *Flower* 3–4 cm across; rhachis, pedicel and ovary pale green to cream, often stained with pink or purple; sepals and petals cream to pale yellow or brownish-pink with a diffuse, central, purple-red stripe to near the apex; lip white with red spots on the mid-lobe and purple-red stripes on the side-lobes, which become confluent at the margin; callus white, sometimes stained pink; column white, stained or lightly spotted purple-pink above, with purple-red dashes below; anther-cap cream. *Pedicel and ovary* 10–37 mm long. *Dorsal sepal* (14)19–26 × (3)4–6 mm, narrowly obovate to ligulate, acute, erect; lateral sepals similar, spreading. *Petals* 14–20 × 5–7 mm, narrowly elliptic, oblique, acute, usually porrect and covering the column. *Lip* 12–17 × 9–11 mm when flattened, almost rhombic in outline, subentire to strongly 3-lobed, minutely papillose; side-lobes erect, obtuse to subacute at the apex; mid-lobe 5–7 × 4–6 mm, triangular to ligulate, acute to obtuse, recurved, the margin usually slightly undulating; callus ridges 2, convergent towards the apex, forming a short tube at the base of the mid-lobe. *Column* 9–13 mm long, slender, arching, narrowly winged towards the apex; pollinia 4, broadly ovate, in two unequal pairs; viscidium broadly crescent-shaped. *Capsule* 30–50 × 9–12 mm, fusiform to ellipsoidal, pedicellate, held erect and parallel to the rhachis, retaining the column which forms a short (5–10 mm) apical beak.

ILLUSTRATIONS. Photographs 141–143; figure 30.7.

DISTRIBUTION. Pakistan (N Punjab), N India (Kashmir, N W Indian Himalaya, Sikkim, Assam, Meghalaya, Nagaland), Nepal, Burma, Thailand, Laos, China (Guizhou, Sichuan, Yunnan), Japan, Taiwan, Ryukyus (map 11E); up to 2500 m (8200 ft).

HABITAT. In broad-leaved and pine forest growing in damp humus, in shade; flowering May–August.

C. macrorhizon is a small, leafless, saprophytic orchid, only visible above ground when the flower spikes emerge from the substrate, or when the erect capsules are developing. The subterranean, branching rhizome is covered in small hairy warts which appear to be linked with the symbiotic fungal growth necessary for its survival. Early descriptions of this species suggested that it was a parasite, but there is no evidence of this, and G. Mann notes on one of his herbarium specimens that attempts to trace the rhizome to its host plant were unsuccessful. The flower spike is usually produced at the tip of the rhizome or of one of its branches, although developmental studies are necessary to determine whether or not it is truly terminally produced. The spike closely resembles that of other species in subgenus *Jensoa*, especially the closely related *C. lancifolium*. The flowers also closely resemble those of *C. lancifolium*. Further evidence of the relationship between these two species may be found in the occasional rhizomatous growth habit of *C. lancifolium*. The rhizomes of both species are segmented, each segment being subtended by a small cataphyll-like bract, although these may be difficult to observe on *C. macrorhizon* as they appear to disintegrate quickly. One specimen from Thailand (*Geesink, Phanichapol & Santisuk* 5536), at present placed in *C. macrorhizon*, appears to be somewhat intermediate between these two species. It lacks leaves, and the basal portion resembles the rhizome of *C. macrorhizon*, but the apical region is fast-growing and covered in cataphylls, and appears similar to an elongated, immature shoot of *C. lancifolium*. This specimen also resembles the type of *C. caulescens* (at present synonymised under *C. lancifolium*), except that it has no true leaves. Further investigation may show *C. caulescens* to be a distinct taxon intermediate between *C. macrorhizon* and *C. lancifolium*, living as a semi-saprophytic plant capable of the production of some green leaves and cataphylls.

C. macrorhizon was described by Lindley in 1833 from a specimen collected by Royle in Kashmir, at about 31°N, and near the extreme north-western edge of the distribution of the genus. Specimens have since been collected in Japan and the Ryukyus in similar northerly latitudes. *C. macrorhizon* seems to be the most westerly representative of the genus *Cymbidium*.

C. nipponicum (Franchet & Savatier) Rolfe is considered by Ohwi (1965) and Garay & Sweet (1974) to be the correct name for the representatives of this species from Japan and the Ryukyus. It was originally described as *Bletia nipponica* in 1879, from a specimen collected by Savatier. Rolfe transferred it to the genus *Cymbidium* in 1895 but noted its similarity to *C. macrorhizon* from northern India. Rolfe (1904) included *C. pedicellatum* Finet (1900) as a synonym of *C. nipponicum*, as both of these are based on the same type specimen. Seidenfaden (1983) compared the type specimens of *C. macrorhizon* and *C. nipponicum*, and concluded that there did not appear to be any justification for recognising them as different taxa.

Finet also published the name *Yoania aberrans* in 1900, illustrating it alongside his *C. pedicellatum*, and noting the similarity of its flowers to those of *Cymbidium*. The illustration shows *Y. aberrans* to be a smaller, fewer-flowered plant than *C. aberrans*, but the flowers are otherwise similar. Schlechter transferred *Y. aberrans* to *Aphyllorchis* in 1911, and in 1919 to *Cymbidium*. Schlechter considered that *C. nipponicum* differed from *C. aberrans* by the larger bracts, the short ovary, the larger flowers and the shorter, broader column of the former. Maekawa (1971) also recognises these two as distinct species, his illustrations indicating the smaller size of *C. aberrans*, its shorter bracts and more pallid, fewer flowers which do not open so widely. It appears probable that *C. aberrans* is simply a poorly growing specimen, and is consequently smaller in size, and the few, poorly coloured flowers do not open so fully. The small bracts are no smaller than those found at the apex of the scape of larger specimens. Therefore, there appears to be only one species of saprophytic *Cymbidium* in Japan.

There have been few comparisons made between the Japanese material and specimens of *C. macrorhizon* from northern India. This may be due to the disjunct distribution of these two regions. The variation in size and flower number already noted from Japan is mirrored in the Indian specimens (see also the discussion of the treatment of Nakai, 1931).

In 1919 Schlechter described *C. aphyllum*, from Sichuan in western China. His description differs from the northern Indian specimens of *C. macrorhizon* only in the obtuse mid-lobe apex. Wu & Chen (1980) in their revision of *Cymbidium* in China place this species as a synonym of *C. macrorhizon*, and their treatment is followed here.

Hu (1973) noted that the name *C. aphyllum* Schltr. was a later homonym of *C. aphyllum* (Roxb.)Sw. (1799) which was used for a species which was later transferred to the genus *Dendrobium*. She therefore published a new name for this taxon, *C. szechuanensis*, basing this species on the type specimen of *C. aphyllum* Schltr.. This name is therefore also treated here as a synonym of *C. macrorhizon*.

The taxa now included in *C. macrorhizon* were studied by Nakai in 1931. He removed them to a new genus, *Pachyrhizanthe*, and recognised five distinct species. These were *P. macrorhizon* from northern India, *P. aphyllum* from China, *P. nipponicum* and *P. aberrans* from Japan, and a third Japanese variant which he named *P. sagamiense*.

This last variant was distinguished by its lack of purple colouring in the flowers, and by its lip which had very poorly defined side-lobes. This is probably an albino variant, lacking the red pigment in the flower, similar to those encountered in many other species such as *C. ensifolium*, *C. sinense*, *C. goeringii*, *C. insigne*, and *C. mastersii*. Variation in the degree of differentiation between the mid-lobe and the side-lobes of the lip can be seen in specimens from both India and Japan. The variation from strongly 3-lobed to entire is continuous and cannot be used as a distinguishing character. This type of variation has also been described in *C. lancifolium*, and several species in sections *Jensoa* and *Maxillarianthe*. Later authors (Ohwi, 1965; Maekawa, 1971; Garay & Sweet, 1974) do not recognise *P. sagamiense* as a distinct taxon, and it is included here in the synonymy of *C. macrorhizon*.

Nakai keys out the remaining four species on the basis of the length of the scape, the colour of the flowers and the degree to which the lip appears to be 3-lobed. These are all continuously variable characters.

The distribution of *C. macrorhizon* is unusual in that it is highly disjunct, and extends over a greater east–west range than any other *Cymbidium* species (map 11E). It occurs as far west as the hills of northern Pakistan and Kashmir. From there it extends east along the Indian Himalaya. It has not been collected in Nepal, but it occurs again in Sikkim and Darjeeling, and in north-eastern India in Meghalaya (Khasia Hills) and Nagaland. The distribution continues south through Burma (including the Chin Hills) into northern Thailand, Laos and the western provinces of China. From there, there is a large jump to Japan, the Ryukyus and Taiwan.

Its altitudinal range varies in the different regions of this distribution. In Japan and the Ryukyus it may be found almost at sea level. In western China, Thailand and Burma it appears to be found between 1000 and 1500 m (3280–4920 ft), in Taiwan at 1800–2200 m (5905–7220 ft) while in northern India the range is between 1000 and 2500 m (3280–8200 ft), although it is found at lower altitudes in the more north-westerly extremes of its range. Flowering time seems constant, between June and August, except in Indo-China where it flowers slightly earlier, during May and June.

In general, *C. macrorhizon* is a forest species, growing in shaded spots, in humus-rich soil. In Japan it has been reported in hills and fields not far from the sea, and from road-sides in China. In Thailand it has been found in open forest, growing in humus-filled rock fissures. In India, it is usually encountered in pine forest, where it thrives in the decaying pine needles, although broad-leaved forest such as Oak and *Shorea* can also support populations of this species.

BIBLIOGRAPHY

Note: bibliographical citations are given on first appearance in the book. For subsequent citations the reader should refer back to earlier chapters.

Chapter 1

Du Puy, D.J. (1986). *A taxonomic revision of the genus* Cymbidium *Sw. (Orchidaceae).* PhD thesis, University of Birmingham and Royal Botanic Gardens, Kew

Seth, C.J. & Cribb, P.J. (1984). *A reassessment of the sectional limits in the genus* Cymbidium. In Arditti, J. (ed.), Orchid Biology, Reviews and Perspectives. Cornell University Press, Ithaca & New York, pp. 283–322

Winter, K., Wallace, B.J., Stocker, G.C. & Roksandic, Z. (1983). *Crassulacean Acid Metabolism in Australian vascular epiphytes and some related species.* Oecologia 57: 129–41

Withner, C.L., Nelson, P.H. & Wejksnora, P.J. (1975). *The anatomy of the orchids.* In Withner, C.L. (ed.), The Orchids, Scientific Studies. J. Wiley & Sons, New York, pp. 267–347

Chapter 2

Ackerman, J.D. & Williams, N.H. (1980). *Pollen morphology of the tribe Neottieae and its impact on the classification of the Orchidaceae.* Grana 19: 7–18

Ackerman, J.D. & Williams, N.H. (1981). *Pollen morphology of the Chloraeinae (Orchidaceae: Diurideae) and related subtribes.* Amer. J. Bot. 68: 1392–402

Arditti, J., Michaud, J.D. & Healey, P.L. (1979). *Morphometry of orchid seeds. I.* Paphiopedilum *and native California and related species of* Cypripedium. Amer. J. Bot. 66: 1128–37

Arditti, J., Michaud J.D. & Healey, P.L. (1980). *Morphometry of orchid seeds. II. Native California and related species of* Calypso, Cephalanthera, Corallorhiza *and* Epipactis. Amer. J. Bot. 67: 508–18

Barthlott, W. (1974). *Morphologie der Samen.* In Senghas, K., Ehler, N., Schill, R. & Barthlott, W. *Neue Untersuchungen und Methoden zur Systematik und Morphologie der Orchideen.* Orchidee 25: 157–69

Barthlott, W. (1976). *Morphologie der Samen von Orchideen im Hinblick auf taxonomische und funtionelle Aspekte.* In Proceedings of the 8th World Orchid Conference, pp. 444–55. Publ: German Orchid Soc. Inc.

Barthlott, W. & Zeigler, B. (1980). *Uber ausziehbare helicale Zellwandverdickungen als Haftapparat der Samenschalen von* Chiloschista lunifera *(Orchidaceae).* Ber. Deutsch Bot. Ges. Bd. 93: 391–403

Barthlott, W. & Zeigler, B. (1981). *Systematic applicability of seed coat micromorphology in orchids.* Ber. Deutsch Bot. Ges. Bd. 94: 267–73

Beer, I.G. (1863). *Biologie und Morphologie der Familie der Orchideen.* Druck und Verlag von Carl Gerold's Sohn, Vienna

Bertsch, K. (1941). *Fruchte und Samen*: 118–21. In *Handbucher der praktischen Vorgeschichtsforschung.* Herausgegeben von Prof. Dr. H. Reinerth, Band 1, Stuttgart

Burgeff, H. (1936). *Samenbeimung der Orchideen und Entwicklung ihrer Keimpflanzen, mit Anhang uber praktishe Orchideenanzucht.* Gustav Fischer Verlag, Jena

Carlson, M.S. (1940). *Formation of the seed of* Cypripedium parviflorum. Bot. Gaz. 102: 295–300

Clifford, H.T. & Smith, W.K. (1969). *Seed morphology and classification of Orchidaceae.* Phytomorphology 19: 133–9

Davis, A. (1946). *Orchid seed and seed germination.* Am. Orch. Soc. Bull. 15: 218–23

Dressler, R.L. (1981). *The Orchids. Natural history and classification.* Harvard University Press, Massachusetts and London. 332pp

Dressler, R.L. & Dodson, C. (1960). *Classification and phylogeny in the Orchidaceae.* Ann. Miss. Bot. Gard. 47: 25–68

Hoehne, F.C. (1949). *Iconographia de orchidaceas de Brasil.* Secretaria de Agricultura, Sao Paolo, Brazil

Newton, G.D. & Williams, N.H. (1978). *Pollen morphology of the Cypripedioideae and the Apostasioideae (Orchidaceae).* Selbyana 2: 169–82

Schill, R. (1974). *Pollenmorphologie.* In Senghas, K., Ehler, N., Schill, R. & Barthlott, W. *Neue Untersuchungen und Methoden zur Systematik und Morphologie der Orchideen.* Orchidee 25: 157–69

Schill, R. & Pfeiffer, W. (1977). *Untersuchungen an Orchideen-pollinien unter besonderer Beruecksichtigung ihrer Feinskulpturen.* Pollen et Spores 19: 5–118

Stoutamire, W.P. (1963). *Terrestrial orchid seedlings.* Australian Plants 2: 119–22

Thomale, H. (1957). *Die Orchideen,* 2nd edn. Eugen Ulmer Verlag, Stuttgart

Tohda, H. (1983). *Seed morphology in the Orchidaceae I.* Sci. Rep. Tohoku Univ. Fourth Ser. (Biol) 38: 253–68

Williams, N.H. & Broome, C.R. (1976). *Scanning electron microscope studies of orchid pollen.* Amer. Orchid. Soc. Bull. 45: 699–707

Zeigler, B. (1981). *Micromorphologie der Orchidaceen-Samen unter beruecksichtigung taxonomisches Aspekte.* Thesis: Ruprecht Karls Universitat, Heidelberg

Chapter 3

Atwood, J.T. (1984). *The relationships of the slipper orchids (subfam. Cypripedioideae).* Selbyana 7: 129–247

Atwood, J.T. & Williams, N.H. (1978). *The utility of epidermal cell features in* Phragmipedium *and* Paphiopedilum *(Orchidaceae) for determining sterile specimens.* Selbyana 2: 356–66

Atwood, J.T. & Williams, N.H. (1979). *Surface features of the adaxial epidermis in the conduplicate-leaved Cypripedioideae (Orchidaceae).* Bot. J. Linn. Soc. 78: 141–56

Dressler, R.L. (1981). *The Orchids. Natural history and classification.* Harvard University Press, Massachusetts and London. 332pp

Dressler, R.L. & Dodson, C. (1960). *Classification and phylogeny in the Orchidaceae.* Ann. Miss. Bot. Gard. 47: 25–68

Kaushik, P. (1983). *Ecological and Anatomical marvels of the Himalayan orchids.* Progress in Ecology, 8. Today and Tomorrow's Printers, New Delhi. 123pp

Lov, L. (1926). *Zur Kenntnis der Entfaltungszellen monokotylen Blatter.* Flora, Jena 120: 283–343

Mobius, M. (1877). *Uber den anatomische Bau der Orchideen – blatter und dessen Bedeutung fur das System dieser Familie.* Jb. Wiss. Bot. 18: 530–607

Pridgeon, A.M. (1982). *Diagnostic anatomical characters in the Pleurothallidinae (Orchidaceae).* Amer. J. Bot. 69: 921–38

Pridgeon, A.M. & Stern, W.L. (1982). *Vegetative anatomy of* Myoxanthus *(Orchidaceae).* Selbyana 7: 55–63

Rasmussen, H. (1981a). *The diversity of stomatal development in the Orchidaceae subfamily Orchidoideae.* Bot. J. Linn. Soc. 82: 381–93

Rasmussen, H. (1981b). *Terminology and classification of stomata and stomatal development – a critical survey.* Bot. J. Linn. Soc. 83: 199–211

Rudall, P. (1983). *Leaf anatomy and relationships of* Dietes *(Iridaceae).* Nordic J. Bot. 3: 471–8

Singh, H. (1981). *Development and organisation of stomata in the Orchidaceae.* Acta. Bot. Indica 9: 94–100

Solereder, H. & Meyer, F.J. (1930). *Systematic anatomy of the Monocotyledons – Microspermae (volume 6).* Israel Program for Scientific Publications (I.P.S.T.) Press, Jerusalem

Wilkinson H.P. (1979). *The plant surface (mainly leaf).Part 1: Stomata.* In C.R. Metcalfe and L. Chalk (eds.), *Anatomy of the Dicotyledons,* 2nd edn, vol. 1. Clarendon Press, Oxford, pp. 97–117

Williams, N.H. (1974). *The value of plant anatomy in orchid taxonomy.* In Proc. 7th. World Orchid Conf., pp. 281–98

Williams, N.H. (1979). *Subsidiary cells in the Orchidaceae: their general distribution with special reference to development in the Orchidieae.* Bot. J. Linn. Soc. 78: 41–66

Winter, K., Wallace, B.J., Stocker, G.C. & Roksandic, Z. (1983). *Crassulacean acid metabolism in Australian vascular epiphytes and some related species.* Oecologia 57: 129–41

Withner, C.L., Nelson, P.H. & Wejksnora, P.J. (1975). *The Anatomy of Orchids,* in Withner, C.L. (ed.), The Orchids, Scientific Studies. John Wiley & Sons, New York, pp. 267–347

Chapter 4

Du Puy, D.J. (1986). *A Taxonomic Revision of the genus* Cymbidium *Sw. (Orchidaceae).* PhD thesis, University of Birmingham and Royal Botanic Gardens, Kew

Leonhardt, K.W. (1979). *Chromosome numbers and cross compatibility in the genus* Cymbidium *and related genera.* The Orchid Advocate 5: 44–51

Tanaka, R. & Kamemoto, H. (1974). *List of chromosome numbers in species of Orchidaceae.* In Withner, C.L. (ed.), The Orchids: Scientific Studies. John Wiley & Sons, New York, pp. 411–83

Wimber, D.E. (1957a). *Cytogenetic studies in the genus* Cymbidium. *I. Chromosome numbers within the genus and related genera.* Amer. Orch. Soc. Bull. 26: 636–9

Wimber, D.E. (1957b). *Cytogenetic studies in the genus* Cymbidium. *II. Pollen formation in the species.* Amer. Orch. Soc. Bull. 26: 700–3

Wimber, D.E. (1957c). *Cytogenetic studies in the genus* Cymbidium. *III. Pollen formation in the hybrids.* Amer. Orch. Soc. Bull. 26: 771–7

Chapter 5

Ackerman, J.D. (1983a). *Euglossine bee pollination of the Orchid* Cochleanthes lipscombiae*: a food source mimic.* Amer. J. Bot. 70: 830–4

Ackerman, J.D. (1983b). *Specificity and mutual dependency of the orchid Euglossine bee interaction.* Biol. Journ. Linn. Soc. 20: 301–14

Bierzychudek, P. (1981). Asclepias, Lantana *and* Epidendrum*: A floral mimicry complex?* Biotropica 13: 54–8

Boyden, T.C. (1980). *Floral mimicry by* Epidendrum ibaguense *(Orchidaceae) in Panama.* Evolution 34: 135–6

Dafni, A. (1983). *Pollination of* Orchis caspia *– a nectarless plant which deceives the pollinators of nectariferous species of other families.* Journ. Ecology 71: 467–74

Dafni, A. & Ivri, Y. (1981a). *The flower biology of* Cephalanthera longifolia *(Orchidaceae) – Pollen imitation and facultative floral mimicry.* Plant Syst. Evol. 137: 229–40

Dafni, A. & Ivri, Y. (1981b). *Floral Mimicry between* Orchis israelitica *Baumann and Dafni (Orchidaceae) and* Bellevalia flexuosa *Boiss. (Liliaceae).* Oecologia 49: 229–32

Dodson, C.H. & Hills, H.G. (1966). *Gas chromatography of orchid fragrances.* Amer. Orch. Soc. Bull. 35: 720–5

Dodson, C.H., Dressler, R.L., Hills, H.G., Adams, R.M. & Williams, N.H. (1969). *Biologically active compounds in orchid fragrances.* Science 164: 1243–9

Frison, T.H. (1934). *Records and descriptions of* Bremus *and* Psithymis *from Formosa and the Asiatic mainland.* Trans. Nat. Hist. Soc. Formosa 24: 150–85

Hills, H.G., Williams, N.H. & Dodson, C.H. (1968). *Identification of some orchid fragrance components.* Amer. Orch. Soc. Bull. 37: 967–71

Hills, H.G., Williams, N.H. & Dodson, C.H. (1972). *Floral fragrances and isolating mechanisms in the genus* Catasetum *(Orchidaceae).* Biotropica 4: 61–76

Kjellsson, G., Rasmussen, F.N. & Du Puy, D.J. (1985). *The pollination of* Dendrobium infundibulum, Cymbidium insigne *(Orchidaceae) and* Rhododendron lyi *(Ericaceae) by* Bombus eximius *(Apidae) in Thailand: a possible case of floral mimicry.* J. Trop. Ecology 1: 289–302

Macpherson, K. & Rupp, H.M.R. (1935). *The pollination of* Cymbidium iridifolium *Cunn.* North Queensland Naturalist 3: 26

Macpherson, K. & Rupp, H.M.R. (1936). *Further notes on orchid pollination.* North Queensland Naturalist 4: 25

Nilsson, L.A. (1978a). *Pollination ecology and adaptation in* Platanthera chlorantha *(Orchidaceae).* Bot. Notiser 131: 35–51

Nilsson, L.A. (1978b). *Pollination ecology of* Epipactis palustris *(Orchidaceae).* Bot. Notiser 131: 355–68

Nilsson, L.A. (1979a). *Anthecological studies on Lady's Slipper,* Cypripedium calceolus *(Orchidaceae).* Bot. Notiser 132: 329–47

Nilsson, L.A. (1979b). *The pollination ecology of* Herminium monorchis *(Orchidaceae).* Bot. Notiser 132: 537–49

Nilsson, L.A. (1980). *The pollination ecology of* Dactylorhiza sambucina *(Orchidaceae).* Bot. Notiser 133: 367–85

Nilsson, L.A. (1981). *The pollination ecology of* Listera ovata (Orchidaceae). Nordic J. Bot. 1: 461–80

Nilsson, L.A. (1983). *Mimesis of bellflower* (Campanula) *by the red helleborine orchid* Cephalanthera rubra. Nature 305: 799–800

Seidenfaden, G. (1984). *Orchid genera in Thailand XI.- Cymbidieae Pfitz.,* Opera Bot. 72: 1–124

Smythe, R. (1970). *Pollination by the Common Native Bee.* Orchadian 3: 149

Chapter 6

Barlow, B.A. (1981). *The Australian Flora: its origin and evolution.* Flora of Australia vol. 1 (Introduction). Canberra (Australian Government Publishing Service), pp. 25–75

Burbridge, N.T. (1960). *The phytogeography of the Australian Region.* Aust. J. Bot. 8: 75–212

Hooker, J.D. (1860). *Introductory Essay, Botany of the Antarctic Voyage of H.M. Discovery ships 'Erebus' and 'Terror' in the years 1839–1843, III.* Flora Tasmainiae, Reeve, London

Hu, S.Y. (1971). *The Orchidaceae of China 2.* Quart. J. Taiwan Mus. 24: 38–112

Lavarack, P.S. (1981). *Origins and Affinities of the Orchid Flora of Cape York Peninsula.* Proceedings of the Orchid Symposium, 13th International Botanical Congress, Spheg. Orchid Society of New South Wales, Sydney, Australia, pp. 17–26

Quisumbing, E.A. (1940). *The genus* Cymbidium *in the Philippines.* Philippine J. Sci. 72: 481–92, pl. 1–8

Specht, R.L. (1981). *Evolution of the Australian flora: some generalisations,* in A. Keast, Ecological Biogeography of Australia. W. Junk, The Hague. pp. 785–805

Wu, Y-S. & Chen, S-C. (1980). *A taxonomic review of the orchid genus* Cymbidium *in China.* Acta Phytototaxonomica Sinica 18: 292–307

Chapter 7

Chen, S-C. & Tang, T. (1982). *A General Review of the Orchid Flora of China.* In Arditti, J. (ed.), Orchid Biology, Reviews and Perspectives, 2, Cornell Univ. Press, Ithaca and London, pp. 39–81

Chow, C. (1979). *Formosan Orchids,* Taichung, Taiwan

Hu, S.Y. (1971). *Orchids in the life and culture of the Chinese people.* Quart. J. Taiwan Mus. 24: 67–103

Miyoshi, M. (1932). *On the manuscripts by Matsuoka Joan.* Honzu 6: 43–55

Nagano, Y. (1952). *Three main species of orchids in Japan.* Amer. Orchid Soc. Bull. 21: 787–9

Nagano, Y. (1953). *History of orchid growing in Japan.* Amer. Orchid Soc. Bull. 22: 331–3

Nagano, Y. (1955). *Miniature Cymbidiums in Japan.* Amer. Orchid Soc. Bull. 24: 735–43

Nagano, Y. (1960). *Orchids in Japan.* In Proc. 3rd World Orchid Conf. Royal Horticultural Soc., London, pp. 50–5

Yen, T-K. (1964). *Icones Cymbidiorum Amoyensium.* Committee Sci. Tech. Amag. Fukien

Chapter 9

Chen, S-C. & Tang, T. (1982). *A general review of the Orchid Flora of China. In Arditti, J. (ed.) Orchid Biology. Reviews & Perspectives, 2. Cornell University Press, Ithaca and London, pp. 39–81*

Hu, S.Y. (1971). The Orchidaceae of China I. Quart. J. Taiwan Mus. 24: 67–103

Lawler, L.J (1984). *Ethnobotany of the Orchidaceae.* In Arditti, J. (ed.) Orchid Biology, Reviews & Perspectives, 3. Cornell University Press, Ithaca and London, pp. 27–149

Chapter 10

Note: bibliographical citations not included below may be found in the synonymy given under each species.

Backer, C.A & Bakhuizen, R.C. (1968). *Orchidaceae.* Flora of Java 3: 215–450

Blume, C.L. (1848). *Orchideae.* Rumphia 4: 38–56

Blume, C.L. (1849). *Cyperorchis.* Mus. Bot. Lugduno-Batavia 1: 48

Blume, C.L. (1858). *Collection des Orchidees les plus remarquables de l'archipel Indien et du Japon* 1: 90–3, t.26

Chow, C. (1979, 1981). *Formosan Orchids*: Taicheng, Taiwan. 39–61

Comber, J.B. (1980). *The species of* Cymbidium *in Java.* Orchid Digest 44: 164–8, + figs.

Dockrill, A.W. (1969). *Australian Indigenous Orchids* 1. The Society for Growing Australian Plants, Sydney, pp. 629–39, + figs.

Dressler, R.L. (1974). *Classification of the orchid family.* Proceedings of the 7th World Orchid Conference: 259–79

Dressler, R.L. (1981). *The Orchids – Natural History and Classification.* Harvard University Press, Cambridge and London, 332pp.

Dressler, R.L. & Dodson, C.H. (1960). *Classification and phylogeny in the Orchidaceae.* Annals of the Missouri Botanical Garden 47: 25–68

Du Puy, D.J. (1983). *The Wildlife Sanctuary of Phu Luang, Thailand and its rich orchid flora.* Orchid Rev. 91: 366–71

Du Puy, D.J. (1984). *Flowers of the Phu Luang Wildlife Sanctuary.* Kew Mag. 1: 75–84

Du Puy, D.J. (1986). *A taxonomic revision of the genus* Cymbidium *Sw. (Orchidaceae).* PhD thesis, University of Birmingham and Royal Botanic Gardens, Kew

Du Puy, D.J., Ford-Lloyd, B.V. & Cribb, P.J. (1984). *A Numerical Taxonomic Analysis of* Cymbidium *section* Iridorchis *(Orchidaceae).* Kew Bulletin 40: 421–34, + figs.

Du Puy, D.J. & Lamb, A. (1984). *The genus* Cymbidium *in Sabah.* Orchid Review 92: 349–58, + figs.

Guillaumin, A. (1932). *Flore Générale de l'Indo-Chine* 6:412.

Guillaumin, A. (1960). *Notules sur quelques Orchidées de l'Indo-Chine.* Bull. Mus. Nat. Hist. (Paris) 2, ser. 32, 1: 115–117.

Guillaumin, A. (1961a). op. cit. 2, ser. 32, 6: 562–565.

Guillaumin, A. (1961b). op. cit. 2, ser. 33, 3: 332–335.

Holttum, R.E. (1953, 1957, 1964). *Orchids of Malaya.* Flora of Malaya 1: 517–26, Government Printing Office, Singapore

Hooker, J.D. (1891). *Flora of British India* 6: 8–15

Hu, S.Y. (1971a). *The Orchidaceae of China, part 1. Orchids in the life and culture of the Chinese people.* Quart. Journ. Taiwan Mus. 24: 67–103

Hu, S.Y. (1971b). *The Orchidaceae of China, part 2. The composition and distribution of orchids in China.* Quart. Journ. Taiwan Mus. 24: 181–255

Hu, S.Y. (1973). *The Orchidaceae of China 5,* Cymbidium *and* Cyperorchis. Quart. Journ. Taiwan Mus. 26: 134–42

Hunt, P.F. (1970). *Notes on Asiatic Orchids* 5. Kew Bulletin 24: 93–4

King, G. & Pantling, R. (1898). *Orchids of the Sikkim – Himalaya.* Annals of the Royal Botanic Gardens, Calcutta 8: 184–96, t.247–262

Lin, T-P. (1977). *Native Orchids of Taiwan* 2: 101–34, + figs., Chong Tao Printing Co. Ltd., Taiwan

Lindley, J. (1830–1840). *The Genera and Species of Orchidaceous Plants,* London, 553pp.

Liu, T-S & Su, H-J. (1978). *Orchidaceae, in the Flora of Taiwan* 5: 937–50

Maekawa, F. (1971). *The Wild Orchids of Japan in Colour.* Japan, 495pp.

Mark, F., Ho, H.S. & Fowlie, J.A. (1986). *An artificial key to the* Cymbidium *species in Taiwan.* Orchid Digest 50: 13–24

Pfitzer, E. (1887). *Entwurf einer naturlichen Anordnung der Orchideen.* Heidelberg

Pradhan, U.C. (1979). *Indian orchids: guide to identification and culture* 2: 465–80. Thomson Press Ltd., India

Reichenbach, H.G. (1852). *Orchidaceae* (Cymbidium *and* Cyperorchis). Walpers Ann. 3: 547–8.

Reichenbach, H.G. (1864). *Orchidaceae* Cymbidium, in C. Mueller, Walpers Ann. 6: 622–6

Schlechter, R. (1924). *Die Gattungen* Cymbidium *Sw. und* Cyperorchis *Bl.* Fedde, Repertorium Species Novarum 20: 96–110

Seidenfaden, G. (1983). *Orchid Genera in Thailand 11, Cymbidieae Pfitz.* Opera Botanica 72: 65–93

Seth, C.J. & Cribb, P.J. (1984). *A Reassessment of the Sectional Limits in the Genus* Cymbidium *Swartz.* In Arditti, J. (ed.) Orchid Biology, Reviews and Perspectives 3. Cornell University Press, Ithaca and London, pp. 283–322

Smith, J.J. (1905, 1911). *Die Orchideen von Java,* Flora von Buitenzorg 6: 475–84, and Figuren-Atlas: t.363–368

Tanaka, R. & Kamemoto, H. (1974). *List of chromosome numbers in species of Orchidaceae.* In Withner, C.L. (ed.), The Orchids, Scientific Studies, John Wiley & Sons, New York, pp. 411–83

Winter, K., Wallace, B.J., Stocker, G.C. & Roksandic, Z. (1983). *Crassulacean acid metabolism in Australian vascular epiphytes and some related species.* Oecologia 57: 129–41

Wu, Y.S. & Chen, S.C. (1980). *A Taxonomic Review of the Orchid genus* Cymbidium *in China.* Acta Phytotaxonomica Sinica 18: 292–307

GLOSSARY

abaxial – the side of an organ away from the axis.
abscission layer – zone of detachment of leaf or other organ.
acuminate – having a gradually diminishing point.
amplexicaul – clasping the stem.
anther – the part of the stamen contaning the pollen or pollinia.
anticlinal – perpendicular to the surface.
apiculate – with a short sharp point.
arcuate – curved.
articulated – jointed.
auricles – small ear-like flaps.
autotrophic – applied to plants which produce their own food by photosynthesis.
axil – angle formed by leaf- or bract-base and stem.
bract – leaf-like organ subtending a flower.
callus – structure on the upper surface of the lip, usually comprising two or three ridges in Cymbidium.
canaliculate – with a longitudinal groove.
capsule – a dry dehiscent seed-vessel.
cataphyll – the early leaf forms of a shoot.
caudicle – the cartilaginous strap that joins the pollinia to the viscidium.
chloroplast – cell-organelle that contains the chlorophyll.
ciliate – bearing hairs along the margin.
clavate – club-shaped.
column – the combination of the stamens and styles into the central organ of the orchid flower.
column-foot – the basal extension of the column.
conduplicate – folded together lengthwise.
convolute – rolled longitudinally with the margins overlapping.
cordate – heart-shaped.
coriaceous – leathery.
corolla – petals.
crenulate – with small teeth.
cucullate – hooded.
cultivar – a cultivated variety; a taxonomic rank used for varieties maintained in cultivation.
cuneate – wedge-shaped.
cuspidate – tipped with a rigid point.
cuticle – the outermost layer.
cymbiform – boat-shaped.
decurved – curved down.
dimorphic – occurring in two forms.
diploid – an organism or cell with twice the haploid number of chromosomes in its nuclei.
disc – the area in the basal part of the lip between the sidelobes and the midlobe.
distichous – borne in two ranks.
duplicate – folded.
emarginate – notched.
epiphyte – growing on a plant.
erose – gnawed.
exserted – protruding beyond.
falcate – sickle-shaped.
filiform – thread-shaped.
fimbriate – with a border of long slender processes.
fractiflex – zig-zag.
fusiform – spindle-shaped.
glabrous – lacking hairs.
haploid – a nucleus or individual containing only one representative of each chromosome of the chromosome complement.
holotype – the one specimen forming the basis for the original description of a new species and designated as such.
homonym – a name rejected because of an earlier application to another taxon.
hyaline – translucent.
hypochile – the basal part of a lip.
indumentum – any covering.
inflorescence – the flowers on the floral axis.
involute – with the edges rolled inwards.
isolectotype – duplicate specimen of lectotype.
isotype – duplicate specimen of the holotype.
lamina – the blade as of a leaf or petal.
lanceolate – narrow and tapering to each end.
lectotype – a specimen chosen from among syntypes.

lignified – converted into wood.
ligulate – tongue-shaped.
lip – the modified third petal of the orchid flower.
lithophyte – plant growing on a rock.
meiotic – applied to the reduction division of chromosomes.
mentum – the chin-like structure formed by a column-foot and the enclosing bases of the lateral sepals.
mesophyll – the interior parenchyma of the leaf.
mitotic – referring to normal cell division.
monopodial – a stem with a single axis.
motile – moveable.
mucilaginous – slimy.
mucro – sharp terminal point.
obcordate – inversely heart-shaped.
obovate – inversely ovate.
ovary – that part of flower that contains the ovules.
ovate – shaped like the longitudinal section of an egg.
palisade cells – perpendicular elongated cells on the surface of most leaves.
paniculate – furnished with a branched raceme.
papillae – soft superficial protuberances.
pedicel – flower stalk.
peduncle – the inflorescence axis below the lowermost flower.
peloric – relating to an irregular flower becoming regular, e.g. when the lip of an orchid becomes petaloid.
perianth – the calyx and corolla.
periclinal – curved in the same direction as the surface.
petiole – the leaf stalk.
phloem – vascular tissue transporting nutrients around the plant.
plicate – pleated.
pollinium (pl. pollinia) – the pollen masses in orchids.
pollinarium – the pollinia, stalk and viscidium in an orchid.
polyad – structure in pollinium of many pollen grains.
polyploid – of organisms or cells with three or more complete sets of chromosomes in their nuclei.
porrect – pointing forwards.
pubescence – hairiness.
pyriform – pear-shaped.
raceme – a simple inflorescence bearing pedicellate flowers.
reclinate – turned or bent downwards.
reniform – kidney-shaped.
rhachis – that part of the inflorescence axis bearing flowers.
rhizome – a horizontal stem.
rhombic – an equilateral oblique-angled figure.
rostellum – a narrow extension of the upper edge of the stigma in orchids.
saccate – pouched.
saprophyte – a plant living upon dead organic matter.
scape – floral axis.
scarious – thin, dry and membranous.
schlerenchyma – thick-walled cells.
sepal – segment of calyx.
serrulate – bearing small saw-like teeth.
sessile – without a stalk.
sigmoid – S-shaped.
sinus – a recess.
s.n. – sine numero (without number).
spathulate – shaped like a spatula.
stigma – the pollen receiving part of the gynoecium.
stipe – stalk joining pollinium and viscidium.
stoma (pl. stomata) – pore on the leaf surface allowing gaseous exchange.
subulate – with a fine sharp point.
sulcate – grooved.
superposed – placed on top of.
sympodium – stem made up of successive growths.
synapomorphy – shared derived character states (term used in cladistics).
syntype – one of two or more specimens cited with the description of a new taxon when none is designated the holotype.
terete – circular in cross-section.
testa – outer covering of seed.
tetrad – a body of four cells as in the formation of pollen.
tetraploid – cell or individual with four times the haploid chromosome number.
triploid – cell or individual with three times the haploid chromosome number.
truncate – cut off.
t.s. – transverse section.
vascular tissue – the cells which transport water (xylem) and nutrients (phloem) around the plant.
velamen – the layer of dead cells that sheaths the root in orchids.
viscidium – the sticky disc to which the pollinia are attached in orchids.
xeromorphic – with adaptations to dry conditions.
zygomorphic – bilaterally symmetrical, of flowers.

INDEX OF SCIENTIFIC NAMES

Accepted names are given in **bold** type.

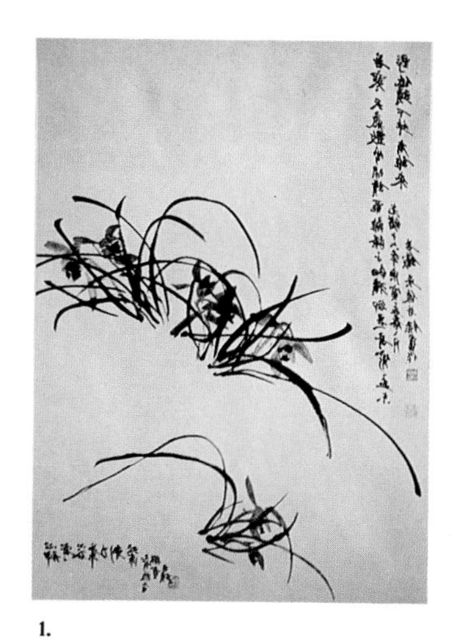

1.

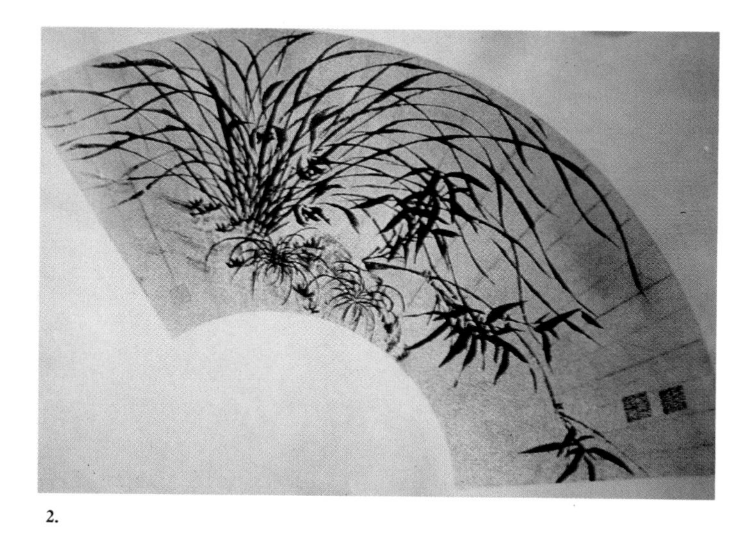

2.

3.

1. Chinese painting from Ming period of *Cymbidium goeringii*

2. Chinese painting of *Cymbidium goeringii* on fan of Ming period

3. *Cymbidium eburneum* being grown in a pot, Dali, Yunnan

4. *Cymbidium ensifolium* subsp. *ensifolium* grown as a pot-plant, Guangzhou, Guangdong, China

5, 6. Selected *Cymbidium goeringii* plants grown in pots, Tokyo, Japan

4.

5.

6.

7. T.s. of leaf of *Cymbidium rectum*

8. T.s. of leaf of *Cymbidium floribundum*

9. T.s. of leaf of *Cymbidium faberi*

10. T.s. of leaf of *Cymbidium ensifolium*

11. T.s. of leaf of *Cymbidium sanderae*

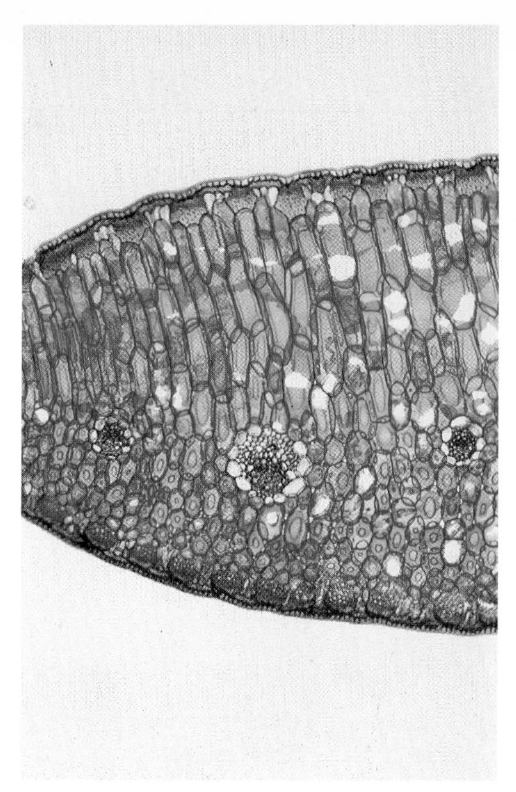

7.

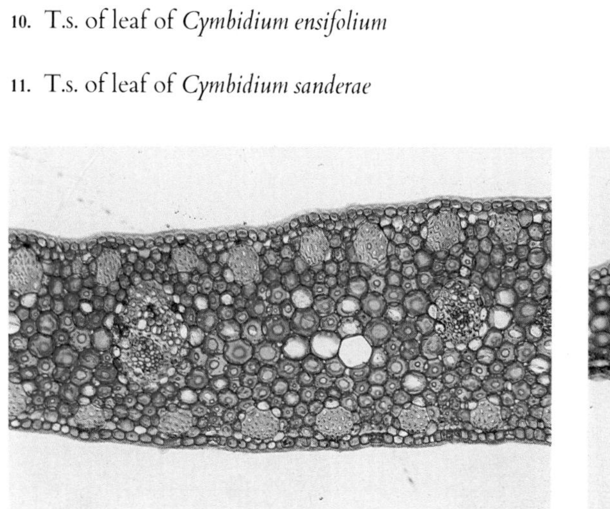

8.

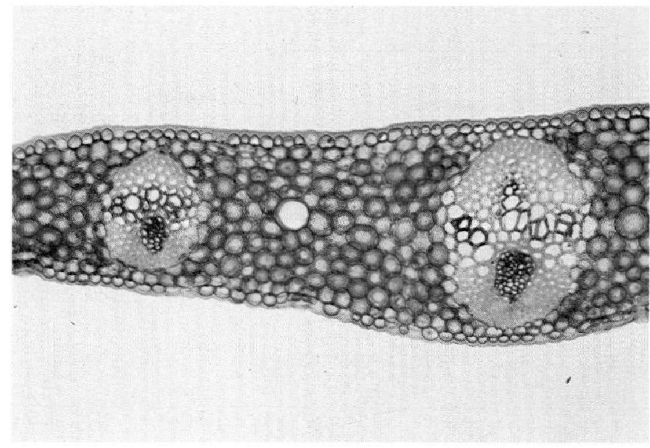

9.

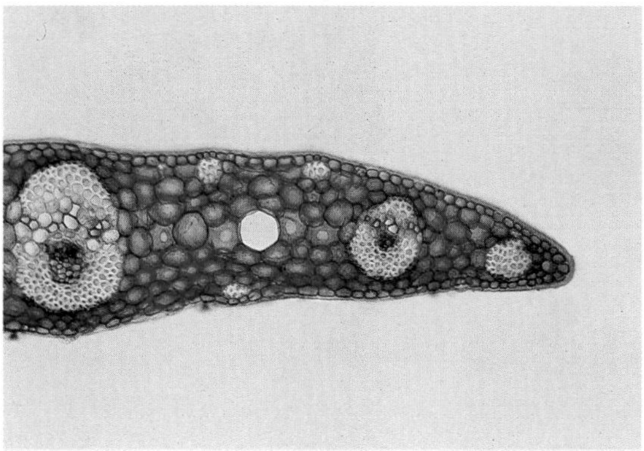

10.

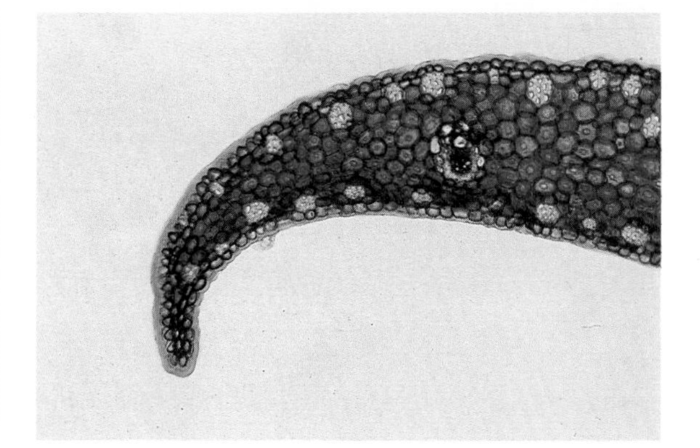

11.

12. *Cymbidium insigne*, albino form

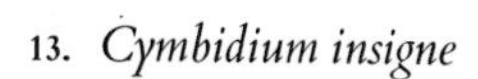

13. *Cymbidium insigne*

14. *Rhododendron lyi* in flower

15. *Rhododendron lyi* and *Cymbidium insigne* flowers

16. Bumble bee pollinating *Cymbidium insigne* flower

17. Bumble bees carrying pollinia of *Dendrobium infundibulum* (centre) and *Cymbidium insigne* (right)

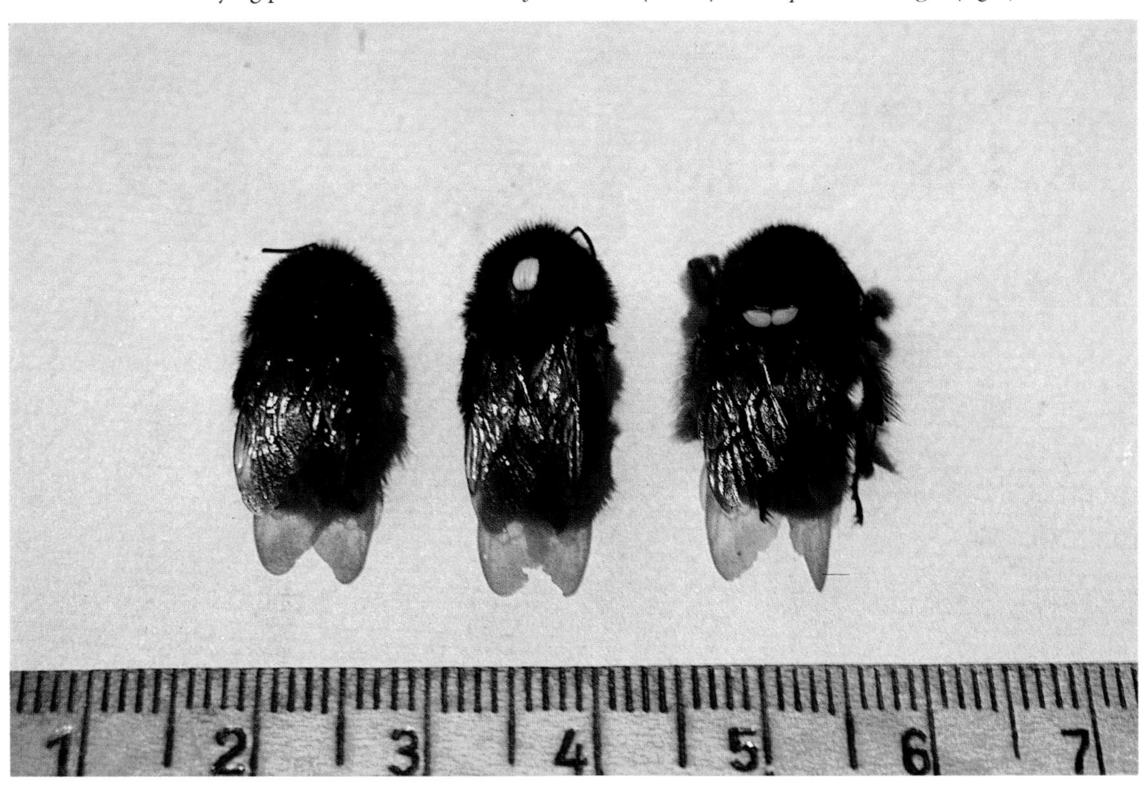

18.

18. *C.* Alexanderi 'Westonbirt'
19. *C.* Early Bird 'Pacific' AM/RHS
20. *C.* Balkis 'Nevada'
21. *C.* Fred Stewart 'Adonis'
22. *C.* Stanley Fouraker 'White Magic'
23. *C.* Rosanna 'Pinkie'
24. *C.* San Miguel 'Limelight'

19.

20.

21.

22.

23.

24.

25.

27.

25. *C.* Coningsbyanum 'Brockhurst'

26. *C.* Louisiana cv.

27. *C.* Liliana × Angelica cv.

28. *C.* Blue Smoke 'Pernod'

29. *C.* Rio Rita 'Radiant'

30. *C. erythrostylum* in its diploid and tetraploid forms

31. *C.* Kiri Te Kanawa cv.

26.

28.

29.

30.

31.

32. *C.* Fancy Free 'Mont Millais'

33. *C.* Via del Playa 'The Globe'

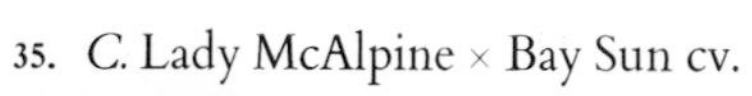

34. *C.* Pontiac 'Grouville'

35. *C.* Lady McAlpine × Bay Sun cv.

36. *C.* Minuet cv.

37. *C.* Mimi 'Safari'

38. *C.* Alison Shaw × Dr. Baker cv.

39. *C.* Mem. Emma Menninger 'Geyserland'

40. *C.* Peter Pan 'Greensleeves'

41. *C.* Beaconfire x tetraploid Peter Pan 'Greensleeves'

42. *C. devonianum* × Piccadilly 'Cardinal'

43. *C.* Jack Hudlow 'Red Velvet'

44. *C.* Tapestry 'Ruby'

45. *C.* Gladys Whitesell 'The Charmer'

46. *C.* Tiger Cub cv.

47. *C.* Tomtit 'Lettuce'

48. *Cymbidium aloifolium* on *Borassus* palm

49. *Cymbidium aloifolium* flower

50. *Cymbidium bicolor* subsp. *bicolor*

51. *Cymbidium bicolor* subsp. *bicolor*

52. *Cymbidium bicolor* subsp. *obtusum*

53. *Cymbidium bicolor* subsp. *obtusum*

54. *Cymbidium bicolor* subsp. *pubescens*

55. *Cymbidium bicolor* subsp. *pubescens*

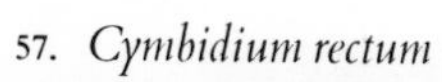

56. *Cymbidium rectum*

57. *Cymbidium rectum*

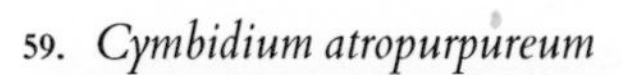

58. *Cymbidium finlaysonianum*

59. *Cymbidium atropurpureum*

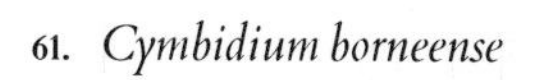

60. *Cymbidium borneense*

62. *Cymbidium borneense*

61. *Cymbidium borneense*

63. *Cymbidium dayanum*

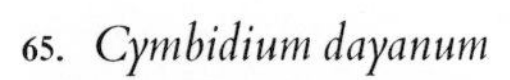

64. *Cymbidium dayanum*

65. *Cymbidium dayanum*

66. *Cymbidium canaliculatum*

67. *Cymbidium canaliculatum*

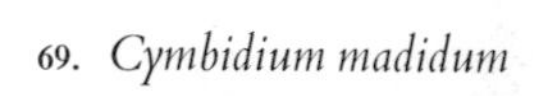

68. *Cymbidium madidum*

70. *Cymbidium suave*

69. *Cymbidium madidum*

71. *Cymbidium suave*

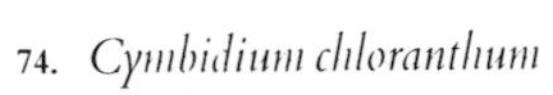

72. *Cymbidium chloranthum*

73. *Cymbidium chloranthum*

74. *Cymbidium chloranthum*

75. *Cymbidium hartinahianum*

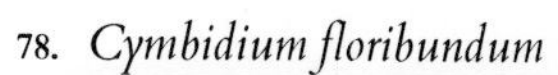

76. *Cymbidium elongatum*

77. *Cymbidium elongatum*

78. *Cymbidium floribundum*

79. *Cymbidium floribundum*, pale flowered form

80. *Cymbidium suavissimum*

81. *Cymbidium suavissimum*

82. Habitat of *Cymbidium devonianum*

83. *Cymbidium devonianum*

84. *Cymbidium tracyanum*

86. *Cymbidium iridioides* seedling on tree

85. *Cymbidium hookerianum*

87. *Cymbidium iridioides*

88. *Cymbidium erythraeum*

89. *Cymbidium erythraeum*

90. *Cymbidium wilsonii*

91. *Cymbidium wilsonii*

92, 93. *Cymbidium lowianum* var. *lowianum*

94. *Cymbidium lowianum* var. *i'ansonii*

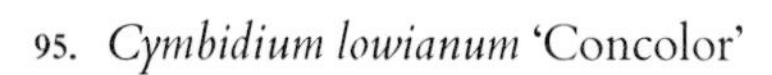

95. *Cymbidium lowianum* 'Concolor'

96. Habitat of *Cymbidium insigne*

97. *Cymbidium insigne*

98. *Cymbidium insigne*

99. *Cymbidium insigne*

100. *Cymbidium insigne*

101. *Cymbidium sanderae*

102. *Cymbidium eburneum*

103. *Cymbidium eburneum*

104. Habitat of *Cymbidium mastersii*

105. *Cymbidium mastersii*

106. *Cymbidium parishii*

107. *Cymbidium roseum*

108. *Cymbidium erythrostylum*

109. *Cymbidium elegans*

110. *Cymbidium elegans*

111. *Cymbidium elegans*

112. *Cymbidium cochleare*

113. *Cymbidium cochleare*

114. *Cymbidium whiteae*

115. *Cymbidium whiteae*

116. *Cymbidium sigmoideum*

117. *Cymbidium sigmoideum*

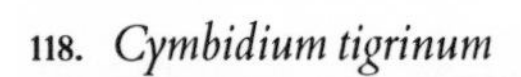

118. *Cymbidium tigrinum*

119. *Cymbidium tigrinum*

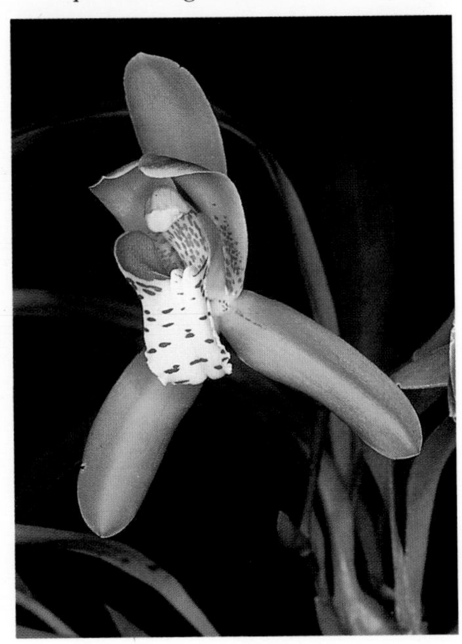

120.

121.

122.

120. *Cymbidium ensifolium* subsp. *ensifolium*

121. *Cymbidium ensifolium* subsp. *ensifolium*

122. Habitat of *Cymbidium ensifolium* subsp. *haematodes*

123. *Cymbidium ensifolium* subsp. *haematodes*

124, 125. *Cymbidium sinense*

123.

124.

125.

127. *Cymbidium kanran*

126. *Cymbidium kanran*

128. *Cymbidium cyperfolium* subsp. *cyperfolium*

129. *Cymbidium cyperfolium* subsp. *cyperfolium*

130.

131.

132.

130. *Cymbidium faberi* var. *faberi*

131. *Cymbidium faberi* var. *szechuanicum*

132. *Cymbidium goeringii* var. *goeringii*

133. *Cymbidium goeringii* var. *goeringii*

134, 135. *Cymbidium goeringii* var. *goeringii*

133.

134.

135.

136. *Cymbidium goeringii* var. *goeringii*

137. *Cymbidium goeringii* var. *serratum*

138. Habitat of *Cymbidium lancifolium*

139. *Cymbidium lancifolium*

140. *Cymbidium lancifolium*

141. *Cymbidium macrorhizon*

142. *Cymbidium macrorhizon*

143. *Cymbidium macrorhizon*